Ford Explorer Automotive Repair Manual

by Jeff Killingsworth
and John H Haynes
Member of the Guild of Motoring Writers

Models covered:
Ford Explorer
2011 through 2017
Does not include information specific to Police Interceptor models

ABCDE
FGHIJ
KLMNO
PQ

Haynes Publishing Group
Sparkford Nr Yeovil
Somerset BA22 7JJ England

Haynes North America, Inc
859 Lawrence Drive
Newbury Park
California 91320 USA
www.haynes.com

Acknowledgements

Technical writers who contributed to this project include Demian Hurst and Scott "Gonzo" Weaver.

© **Haynes North America, Inc. 2017**

With permission from J.H. Haynes & Co. Ltd.

A book in the Haynes Automotive Repair Manual Series

Printed in Malaysia

All rights reserved. No part of this book may be reproduced or transmitted in any form or by any means, electronic or mechanical, including photocopying, recording or by any information storage or retrieval system, without permission in writing from the copyright holder.

ISBN-13: 978-1-62092-285-9
ISBN-10: 1-62092-285-1

Library of Congress Control Number: 2017958663

While every attempt is made to ensure that the information in this manual is correct, no liability can be accepted by the authors or publishers for loss, damage or injury caused by any errors in, or omissions from, the information given.

"Ford" and the Ford logo are registered trademarks of Ford Motor Company. Ford Motor Company is not a sponsor or affiliate of Haynes Publishing Group or Haynes North America, Inc. and is not a contributor to the content of this manual.

Contents

Introductory pages

About this manual	0-5
Introduction	0-5
Vehicle identification numbers	0-6
Recall information	0-7
Buying parts	0-9
Maintenance techniques, tools and working facilities	0-9
Jacking and towing	0-16
Booster battery (jump) starting	0-17
Automotive chemicals and lubricants	0-18
Conversion factors	0-19
Fraction/decimal/millimeter equivalents	0-20
Safety first!	0-21
Troubleshooting	0-22

Chapter 1
Tune-up and routine maintenance — 1-1

Chapter 2 Part A
Four-cylinder engines — 2A-1

Chapter 2 Part B
V6 engines — 2B-1

Chapter 2 Part C
General engine overhaul procedures — 2C-1

Chapter 3
Cooling, heating and air conditioning systems — 3-1

Chapter 4
Fuel and exhaust systems — 4-1

Chapter 5
Engine electrical systems — 5-1

Chapter 6
Emissions and engine control systems — 6-1

Chapter 7 Part A
Automatic transaxle — 7A-1

Chapter 7 Part B
Transfer case — 7B-1

Chapter 8
Driveline — 8-1

Chapter 9
Brakes — 9-1

Chapter 10
Suspension and steering systems — 10-1

Chapter 11
Body — 11-1

Chapter 12
Chassis electrical system — 12-1

Wiring diagrams — 12-25

Index — IND-1

Haynes mechanic and photographer with a 2014 Ford Explorer

About this manual

Its purpose

The purpose of this manual is to help you get the best value from your vehicle. It can do so in several ways. It can help you decide what work must be done, even if you choose to have it done by a dealer service department or a repair shop; it provides information and procedures for routine maintenance and servicing; and it offers diagnostic and repair procedures to follow when trouble occurs.

We hope you use the manual to tackle the work yourself. For many simpler jobs, doing it yourself may be quicker than arranging an appointment to get the vehicle into a shop and making the trips to leave it and pick it up. More importantly, a lot of money can be saved by avoiding the expense the shop must pass on to you to cover its labor and overhead costs. An added benefit is the sense of satisfaction and accomplishment that you feel after doing the job yourself.

Using the manual

The manual is divided into Chapters. Each Chapter is divided into numbered Sections, which are headed in bold type between horizontal lines. Each Section consists of consecutively numbered paragraphs.

The reference numbers used in illustration captions pinpoint the pertinent Section and the Step within that Section. That is, illustration 3.2 means the illustration refers to Section 3 and Step (or paragraph) 2 within that Section.

Procedures, once described in the text, are not normally repeated. When it's necessary to refer to another Chapter, the reference will be given as Chapter and Section number. Cross references given without use of the word "Chapter" apply to Sections and/or paragraphs in the same Chapter. For example, "see Section 8" means in the same Chapter.

References to the left or right side of the vehicle assume you are sitting in the driver's seat, facing forward.

Even though we have prepared this manual with extreme care, neither the publisher nor the author can accept responsibility for any errors in, or omissions from, the information given.

NOTE

A **Note** provides information necessary to properly complete a procedure or information which will make the procedure easier to understand.

CAUTION

A **Caution** provides a special procedure or special steps which must be taken while completing the procedure where the Caution is found. Not heeding a Caution can result in damage to the assembly being worked on.

WARNING

A **Warning** provides a special procedure or special steps which must be taken while completing the procedure where the Warning is found. Not heeding a Warning can result in personal injury.

Introduction

This manual covers the Ford Explorer. The available engines are a 2.0L turbocharged in-line four-cylinder, a 2.3L turbocharged in-line four-cylinder, a 3.5L V6 and a 3.5L twin turbocharged V6.

The engine drives the front wheels through a six-speed automatic transaxle via independent driveaxles. On All-Wheel Drive (AWD) models, the rear wheels are also propelled via a driveshaft, rear differential, and two rear driveaxles.

Suspension is independent at all four wheels. The front suspension uses strut/coil spring assemblies with lower control arms. The rear uses a multi-link design with shock absorbers and coil springs. The electrically assisted rack-and-pinion steering unit is mounted on the front subframe.

The brakes are disc at the front and rear, with power assist standard. An Anti-lock Brake System (ABS) is standard equipment.

Vehicle identification numbers

1 Modifications are a continuing and unpublicized process in vehicle manufacturing. Since spare parts manuals and lists are compiled on a numerical basis, the individual vehicle numbers are essential to correctly identify the component required.

Vehicle Identification Number (VIN)

2 This very important identification number is stamped on a plate attached to the dashboard inside the windshield on the driver's side of the vehicle (see illustration). The VIN also appears on the Vehicle Certificate of Title and Registration. It contains information such as where and when the vehicle was manufactured, the model year and the body style.

Manufacturer's Certification Regulation label

3 The Manufacturer's Certification Regulation label is attached to the driver's side door opening (see illustration). The label contains the name of the manufacturer, the month and year of production, the Gross Vehicle Weight Rating (GVWR), the Gross Axle Weight Rating (GAWR) and the certification statement.

VIN engine code

4 Counting from the left, the engine code letter designation is the 8th character. On all models covered by this manual the engine codes are:

9 .... 2.0L four-cylinder engine
H .... 2.3L four-cylinder engine
8 .... 3.5L V6 engine
T .... 3.5L twin-turbo V6 engine

VIN model year code

5 Counting from the left, the model year code letter designation is the 10th character. On all models covered by this manual the model year codes are:

B 2011
C 2012
D 2013
E 2014
F 2015
G 2016
H 2017

Engine number

6 The engine identification numbers are on a sticker attached to the front of the valve cover (see illustrations).

Transmission identification

7 The transmission identification code is listed at the bottom of the Manufacturer's Certification Regulation label, under the heading "TR."

8 The codes are as follows:

6 .... 6-speed automatic transmission (6F) mid-range
C .... 6-speed automatic transmission (6F55)
J .... 6-speed automatic transmission (6F)

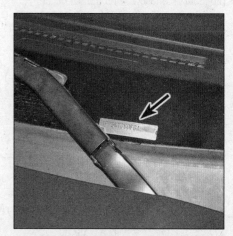

3.2 The Vehicle Identification Number (VIN) is visible through the driver's side of the windshield

3.3 The Manufacturer's Certification Regulation label is located on the driver's door opening

3.6a Engine identification label - four-cylinder engines

3.6b On V6 engines, the engine identification label is affixed to the left (driver's side) valve cover

Recall information

Vehicle recalls are carried out by the manufacturer in the rare event of a possible safety-related defect. The vehicle's registered owner is contacted at the address on file at the Department of Motor Vehicles and given the details of the recall. Remedial work is carried out free of charge at a dealer service department.

If you are the new owner of a used vehicle which was subject to a recall and you want to be sure that the work has been carried out, it's best to contact a dealer service department and ask about your individual vehicle - you'll need to furnish them your Vehicle Identification Number (VIN).

The table below is based on information provided by the National Highway Traffic Safety Administration (NHTSA), the body which oversees vehicle recalls in the United States. The recall database is updated constantly. For the latest information on vehicle recalls, check the NHTSA website at www.nhtsa.gov, www.safercar.gov, or call the NHTSA hotline at 1-888-327-4236.

Recall date	Recall campaign number	Model(s) affected	Concern
FEB 02, 2011	11V063000	2011 Explorer	Some models equipped with second row 60-percent type seats with manual recliner mechanisms may have components that are out of dimensional specification. In the event of a crash, the seat back may not provide the required strength, increasing the risk of injury.
MAY 31, 2013	13V227000	2013 Explorer	On some models, the fuel pump module may develop a crack, allowing fuel to leak. A fuel leak in the presence of an ignition source may result in a fire.
JUNE 26, 2013	13V270000	2013 Explorer	On some models, with sufficient door openings and closings, the child safety locks may change from an activated position to a deactivated position. If the child lock is deactivated, the door could be unlocked and opened from the inside which could lead to injury to an unrestrained child.
JAN 06, 2014	14E001000	2011, 2012 Explorer	Ford is recalling certain replacement steering gears installed on model year 2011 and 2012 Explorers as service parts in later years. The affected gears may lock, preventing the driver from being able to steer the vehicle, which could result in a crash.
MAY 29, 2014	14V286000	2011, 2012 and 2013 Explorer	Some models may experience an intermittent connection in the electric power steering gear, which can cause a loss of the motor position sensor signal resulting, in a shut-down of the power steering assist. If the vehicle experiences a loss of power steering assist it will require extra steering effort at lower speeds, increasing the risk of a crash.
MAR 24, 2015	15V171000	2011, 2012, 2013 Explorer	On some models, the interior door handle return spring may unseat, resulting in interior door handle that does not return to the fully stowed position after actuation. If the interior door handle return spring is unseated, the door may unlatch in the event of a side impact crash, increasing the risk of injury.

Recall information

Recall date	Recall campaign number	Model(s) affected	Concern
JULY 24, 2015	15V464000	2015, 2016 Explorer	Some models might have a parking brake that may not fully engage when applied. If the parking brake does not fully engage and the transmission is left in a gear other than "Park" while on a slope, the vehicle may roll away, increasing the risk of a crash.
SEPT 28, 2015	15V605000	2016 Explorer	On some models, the fuel tank attachment bolts may not have been properly tightened. As a result, the fuel tank straps may fracture and the fuel tank could separate from the vehicle. If the fuel tank separates from the vehicle, a fuel leak may occur, increasing the risk of a fire.
DEC 02, 2015	15V812000	2013 Explorer	On some models, the fuel pump module may crack, allowing fuel to leak. A fuel leak in the presence of an ignition source may result in a fire.
MAR 31, 2016	16V183000	2016 Explorer	Ford is recalling certain vehicles are equipped with 2.3L GTDI engines and engine block heaters. These engine block heaters have elements that may overheat while plugged in. Overheating of the engine block increase the risk of a fire.
APR 26, 2016	16V245000	2014, 2015 Explorer	Some models may have improperly welded rear suspension toe links that may fracture. A fracture of the rear suspension toe link may result in a loss of steering control, increasing the risk of a crash.
JUNE 28, 2016	16V475000	2016 Explorer	On some models equipped with a manual recline driver's seat, the seat back frame may have insufficient welds. The seat back may not adequately restrain the occupant during a crash, increasing the risk of injury.
DEC 22, 2016	16V925000	2016, 2017 Explorer	Ford is recalling certain vehicles equipped with 3.5L GTDI engines. Improperly brazed turbocharger oil supply tubes may leak oil on engine components. An oil leak, in the presence of an ignition source increases the risk of a fire.
MAY 22, 2017	17V332000	2017 Explorer	Ford is recalling certain vehicles equipped with a manual driver's seat back recliner mechanism. In the event of a crash, the seat back frame may not restrain the occupant due to having inadequate welds. If the occupant is not adequately restrained in the event of a crash, they have an increased risk of injury.
JUNE 26, 2017	17V401000	2017 Explorer	Some models may be missing the two front inboard attachments for the second row seats. Seats with missing inboard attachments may not adequately restrain an occupant in a crash, increasing the risk of injury.

Buying parts

Replacement parts are available from many sources, which generally fall into one of two categories - authorized dealer parts departments and independent retail auto parts stores. Our advice concerning these parts is as follows:

Retail auto parts stores: Good auto parts stores will stock frequently needed components which wear out relatively fast, such as clutch components, exhaust systems, brake parts, tune-up parts, etc. These stores often supply new or reconditioned parts on an exchange basis, which can save a considerable amount of money. Discount auto parts stores are often very good places to buy materials and parts needed for general vehicle maintenance such as oil, grease, filters, spark plugs, belts, touch-up paint, bulbs, etc. They also usually sell tools and general accessories, have convenient hours, charge lower prices and can often be found not far from home.

Authorized dealer parts department: This is the best source for parts which are unique to the vehicle and not generally available elsewhere (such as major engine parts, transmission parts, trim pieces, etc.).

Warranty information: If the vehicle is still covered under warranty, be sure that any replacement parts purchased - regardless of the source - do not invalidate the warranty!

To be sure of obtaining the correct parts, have engine and chassis numbers available and, if possible, take the old parts along for positive identification.

Maintenance techniques, tools and working facilities

Maintenance techniques

There are a number of techniques involved in maintenance and repair that will be referred to throughout this manual. Application of these techniques will enable the home mechanic to be more efficient, better organized and capable of performing the various tasks properly, which will ensure that the repair job is thorough and complete.

Fasteners

Fasteners are nuts, bolts, studs and screws used to hold two or more parts together. There are a few things to keep in mind when working with fasteners. Almost all of them use a locking device of some type, either a lockwasher, locknut, locking tab or thread adhesive. All threaded fasteners should be clean and straight, with undamaged threads and undamaged corners on the hex head where the wrench fits. Develop the habit of replacing all damaged nuts and bolts with new ones. Special locknuts with nylon or fiber inserts can only be used once. If they are removed, they lose their locking ability and must be replaced with new ones.

Rusted nuts and bolts should be treated with a penetrating fluid to ease removal and prevent breakage. Some mechanics use turpentine in a spout-type oil can, which works quite well. After applying the rust penetrant, let it work for a few minutes before trying to loosen the nut or bolt. Badly rusted fasteners may have to be chiseled or sawed off or removed with a special nut breaker, available at tool stores.

If a bolt or stud breaks off in an assembly, it can be drilled and removed with a special tool commonly available for this purpose. Most automotive machine shops can perform this task, as well as other repair procedures, such as the repair of threaded holes that have been stripped out.

Flat washers and lockwashers, when removed from an assembly, should always be replaced exactly as removed. Replace any damaged washers with new ones. Never use a lockwasher on any soft metal surface (such as aluminum), thin sheet metal or plastic.

Fastener sizes

For a number of reasons, automobile manufacturers are making wider and wider use of metric fasteners. Therefore, it is important to be able to tell the difference between standard (sometimes called U.S. or SAE) and metric hardware, since they cannot be interchanged.

All bolts, whether standard or metric, are sized according to diameter, thread pitch and length. For example, a standard 1/2 - 13 x 1 bolt is 1/2 inch in diameter, has 13 threads per inch and is 1 inch long. An M12 - 1.75 x 25 metric bolt is 12 mm in diameter, has a thread pitch of 1.75 mm (the distance between threads) and is 25 mm long. The two bolts are nearly identical, and easily confused, but they are not interchangeable.

In addition to the differences in diameter, thread pitch and length, metric and standard bolts can also be distinguished by examining the bolt heads. To begin with, the distance across the flats on a standard bolt head is measured in inches, while the same dimension on a metric bolt is sized in millimeters

Maintenance techniques, tools and working facilities

(the same is true for nuts). As a result, a standard wrench should not be used on a metric bolt and a metric wrench should not be used on a standard bolt. Also, most standard bolts have slashes radiating out from the center of the head to denote the grade or strength of the bolt, which is an indication of the amount of torque that can be applied to it. The greater the number of slashes, the greater the strength of the bolt. Grades 0 through 5 are commonly used on automobiles. Metric bolts have a property class (grade) number, rather than a slash, molded into their heads to indicate bolt strength. In this case, the higher the number, the stronger the bolt. Property class numbers 8.8, 9.8 and 10.9 are commonly used on automobiles.

Strength markings can also be used to distinguish standard hex nuts from metric hex nuts. Many standard nuts have dots stamped into one side, while metric nuts are marked with a number. The greater the number of dots, or the higher the number, the greater the strength of the nut.

Metric studs are also marked on their ends according to property class (grade). Larger studs are numbered (the same as metric bolts), while smaller studs carry a geometric code to denote grade.

It should be noted that many fasteners, especially Grades 0 through 2, have no distinguishing marks on them. When such is the case, the only way to determine whether it is standard or metric is to measure the thread pitch or compare it to a known fastener of the same size.

Standard fasteners are often referred to as SAE, as opposed to metric. However, it should be noted that SAE technically refers to a non-metric fine thread fastener only. Coarse thread non-metric fasteners are referred to as USS sizes.

Since fasteners of the same size (both standard and metric) may have different strength ratings, be sure to reinstall any bolts, studs or nuts removed from your vehicle in their original locations. Also, when replacing a fastener with a new one, make sure that the new one has a strength rating equal to or greater than the original.

Tightening sequences and procedures

Most threaded fasteners should be tightened to a specific torque value (torque is the twisting force applied to a threaded component such as a nut or bolt). Overtightening the fastener can weaken it and cause it to break, while undertightening can cause it to eventually come loose. Bolts, screws and studs, depending on the material they are made of and their thread diameters, have specific torque values, many of which are noted in the Specifications at the beginning of each Chapter. Be sure to follow the torque recommen-

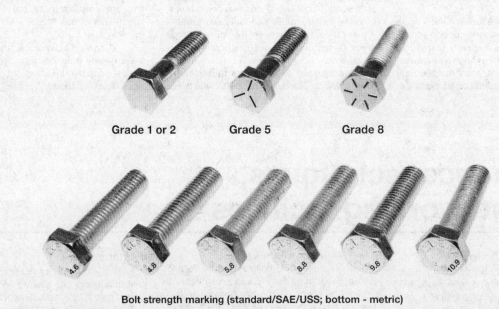

Maintenance techniques, tools and working facilities

dations closely. For fasteners not assigned a specific torque, a general torque value chart is presented here as a guide. These torque values are for dry (unlubricated) fasteners threaded into steel or cast iron (not aluminum). As was previously mentioned, the size and grade of a fastener determine the amount of torque that can safely be applied to it. The figures listed here are approximate for Grade 2 and Grade 3 fasteners. Higher grades can tolerate higher torque values.

Fasteners laid out in a pattern, such as cylinder head bolts, oil pan bolts, differential cover bolts, etc., must be loosened or tightened in sequence to avoid warping the component. This sequence will normally be shown in the appropriate Chapter. If a specific pattern is not given, the following procedures can be used to prevent warping.

Initially, the bolts or nuts should be assembled finger-tight only. Next, they should be tightened one full turn each, in a criss-cross or diagonal pattern. After each one has been tightened one full turn, return to the first one and tighten them all one-half turn, following the same pattern. Finally, tighten each of them one-quarter turn at a time until each fastener has been tightened to the proper torque. To loosen and remove the fasteners, the procedure would be reversed.

	Ft-lbs	Nm
Metric thread sizes		
M-6	6 to 9	9 to 12
M-8	14 to 21	19 to 28
M-10	28 to 40	38 to 54
M-12	50 to 71	68 to 96
M-14	80 to 140	109 to 154
Pipe thread sizes		
1/8	5 to 8	7 to 10
1/4	12 to 18	17 to 24
3/8	22 to 33	30 to 44
1/2	25 to 35	34 to 47
U.S. thread sizes		
1/4 - 20	6 to 9	9 to 12
5/16 - 18	12 to 18	17 to 24
5/16 - 24	14 to 20	19 to 27
3/8 - 16	22 to 32	30 to 43
3/8 - 24	27 to 38	37 to 51
7/16 - 14	40 to 55	55 to 74
7/16 - 20	40 to 60	55 to 81
1/2 - 13	55 to 80	75 to 108

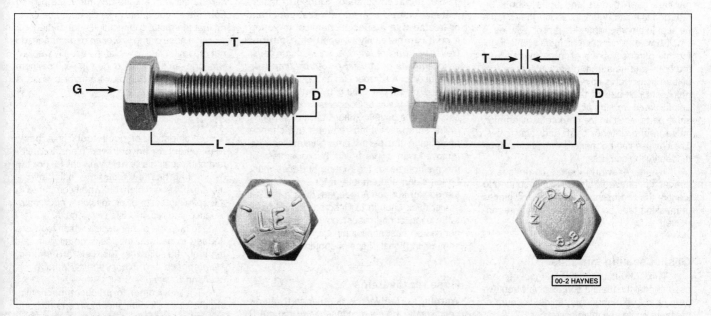

Standard (SAE and USS) bolt dimensions/grade marks

- G Grade marks (bolt strength)
- L Length (in inches)
- T Thread pitch (number of threads per inch)
- D Nominal diameter (in inches)

Metric bolt dimensions/grade marks

- P Property class (bolt strength)
- L Length (in millimeters)
- T Thread pitch (distance between threads in millimeters)
- D Diameter

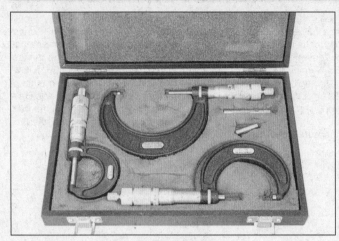
Micrometer set

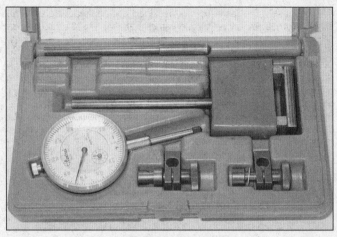
Dial indicator set

Component disassembly

Component disassembly should be done with care and purpose to help ensure that the parts go back together properly. Always keep track of the sequence in which parts are removed. Make note of special characteristics or marks on parts that can be installed more than one way, such as a grooved thrust washer on a shaft. It is a good idea to lay the disassembled parts out on a clean surface in the order that they were removed. It may also be helpful to make sketches or take instant photos of components before removal.

When removing fasteners from a component, keep track of their locations. Sometimes threading a bolt back in a part, or putting the washers and nut back on a stud, can prevent mix-ups later. If nuts and bolts cannot be returned to their original locations, they should be kept in a compartmented box or a series of small boxes. A cupcake or muffin tin is ideal for this purpose, since each cavity can hold the bolts and nuts from a particular area (i.e. oil pan bolts, valve cover bolts, engine mount bolts, etc.). A pan of this type is especially helpful when working on assemblies with very small parts, such as the carburetor, alternator, valve train or interior dash and trim pieces. The cavities can be marked with paint or tape to identify the contents.

Whenever wiring looms, harnesses or connectors are separated, it is a good idea to identify the two halves with numbered pieces of masking tape so they can be easily reconnected.

Gasket sealing surfaces

Throughout any vehicle, gaskets are used to seal the mating surfaces between two parts and keep lubricants, fluids, vacuum or pressure contained in an assembly.

Many times these gaskets are coated with a liquid or paste-type gasket sealing compound before assembly. Age, heat and pressure can sometimes cause the two parts to stick together so tightly that they are very difficult to separate. Often, the assembly can be loosened by striking it with a soft-face hammer near the mating surfaces. A regular hammer can be used if a block of wood is placed between the hammer and the part. Do not hammer on cast parts or parts that could be easily damaged. With any particularly stubborn part, always recheck to make sure that every fastener has been removed.

Avoid using a screwdriver or bar to pry apart an assembly, as they can easily mar the gasket sealing surfaces of the parts, which must remain smooth. If prying is absolutely necessary, use an old broom handle, but keep in mind that extra clean up will be necessary if the wood splinters.

After the parts are separated, the old gasket must be carefully scraped off and the gasket surfaces cleaned. Stubborn gasket material can be soaked with rust penetrant or treated with a special chemical to soften it so it can be easily scraped off. **Caution:** *Never use gasket removal solutions or caustic chemicals on plastic or other composite components.* A scraper can be fashioned from a piece of copper tubing by flattening and sharpening one end. Copper is recommended because it is usually softer than the surfaces to be scraped, which reduces the chance of gouging the part. Some gaskets can be removed with a wire brush, but regardless of the method used, the mating surfaces must be left clean and smooth. If for some reason the gasket surface is gouged, then a gasket sealer thick enough to fill scratches will have to be used during reassembly of the components. For most applications, a non-drying (or semi-drying) gasket sealer should be used.

Hose removal tips

Warning: *If the vehicle is equipped with air conditioning, do not disconnect any of the A/C hoses without first having the system depressurized by a dealer service department or a service station.*

Hose removal precautions closely parallel gasket removal precautions. Avoid scratching or gouging the surface that the hose mates against or the connection may leak. This is especially true for radiator hoses. Because of various chemical reactions, the rubber in hoses can bond itself to the metal spigot that the hose fits over. To remove a hose, first loosen the hose clamps that secure it to the spigot. Then, with slip-joint pliers, grab the hose at the clamp and rotate it around the spigot. Work it back and forth until it is completely free, then pull it off. Silicone or other lubricants will ease removal if they can be applied between the hose and the outside of the spigot. Apply the same lubricant to the inside of the hose and the outside of the spigot to simplify installation.

As a last resort (and if the hose is to be replaced with a new one anyway), the rubber can be slit with a knife and the hose peeled from the spigot. If this must be done, be careful that the metal connection is not damaged.

If a hose clamp is broken or damaged, do not reuse it. Wire-type clamps usually weaken with age, so it is a good idea to replace them with screw-type clamps whenever a hose is removed.

Tools

A selection of good tools is a basic requirement for anyone who plans to maintain and repair his or her own vehicle. For the owner who has few tools, the initial investment might seem high, but when compared to the spiraling costs of professional auto maintenance and repair, it is a wise one.

To help the owner decide which tools are needed to perform the tasks detailed in this manual, the following tool lists are offered: *Maintenance and minor repair*, *Repair/overhaul* and *Special*.

The newcomer to practical mechanics should start off with the *maintenance and minor repair* tool kit, which is adequate for the simpler jobs performed on a vehicle. Then, as confidence and experience grow, the owner can tackle more difficult tasks, buying additional tools as they are needed. Eventually the basic kit will be expanded into the *repair and overhaul* tool set. Over a period of time, the

Maintenance techniques, tools and working facilities

Dial caliper

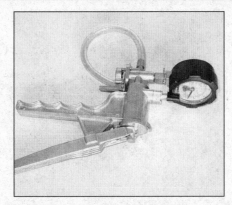
Hand-operated vacuum pump

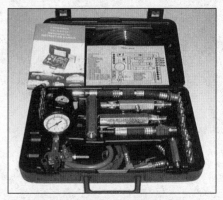
Fuel pressure gauge set

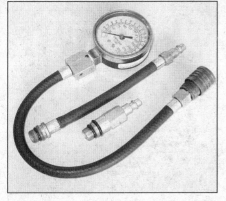
Compression gauge with spark plug hole adapter

Damper/steering wheel puller

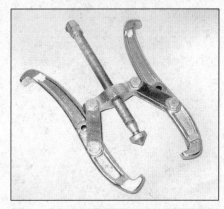
General purpose puller

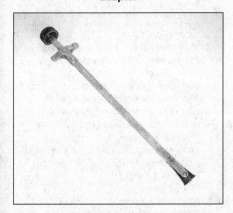
Hydraulic lifter removal tool

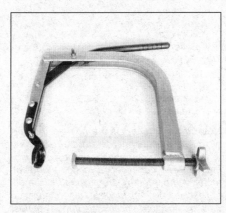
Valve spring compressor

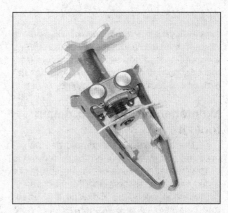
Valve spring compressor

Ridge reamer

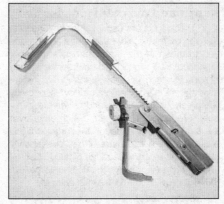
Piston ring groove cleaning tool

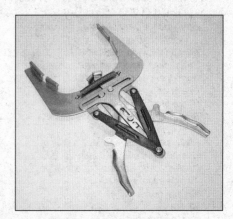
Ring removal/installation tool

Ring compressor

Cylinder hone

Brake hold-down spring tool

Torque angle gauge

Clutch plate alignment tool

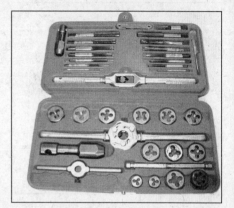
Tap and die set

experienced do-it-yourselfer will assemble a tool set complete enough for most repair and overhaul procedures and will add tools from the special category when it is felt that the expense is justified by the frequency of use.

Maintenance and minor repair tool kit

The tools in this list should be considered the minimum required for performance of routine maintenance, servicing and minor repair work. We recommend the purchase of combination wrenches (box-end and open-end combined in one wrench). While more expensive than open end wrenches, they offer the advantages of both types of wrench.

Combination wrench set (1/4-inch to 1 inch or 6 mm to 19 mm)
Adjustable wrench, 8 inch
Spark plug wrench with rubber insert
Spark plug gap adjusting tool
Feeler gauge set
Brake bleeder wrench
Standard screwdriver (5/16-inch x 6 inch)
Phillips screwdriver (No. 2 x 6 inch)
Combination pliers - 6 inch
Hacksaw and assortment of blades
Tire pressure gauge
Grease gun
Oil can
Fine emery cloth
Wire brush
Battery post and cable cleaning tool
Oil filter wrench
Funnel (medium size)
Safety goggles
Jackstands (2)
Drain pan

Note: *If basic tune-ups are going to be part of routine maintenance, it will be necessary to purchase a good quality stroboscopic timing light and combination tachometer/dwell meter. Although they are included in the list of special tools, it is mentioned here because they are absolutely necessary for tuning most vehicles properly.*

Repair and overhaul tool set

These tools are essential for anyone who plans to perform major repairs and are in addition to those in the maintenance and minor repair tool kit. Included is a comprehensive set of sockets which, though expensive, are invaluable because of their versatility, especially when various extensions and drives are available. We recommend the 1/2-inch drive over the 3/8-inch drive. Although the larger drive is bulky and more expensive, it has the capacity of accepting a very wide range of large sockets. Ideally, however, the mechanic should have a 3/8-inch drive set and a 1/2-inch drive set.

Socket set(s)
Reversible ratchet
Extension - 10 inch
Universal joint
Torque wrench (same size drive as sockets)
Ball peen hammer - 8 ounce
Soft-face hammer (plastic/rubber)
Standard screwdriver (1/4-inch x 6 inch)
Standard screwdriver (stubby - 5/16-Inch)
Phillips screwdriver (No. 3 x 8 inch)
Phillips screwdriver (stubby - No. 2)
Pliers - vise grip
Pliers - lineman's
Pliers - needle nose
Pliers - snap-ring (internal and external)
Cold chisel - 1/2-inch
Scribe
Scraper (made from flattened copper tubing)
Centerpunch
Pin punches (1/16, 1/8, 3/16-inch)
Steel rule/straightedge - 12 inch
Allen wrench set (1/8 to 3/8-inch or 4 mm to 10 mm)
A selection of files
Wire brush (large)
Jackstands (second set)
Jack (scissor or hydraulic type)

Note: *Another tool which is often useful is an electric drill with a chuck capacity of 3/8-inch and a set of good quality drill bits.*

Special tools

The tools in this list include those which are not used regularly, are expensive to buy, or which need to be used in accordance with their manufacturer's instructions. Unless these tools will be used frequently, it is not very economical to purchase many of them. A consideration would be to split the cost and use between yourself and a friend or friends. In addition, most of these tools can be obtained from a tool rental shop on a temporary basis.

This list primarily contains only those tools and instruments widely available to the public, and not those special tools produced by the vehicle manufacturer for distribution to dealer service departments. Occasionally, references to the manufacturer's special tools are included in the text of this manual. Generally, an alternative method of doing the job without the special tool is offered. However, sometimes there is no alternative to their use. Where this is the case, and the tool cannot be purchased or borrowed, the work should be turned over to the dealer service department or an automotive repair shop.

Valve spring compressor
Piston ring groove cleaning tool
Piston ring compressor
Piston ring installation tool
Cylinder compression gauge
Cylinder ridge reamer
Cylinder surfacing hone
Cylinder bore gauge
Micrometers and/or dial calipers
Hydraulic lifter removal tool
Balljoint separator
Universal-type puller
Impact screwdriver
Dial indicator set
Stroboscopic timing light (inductive pick-up)
Hand operated vacuum/pressure pump
Tachometer/dwell meter
Universal electrical multimeter
Cable hoist
Brake spring removal and installation tools
Floor jack

Buying tools

For the do-it-yourselfer who is just starting to get involved in vehicle maintenance and repair, there are a number of options available when purchasing tools. If maintenance and minor repair is the extent of the work to be done, the purchase of individual tools is satisfactory. If, on the other hand, extensive work is planned, it would be a good idea to purchase a modest tool set from one of the large retail chain stores. A set can usually be bought at a substantial savings over the individual tool prices, and they often come with a tool box. As additional tools are needed, add-on sets, individual tools and a larger tool box can be purchased to expand the tool selection. Building a tool set gradually allows the cost of the tools to be spread over a longer period of time and gives the mechanic the freedom to choose only those tools that will actually be used.

Tool stores will often be the only source of some of the special tools that are needed, but regardless of where tools are bought, try to avoid cheap ones, especially when buying screwdrivers and sockets, because they won't last very long. The expense involved in replacing cheap tools will eventually be greater than the initial cost of quality tools.

Care and maintenance of tools

Good tools are expensive, so it makes sense to treat them with respect. Keep them clean and in usable condition and store them properly when not in use. Always wipe off any dirt, grease or metal chips before putting them away. Never leave tools lying around in the work area. Upon completion of a job, always check closely under the hood for tools that may have been left there so they won't get lost during a test drive.

Some tools, such as screwdrivers, pliers, wrenches and sockets, can be hung on a panel mounted on the garage or workshop wall, while others should be kept in a tool box or tray. Measuring instruments, gauges, meters, etc. must be carefully stored where they cannot be damaged by weather or impact from other tools.

When tools are used with care and stored properly, they will last a very long time. Even with the best of care, though, tools will wear out if used frequently. When a tool is damaged or worn out, replace it. Subsequent jobs will be safer and more enjoyable if you do.

How to repair damaged threads

Sometimes, the internal threads of a nut or bolt hole can become stripped, usually from overtightening. Stripping threads is an all-too-common occurrence, especially when working with aluminum parts, because aluminum is so soft that it easily strips out.

Usually, external or internal threads are only partially stripped. After they've been cleaned up with a tap or die, they'll still work. Sometimes, however, threads are badly damaged. When this happens, you've got three choices:

1) *Drill and tap the hole to the next suitable oversize and install a larger diameter bolt, screw or stud.*
2) *Drill and tap the hole to accept a threaded plug, then drill and tap the plug to the original screw size. You can also buy a plug already threaded to the original size. Then you simply drill a hole to the specified size, then run the threaded plug into the hole with a bolt and jam nut. Once the plug is fully seated, remove the jam nut and bolt.*
3) *The third method uses a patented thread repair kit like Heli-Coil or Slimsert. These easy-to-use kits are designed to repair damaged threads in straight-through holes and blind holes. Both are available as kits which can handle a variety of sizes and thread patterns. Drill the hole, then tap it with the special included tap. Install the Heli-Coil and the hole is back to its original diameter and thread pitch.*

Regardless of which method you use, be sure to proceed calmly and carefully. A little impatience or carelessness during one of these relatively simple procedures can ruin your whole day's work and cost you a bundle if you wreck an expensive part.

Working facilities

Not to be overlooked when discussing tools is the workshop. If anything more than routine maintenance is to be carried out, some sort of suitable work area is essential.

It is understood, and appreciated, that many home mechanics do not have a good workshop or garage available, and end up removing an engine or doing major repairs outside. It is recommended, however, that the overhaul or repair be completed under the cover of a roof.

A clean, flat workbench or table of comfortable working height is an absolute necessity. The workbench should be equipped with a vise that has a jaw opening of at least four inches.

As mentioned previously, some clean, dry storage space is also required for tools, as well as the lubricants, fluids, cleaning solvents, etc. which soon become necessary.

Sometimes waste oil and fluids, drained from the engine or cooling system during normal maintenance or repairs, present a disposal problem. To avoid pouring them on the ground or into a sewage system, pour the used fluids into large containers, seal them with caps and take them to an authorized disposal site or recycling center. Plastic jugs, such as old antifreeze containers, are ideal for this purpose.

Always keep a supply of old newspapers and clean rags available. Old towels are excellent for mopping up spills. Many mechanics use rolls of paper towels for most work because they are readily available and disposable. To help keep the area under the vehicle clean, a large cardboard box can be cut open and flattened to protect the garage or shop floor.

Whenever working over a painted surface, such as when leaning over a fender to service something under the hood, always cover it with an old blanket or bedspread to protect the finish. Vinyl covered pads, made especially for this purpose, are available at auto parts stores.

Jacking and towing

Warning: *The jack supplied with the vehicle should only be used for changing a tire or placing jackstands under the frame. Never work under the vehicle or start the engine while this jack is being used as the only means of support.*

1 The vehicle should be on level ground. Place the shift lever in Park. Block the wheel diagonally opposite the wheel being changed. Set the parking brake.

2 Remove the spare tire and jack from stowage. Remove the wheel cover and trim ring (if so equipped) with the tapered end of the lug nut wrench by inserting and twisting the handle and then prying against the back of the wheel cover. Loosen, but do not remove, the lug nuts (one-half turn is sufficient).

3 Place the scissors-type jack under the vehicle and adjust the jack height until it engages with the proper jacking point. There is a front and rear jacking point on each side of the vehicle (see illustrations).

4 Turn the jack handle clockwise until the tire clears the ground. Remove the lug nuts and pull the wheel off, then install the spare.

5 Install the lug nuts with the beveled edges facing in. Tighten them snugly. Don't attempt to tighten them completely until the vehicle is lowered or it could slip off the jack. Turn the jack handle counterclockwise to lower the vehicle. Remove the jack and tighten the lug nuts in a diagonal pattern.

6 Stow the tire, jack and wrench. Unblock the wheels.

Towing

7 Two-wheel drive models can be towed from the front with the front wheels off the ground, using a wheel lift type tow truck. If towed from the rear, the front wheels must be placed on a dolly. All-wheel drive models must be towed with all four wheels off the ground. A sling-type tow truck cannot be used, as body damage will result. The best way to tow the vehicle is with a flat-bed car carrier.

8 In an emergency the vehicle can be towed a short distance with a cable or chain attached to one of the towing eyelets located under the front or rear bumpers. The driver must remain in the vehicle to operate the steering and brakes (remember that power steering and power brakes will not work with the engine off).

Place the jack in the notched area of the rocker panel flange

Front and rear jacking points

Booster battery (jump) starting

1 Observe these precautions when using a booster battery to start a vehicle:

a) *Before connecting the booster battery, make sure the ignition switch is in the Off position.*
b) *Turn off the lights, heater and other electrical loads.*
c) *Your eyes should be shielded. Safety goggles are a good idea.*
d) *Make sure the booster battery is the same voltage as the dead one in the vehicle.*
e) *The two vehicles MUST NOT TOUCH each other!*
f) *Make sure the transaxle is in Neutral (manual) or Park (automatic).*
g) *If the booster battery is not a maintenance-free type, remove the vent caps and lay a cloth over the vent holes.*

2 Connect the red jumper cable to the positive (+) terminals of each battery (see illustration).

3 Connect one end of the black jumper cable to the negative (-) terminal of the booster battery. The other end of this cable should be connected to a good ground on the vehicle to be started, such as a bolt or bracket on the body.

4 Start the engine using the booster battery, then, with the engine running at idle speed, disconnect the jumper cables in the reverse order of connection.

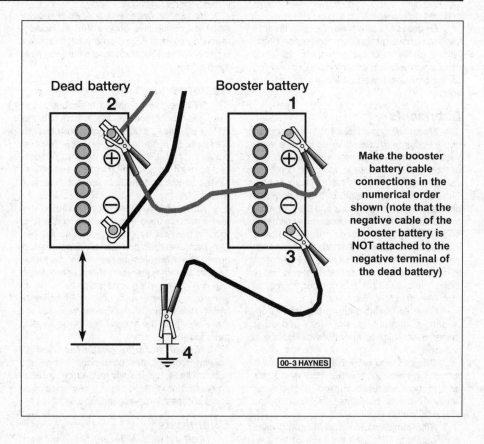

Make the booster battery cable connections in the numerical order shown (note that the negative cable of the booster battery is NOT attached to the negative terminal of the dead battery)

Automotive chemicals and lubricants

A number of automotive chemicals and lubricants are available for use during vehicle maintenance and repair. They include a wide variety of products ranging from cleaning solvents and degreasers to lubricants and protective sprays for rubber, plastic and vinyl.

Cleaners

Carburetor cleaner and choke cleaner is a strong solvent for gum, varnish and carbon. Most carburetor cleaners leave a dry-type lubricant film which will not harden or gum up. Because of this film it is not recommended for use on electrical components.

Brake system cleaner is used to remove brake dust, grease and brake fluid from the brake system, where clean surfaces are absolutely necessary. It leaves no residue and often eliminates brake squeal caused by contaminants.

Electrical cleaner removes oxidation, corrosion and carbon deposits from electrical contacts, restoring full current flow. It can also be used to clean spark plugs, carburetor jets, voltage regulators and other parts where an oil-free surface is desired.

Demoisturants remove water and moisture from electrical components such as alternators, voltage regulators, electrical connectors and fuse blocks. They are non-conductive and non-corrosive.

Degreasers are heavy-duty solvents used to remove grease from the outside of the engine and from chassis components. They can be sprayed or brushed on and, depending on the type, are rinsed off either with water or solvent.

Lubricants

Motor oil is the lubricant formulated for use in engines. It normally contains a wide variety of additives to prevent corrosion and reduce foaming and wear. Motor oil comes in various weights (viscosity ratings) from 0 to 50. The recommended weight of the oil depends on the season, temperature and the demands on the engine. Light oil is used in cold climates and under light load conditions. Heavy oil is used in hot climates and where high loads are encountered. Multi-viscosity oils are designed to have characteristics of both light and heavy oils and are available in a number of weights from 0W-20 to 20W-50.

Gear oil is designed to be used in differentials, manual transmissions and other areas where high-temperature lubrication is required.

Chassis and wheel bearing grease is a heavy grease used where increased loads and friction are encountered, such as for wheel bearings, balljoints, tie-rod ends and universal joints.

High-temperature wheel bearing grease is designed to withstand the extreme temperatures encountered by wheel bearings in disc brake equipped vehicles. It usually contains molybdenum disulfide (moly), which is a dry-type lubricant.

White grease is a heavy grease for metal-to-metal applications where water is a problem. White grease stays soft under both low and high temperatures (usually from -100 to +190-degrees F), and will not wash off or dilute in the presence of water.

Assembly lube is a special extreme pressure lubricant, usually containing moly, used to lubricate high-load parts (such as main and rod bearings and cam lobes) for initial start-up of a new engine. The assembly lube lubricates the parts without being squeezed out or washed away until the engine oiling system begins to function.

Silicone lubricants are used to protect rubber, plastic, vinyl and nylon parts.

Graphite lubricants are used where oils cannot be used due to contamination problems, such as in locks. The dry graphite will lubricate metal parts while remaining uncontaminated by dirt, water, oil or acids. It is electrically conductive and will not foul electrical contacts in locks such as the ignition switch.

Moly penetrants loosen and lubricate frozen, rusted and corroded fasteners and prevent future rusting or freezing.

Heat-sink grease is a special electrically non-conductive grease that is used for mounting electronic ignition modules where it is essential that heat is transferred away from the module.

Sealants

RTV sealant is one of the most widely used gasket compounds. Made from silicone, RTV is air curing, it seals, bonds, waterproofs, fills surface irregularities, remains flexible, doesn't shrink, is relatively easy to remove, and is used as a supplementary sealer with almost all low and medium temperature gaskets.

Anaerobic sealant is much like RTV in that it can be used either to seal gaskets or to form gaskets by itself. It remains flexible, is solvent resistant and fills surface imperfections. The difference between an anaerobic sealant and an RTV-type sealant is in the curing. RTV cures when exposed to air, while an anaerobic sealant cures only in the absence of air. This means that an anaerobic sealant cures only after the assembly of parts, sealing them together.

Thread and pipe sealant is used for sealing hydraulic and pneumatic fittings and vacuum lines. It is usually made from a Teflon compound, and comes in a spray, a paint-on liquid and as a wrap-around tape.

Chemicals

Anti-seize compound prevents seizing, galling, cold welding, rust and corrosion in fasteners. High-temperature ant-seize, usually made with copper and graphite lubricants, is used for exhaust system and exhaust manifold bolts.

Anaerobic locking compounds are used to keep fasteners from vibrating or working loose and cure only after installation, in the absence of air. Medium strength locking compound is used for small nuts, bolts and screws that may be removed later. High-strength locking compound is for large nuts, bolts and studs which aren't removed on a regular basis.

Oil additives range from viscosity index improvers to chemical treatments that claim to reduce internal engine friction. It should be noted that most oil manufacturers caution against using additives with their oils.

Gas additives perform several functions, depending on their chemical makeup. They usually contain solvents that help dissolve gum and varnish that build up on carburetor, fuel injection and intake parts. They also serve to break down carbon deposits that form on the inside surfaces of the combustion chambers. Some additives contain upper cylinder lubricants for valves and piston rings, and others contain chemicals to remove condensation from the gas tank.

Miscellaneous

Brake fluid is specially formulated hydraulic fluid that can withstand the heat and pressure encountered in brake systems. Care must be taken so this fluid does not come in contact with painted surfaces or plastics. An opened container should always be resealed to prevent contamination by water or dirt.

Weatherstrip adhesive is used to bond weatherstripping around doors, windows and trunk lids. It is sometimes used to attach trim pieces.

Undercoating is a petroleum-based, tar-like substance that is designed to protect metal surfaces on the underside of the vehicle from corrosion. It also acts as a sound-deadening agent by insulating the bottom of the vehicle.

Waxes and polishes are used to help protect painted and plated surfaces from the weather. Different types of paint may require the use of different types of wax and polish. Some polishes utilize a chemical or abrasive cleaner to help remove the top layer of oxidized (dull) paint on older vehicles. In recent years many non-wax polishes that contain a wide variety of chemicals such as polymers and silicones have been introduced. These non-wax polishes are usually easier to apply and last longer than conventional waxes and polishes.

Conversion factors

Length (distance)
Inches (in)	X 25.4	= Millimeters (mm)	X 0.0394	= Inches (in)
Feet (ft)	X 0.305	= Meters (m)	X 3.281	= Feet (ft)
Miles	X 1.609	= Kilometers (km)	X 0.621	= Miles

Volume (capacity)
Cubic inches (cu in; in^3)	X 16.387	= Cubic centimeters (cc; cm^3)	X 0.061	= Cubic inches (cu in; in^3)
Imperial pints (Imp pt)	X 0.568	= Liters (l)	X 1.76	= Imperial pints (Imp pt)
Imperial quarts (Imp qt)	X 1.137	= Liters (l)	X 0.88	= Imperial quarts (Imp qt)
Imperial quarts (Imp qt)	X 1.201	= US quarts (US qt)	X 0.833	= Imperial quarts (Imp qt)
US quarts (US qt)	X 0.946	= Liters (l)	X 1.057	= US quarts (US qt)
Imperial gallons (Imp gal)	X 4.546	= Liters (l)	X 0.22	= Imperial gallons (Imp gal)
Imperial gallons (Imp gal)	X 1.201	= US gallons (US gal)	X 0.833	= Imperial gallons (Imp gal)
US gallons (US gal)	X 3.785	= Liters (l)	X 0.264	= US gallons (US gal)

Mass (weight)
Ounces (oz)	X 28.35	= Grams (g)	X 0.035	= Ounces (oz)
Pounds (lb)	X 0.454	= Kilograms (kg)	X 2.205	= Pounds (lb)

Force
Ounces-force (ozf; oz)	X 0.278	= Newtons (N)	X 3.6	= Ounces-force (ozf; oz)
Pounds-force (lbf; lb)	X 4.448	= Newtons (N)	X 0.225	= Pounds-force (lbf; lb)
Newtons (N)	X 0.1	= Kilograms-force (kgf; kg)	X 9.81	= Newtons (N)

Pressure
Pounds-force per square inch (psi; lbf/in^2; lb/in^2)	X 0.070	= Kilograms-force per square centimeter (kgf/cm^2; kg/cm^2)	X 14.223	= Pounds-force per square inch (psi; lbf/in^2; lb/in^2)
Pounds-force per square inch (psi; lbf/in^2; lb/in^2)	X 0.068	= Atmospheres (atm)	X 14.696	= Pounds-force per square inch (psi; lbf/in^2; lb/in^2)
Pounds-force per square inch (psi; lbf/in^2; lb/in^2)	X 0.069	= Bars	X 14.5	= Pounds-force per square inch (psi; lbf/in^2; lb/in^2)
Pounds-force per square inch (psi; lbf/in^2; lb/in^2)	X 6.895	= Kilopascals (kPa)	X 0.145	= Pounds-force per square inch (psi; lbf/in^2; lb/in^2)
Kilopascals (kPa)	X 0.01	= Kilograms-force per square centimeter (kgf/cm^2; kg/cm^2)	X 98.1	= Kilopascals (kPa)

Torque (moment of force)
Pounds-force inches (lbf in; lb in)	X 1.152	= Kilograms-force centimeter (kgf cm; kg cm)	X 0.868	= Pounds-force inches (lbf in; lb in)
Pounds-force inches (lbf in; lb in)	X 0.113	= Newton meters (Nm)	X 8.85	= Pounds-force inches (lbf in; lb in)
Pounds-force inches (lbf in; lb in)	X 0.083	= Pounds-force feet (lbf ft; lb ft)	X 12	= Pounds-force inches (lbf in; lb in)
Pounds-force feet (lbf ft; lb ft)	X 0.138	= Kilograms-force meters (kgf m; kg m)	X 7.233	= Pounds-force feet (lbf ft; lb ft)
Pounds-force feet (lbf ft; lb ft)	X 1.356	= Newton meters (Nm)	X 0.738	= Pounds-force feet (lbf ft; lb ft)
Newton meters (Nm)	X 0.102	= Kilograms-force meters (kgf m; kg m)	X 9.804	= Newton meters (Nm)

Vacuum
Inches mercury (in. Hg)	X 3.377	= Kilopascals (kPa)	X 0.2961	= Inches mercury
Inches mercury (in. Hg)	X 25.4	= Millimeters mercury (mm Hg)	X 0.0394	= Inches mercury

Power
Horsepower (hp)	X 745.7	= Watts (W)	X 0.0013	= Horsepower (hp)

Velocity (speed)
Miles per hour (miles/hr; mph)	X 1.609	= Kilometers per hour (km/hr; kph)	X 0.621	= Miles per hour (miles/hr; mph)

*Fuel consumption**
Miles per gallon, Imperial (mpg)	X 0.354	= Kilometers per liter (km/l)	X 2.825	= Miles per gallon, Imperial (mpg)
Miles per gallon, US (mpg)	X 0.425	= Kilometers per liter (km/l)	X 2.352	= Miles per gallon, US (mpg)

Temperature
Degrees Fahrenheit = (°C x 1.8) + 32   Degrees Celsius (Degrees Centigrade; °C) = (°F - 32) x 0.56

*It is common practice to convert from miles per gallon (mpg) to liters/100 kilometers (l/100km), where mpg (Imperial) x l/100 km = 282 and mpg (US) x l/100 km = 235

DECIMALS to MILLIMETERS

Decimal	mm	Decimal	mm
0.001	0.0254	0.500	12.7000
0.002	0.0508	0.510	12.9540
0.003	0.0762	0.520	13.2080
0.004	0.1016	0.530	13.4620
0.005	0.1270	0.540	13.7160
0.006	0.1524	0.550	13.9700
0.007	0.1778	0.560	14.2240
0.008	0.2032	0.570	14.4780
0.009	0.2286	0.580	14.7320
		0.590	14.9860
0.010	0.2540		
0.020	0.5080		
0.030	0.7620		
0.040	1.0160	0.600	15.2400
0.050	1.2700	0.610	15.4940
0.060	1.5240	0.620	15.7480
0.070	1.7780	0.630	16.0020
0.080	2.0320	0.640	16.2560
0.090	2.2860	0.650	16.5100
		0.660	16.7640
0.100	2.5400	0.670	17.0180
0.110	2.7940	0.680	17.2720
0.120	3.0480	0.690	17.5260
0.130	3.3020		
0.140	3.5560		
0.150	3.8100		
0.160	4.0640	0.700	17.7800
0.170	4.3180	0.710	18.0340
0.180	4.5720	0.720	18.2880
0.190	4.8260	0.730	18.5420
		0.740	18.7960
0.200	5.0800	0.750	19.0500
0.210	5.3340	0.760	19.3040
0.220	5.5880	0.770	19.5580
0.230	5.8420	0.780	19.8120
0.240	6.0960	0.790	20.0660
0.250	6.3500		
0.260	6.6040		
0.270	6.8580	0.800	20.3200
0.280	7.1120	0.810	20.5740
0.290	7.3660	0.820	21.8280
		0.830	21.0820
0.300	7.6200	0.840	21.3360
0.310	7.8740	0.850	21.5900
0.320	8.1280	0.860	21.8440
0.330	8.3820	0.870	22.0980
0.340	8.6360	0.880	22.3520
0.350	8.8900	0.890	22.6060
0.360	9.1440		
0.370	9.3980		
0.380	9.6520		
0.390	9.9060		
		0.900	22.8600
0.400	10.1600	0.910	23.1140
0.410	10.4140	0.920	23.3680
0.420	10.6680	0.930	23.6220
0.430	10.9220	0.940	23.8760
0.440	11.1760	0.950	24.1300
0.450	11.4300	0.960	24.3840
0.460	11.6840	0.970	24.6380
0.470	11.9380	0.980	24.8920
0.480	12.1920	0.990	25.1460
0.490	12.4460	1.000	25.4000

FRACTIONS to DECIMALS to MILLIMETERS

Fraction	Decimal	mm	Fraction	Decimal	mm
1/64	0.0156	0.3969	33/64	0.5156	13.0969
1/32	0.0312	0.7938	17/32	0.5312	13.4938
3/64	0.0469	1.1906	35/64	0.5469	13.8906
1/16	0.0625	1.5875	9/16	0.5625	14.2875
5/64	0.0781	1.9844	37/64	0.5781	14.6844
3/32	0.0938	2.3812	19/32	0.5938	15.0812
7/64	0.1094	2.7781	39/64	0.6094	15.4781
1/8	0.1250	3.1750	5/8	0.6250	15.8750
9/64	0.1406	3.5719	41/64	0.6406	16.2719
5/32	0.1562	3.9688	21/32	0.6562	16.6688
11/64	0.1719	4.3656	43/64	0.6719	17.0656
3/16	0.1875	4.7625	11/16	0.6875	17.4625
13/64	0.2031	5.1594	45/64	0.7031	17.8594
7/32	0.2188	5.5562	23/32	0.7188	18.2562
15/64	0.2344	5.9531	47/64	0.7344	18.6531
1/4	0.2500	6.3500	3/4	0.7500	19.0500
17/64	0.2656	6.7469	49/64	0.7656	19.4469
9/32	0.2812	7.1438	25/32	0.7812	19.8438
19/64	0.2969	7.5406	51/64	0.7969	20.2406
5/16	0.3125	7.9375	13/16	0.8125	20.6375
21/64	0.3281	8.3344	53/64	0.8281	21.0344
11/32	0.3438	8.7312	27/32	0.8438	21.4312
23/64	0.3594	9.1281	55/64	0.8594	21.8281
3/8	0.3750	9.5250	7/8	0.8750	22.2250
25/64	0.3906	9.9219	57/64	0.8906	22.6219
13/32	0.4062	10.3188	29/32	0.9062	23.0188
27/64	0.4219	10.7156	59/64	0.9219	23.4156
7/16	0.4375	11.1125	15/16	0.9375	23.8125
29/64	0.4531	11.5094	61/64	0.9531	24.2094
15/32	0.4688	11.9062	31/32	0.9688	24.6062
31/64	0.4844	12.3031	63/64	0.9844	25.0031
1/2	0.5000	12.7000	1	1.0000	25.4000

Safety first!

Regardless of how enthusiastic you may be about getting on with the job at hand, take the time to ensure that your safety is not jeopardized. A moment's lack of attention can result in an accident, as can failure to observe certain simple safety precautions. The possibility of an accident will always exist, and the following points should not be considered a comprehensive list of all dangers. Rather, they are intended to make you aware of the risks and to encourage a safety conscious approach to all work you carry out on your vehicle.

Essential DOs and DON'Ts

DON'T rely on a jack when working under the vehicle. Always use approved jackstands to support the weight of the vehicle and place them under the recommended lift or support points.
DON'T attempt to loosen extremely tight fasteners (i.e. wheel lug nuts) while the vehicle is on a jack - it may fall.
DON'T start the engine without first making sure that the transmission is in Neutral (or Park where applicable) and the parking brake is set.
DON'T remove the radiator cap from a hot cooling system - let it cool or cover it with a cloth and release the pressure gradually.
DON'T attempt to drain the engine oil until you are sure it has cooled to the point that it will not burn you.
DON'T touch any part of the engine or exhaust system until it has cooled sufficiently to avoid burns.
DON'T siphon toxic liquids such as gasoline, antifreeze and brake fluid by mouth, or allow them to remain on your skin.
DON'T inhale brake lining dust - it is potentially hazardous (see *Asbestos* below).
DON'T allow spilled oil or grease to remain on the floor - wipe it up before someone slips on it.
DON'T use loose fitting wrenches or other tools which may slip and cause injury.
DON'T push on wrenches when loosening or tightening nuts or bolts. Always try to pull the wrench toward you. If the situation calls for pushing the wrench away, push with an open hand to avoid scraped knuckles if the wrench should slip.
DON'T attempt to lift a heavy component alone - get someone to help you.
DON'T rush or take unsafe shortcuts to finish a job.
DON'T allow children or animals in or around the vehicle while you are working on it.
DO wear eye protection when using power tools such as a drill, sander, bench grinder, etc. and when working under a vehicle.
DO keep loose clothing and long hair well out of the way of moving parts.
DO make sure that any hoist used has a safe working load rating adequate for the job.
DO get someone to check on you periodically when working alone on a vehicle.
DO carry out work in a logical sequence and make sure that everything is correctly assembled and tightened.
DO keep chemicals and fluids tightly capped and out of the reach of children and pets.
DO remember that your vehicle's safety affects that of yourself and others. If in doubt on any point, get professional advice.

Steering, suspension and brakes

These systems are essential to driving safety, so make sure you have a qualified shop or individual check your work. Also, compressed suspension springs can cause injury if released suddenly - be sure to use a spring compressor.

Airbags

Airbags are explosive devices that can **CAUSE** injury if they deploy while you're working on the vehicle. Follow the manufacturer's instructions to disable the airbag whenever you're working in the vicinity of airbag components.

Asbestos

Certain friction, insulating, sealing, and other products - such as brake linings, brake bands, clutch linings, torque converters, gaskets, etc. - may contain asbestos or other hazardous friction material. Extreme care must be taken to avoid inhalation of dust from such products, since it is hazardous to health. If in doubt, assume that they do contain asbestos.

Fire

Remember at all times that gasoline is highly flammable. Never smoke or have any kind of open flame around when working on a vehicle. But the risk does not end there. A spark caused by an electrical short circuit, by two metal surfaces contacting each other, or even by static electricity built up in your body under certain conditions, can ignite gasoline vapors, which in a confined space are highly explosive. Do not, under any circumstances, use gasoline for cleaning parts. Use an approved safety solvent.
Always disconnect the battery ground (-) cable at the battery before working on any part of the fuel system or electrical system. Never risk spilling fuel on a hot engine or exhaust component. It is strongly recommended that a fire extinguisher suitable for use on fuel and electrical fires be kept handy in the garage or workshop at all times. Never try to extinguish a fuel or electrical fire with water.

Fumes

Certain fumes are highly toxic and can quickly cause unconsciousness and even death if inhaled to any extent. Gasoline vapor falls into this category, as do the vapors from some cleaning solvents. Any draining or pouring of such volatile fluids should be done in a well ventilated area.

When using cleaning fluids and solvents, read the instructions on the container carefully. Never use materials from unmarked containers.
Never run the engine in an enclosed space, such as a garage. Exhaust fumes contain carbon monoxide, which is extremely poisonous. If you need to run the engine, always do so in the open air, or at least have the rear of the vehicle outside the work area.

The battery

Never create a spark or allow a bare light bulb near a battery. They normally give off a certain amount of hydrogen gas, which is highly explosive.
Always disconnect the battery ground (-) cable at the battery before working on the fuel or electrical systems.
If possible, loosen the filler caps or cover when charging the battery from an external source (this does not apply to sealed or maintenance-free batteries). Do not charge at an excessive rate or the battery may burst.
Take care when adding water to a non maintenance-free battery and when carrying a battery. The electrolyte, even when diluted, is very corrosive and should not be allowed to contact clothing or skin.
Always wear eye protection when cleaning the battery to prevent the caustic deposits from entering your eyes.

Household current

When using an electric power tool, inspection light, etc., which operates on household current, always make sure that the tool is correctly connected to its plug and that, where necessary, it is properly grounded. Do not use such items in damp conditions and, again, do not create a spark or apply excessive heat in the vicinity of fuel or fuel vapor.

Secondary ignition system voltage

A severe electric shock can result from touching certain parts of the ignition system (such as the spark plug wires) when the engine is running or being cranked, particularly if components are damp or the insulation is defective. In the case of an electronic ignition system, the secondary system voltage is much higher and could prove fatal.

Hydrofluoric acid

This extremely corrosive acid is formed when certain types of synthetic rubber, found in some O-rings, oil seals, fuel hoses, etc. are exposed to temperatures above 750-degrees F (400-degrees C). The rubber changes into a charred or sticky substance containing the acid. *Once formed, the acid remains dangerous for years. If it gets onto the skin, it may be necessary to amputate the limb concerned.*
When dealing with a vehicle which has suffered a fire, or with components salvaged from such a vehicle, wear protective gloves and discard them after use.

Troubleshooting

Contents

Symptom	Section
Engine	
Engine backfires	15
Engine continues to run after switching off	18
Engine hard to start when cold	3
Engine hard to start when hot	4
Engine lacks power	14
Engine lopes while idling or idles erratically	8
Engine misses at idle speed	9
Engine misses throughout driving speed range	10
Engine rotates but will not start	2
Engine runs with oil pressure light on	17
Engine stalls	13
Engine starts but stops immediately	6
Engine stumbles on acceleration	11
Engine surges while holding accelerator steady	12
Engine will not rotate when attempting to start	1
Oil puddle under engine	7
Pinging or knocking engine sounds during acceleration or uphill	16
Starter motor noisy or excessively rough in engagement	5
Engine electrical systems	
Alternator light fails to come on when key is turned on	21
Alternator light fails to go out	20
Battery will not hold a charge	19
Fuel system	
Excessive fuel consumption	22
Fuel leakage and/or fuel odor	23
Cooling system	
Coolant loss	28
External coolant leakage	26
Internal coolant leakage	27
Overcooling	25
Overheating	24
Poor coolant circulation	29
Automatic transaxle	
Fluid leakage	30
General shift mechanism problems	32
Transaxle fluid brown or has a burned smell	31
Transaxle slips, shifts roughly, is noisy or has no drive in forward or reverse gears	33

Symptom	Section
Driveaxles	
Clicking noise in turns	34
Shudder or vibration during acceleration	35
Vibration at highway speeds	36
Brakes	
Brake pedal feels spongy when depressed	44
Brake pedal travels to the floor with little resistance	45
Brake roughness or chatter (pedal pulsates)	39
Dragging brakes	42
Excessive brake pedal effort required to stop vehicle	40
Excessive brake pedal travel	41
Grabbing or uneven braking action	43
Noise (high-pitched squeal when the brakes are applied) during braking	38
Parking brake does not hold	46
Vehicle pulls to one side	37
Suspension and steering systems	
Abnormal noise at the front end	53
Abnormal or excessive tire wear	48
Cupped tires	58
Erratic steering when braking	55
Excessive pitching and/or rolling around corners or during braking	56
Excessive play or looseness in steering system	62
Excessive tire wear on inside edge	60
Excessive tire wear on outside edge	59
Hard steering	51
Poor returnability of steering to center	52
Rattling or clicking noise in steering gear	63
Shimmy, shake or vibration	50
Suspension bottoms	57
Tire tread worn in one place	61
Vehicle pulls to one side	47
Wander or poor steering stability	54
Wheel makes a thumping noise	49

Troubleshooting

1 This section provides an easy reference guide to the more common problems which may occur during the operation of your vehicle. These problems and their possible causes are grouped under headings denoting various components or systems, such as Engine, Cooling system, etc. They also refer you to the chapter and/or section which deals with the problem.

2 Remember that successful troubleshooting is not a mysterious black art practiced only by professional mechanics. It is simply the result of the right knowledge combined with an intelligent, systematic approach to the problem. Always work by a process of elimination, starting with the simplest solution and working through to the most complex - and never overlook the obvious. Anyone can run the gas tank dry or leave the lights on overnight, so don't assume that you are exempt from such oversights.

3 Finally, always establish a clear idea of why a problem has occurred and take steps to ensure that it doesn't happen again. If the electrical system fails because of a poor connection, check the other connections in the system to make sure that they don't fail as well. If a particular fuse continues to blow, find out why - don't just replace one fuse after another. Remember, failure of a small component can often be indicative of potential failure or incorrect functioning of a more important component or system.

Engine

1 Engine will not rotate when attempting to start

1 Battery terminal connections loose or corroded (Chapter 1).
2 Battery discharged or faulty (Chapters 1 and 5).
3 Automatic transaxle not completely engaged in Park (Chapter 7A).
4 Broken, loose or disconnected wiring in the starting circuit (Chapters 5 and 12).
5 Starter motor pinion jammed in flywheel ring gear (Chapter 5).
6 Starter solenoid faulty (Chapter 5).
7 Starter motor faulty (Chapter 5).
8 Ignition switch faulty (Chapter 12).
9 Starter pinion or flywheel teeth worn or broken (Chapter 5).

2 Engine rotates but will not start

1 Fuel tank empty.
2 Battery discharged (engine rotates slowly) (Chapter 5).
3 Battery terminal connections loose or corroded (Chapter 1).
4 Leaking fuel injector(s), faulty fuel pump, pressure regulator, etc. (Chapter 4).
5 Broken timing chain (Chapter 2A or 2B).
6 Ignition components damp or damaged (Chapter 5).
7 Worn, faulty or incorrectly-gapped spark plugs (Chapter 1).
8 Broken, loose or disconnected wiring in the starting circuit (Chapter 5).
9 Broken, loose or disconnected wires at the ignition coil or faulty coil (Chapter 5).
10 Defective crankshaft or camshaft sensor (Chapter 6).

3 Engine hard to start when cold

1 Battery discharged or low (Chapter 1).
2 Malfunctioning fuel system (Chapter 4).
3 Faulty coolant temperature sensor or intake air temperature sensor (Chapter 6).
4 Faulty ignition system (Chapter 5).

4 Engine hard to start when hot

1 Air filter clogged (Chapter 1).
2 Fuel not reaching the fuel injection system (Chapter 4).
3 Corroded battery connections (Chapter 1).
4 Faulty coolant temperature sensor or intake air temperature sensor (Chapter 6).

5 Starter motor noisy or excessively rough in engagement

1 Pinion or flywheel gear teeth worn or broken (Chapter 5).
2 Starter motor mounting bolts loose or missing (Chapter 5).

6 Engine starts but stops immediately

1 Insufficient fuel reaching the fuel injector(s) (Chapter 4).
2 Vacuum leak at the gasket between the intake manifold/plenum and throttle body (Chapter 4).

7 Oil puddle under engine

1 Oil pan gasket and/or oil pan drain bolt washer leaking (Chapter 2A).
2 Oil pressure sending unit leaking (Chapter 2A).
3 Valve cover leaking (Chapter 2A).
4 Engine oil seals leaking (Chapter 2A).
5 Oil pump housing leaking (Chapter 2A).

8 Engine lopes while idling or idles erratically

1 Vacuum leakage (Chapters 2A, 2B, and 4).
2 Leaking EGR valve (Chapter 6).
3 Air filter clogged (Chapter 1).
4 Malfunction in the fuel injection or engine control system (Chapter 4 and 6).
5 Leaking head gasket (Chapter 2A).
6 Timing chain and/or sprockets worn (Chapter 2A).
7 Camshaft lobes worn (Chapter 2A).

9 Engine misses at idle speed

1 Spark plugs worn or not gapped properly (Chapter 1).
2 Faulty coil(s) (Chapter 1).
3 Vacuum leaks (Chapter 1).
4 Uneven or low compression (Chapter 2A).
5 Problem with the fuel injection system (Chapter 4).

10 Engine misses throughout driving speed range

1 Fuel filter clogged and/or impurities in the fuel system (Chapters 1 and 4).
2 Low fuel pressure (Chapter 4).
3 Faulty or incorrectly gapped spark plugs (Chapter 1).
4 Faulty emission system components (Chapter 6).
5 Low or uneven cylinder compression pressures (Chapter 2A).
6 Weak or faulty ignition system (Chapter 5).
7 Vacuum leak in fuel injection system, intake manifold or vacuum hoses (Chapters 1 and 4).

11 Engine stumbles on acceleration

1 Spark plugs fouled (Chapter 1).
2 Problem with fuel injection or engine control system (Chapters 4 and 6).
3 Fuel filter clogged (Chapters 1 and 4).
4 Intake manifold air leak (Chapters 2A, 2B, and 4).
5 Problem with the emissions control system (Chapter 6).

12 Engine surges while holding accelerator steady

1 Intake air leak (Chapter 4).
2 Fuel pump or fuel pressure regulator faulty (Chapter 4).
3 Problem with the fuel injection system (Chapter 4).
4 Problem with the emissions control system (Chapter 6).

Troubleshooting

13 Engine stalls

1 Idle speed incorrect (Chapter 1).
2 Fuel filter clogged and/or water and impurities in the fuel system (Chapters 1 and 4).
3 Faulty emissions system components (Chapter 6).
4 Faulty or incorrectly-gapped spark plugs (Chapter 1).
5 Vacuum leak in the fuel injection system, intake manifold or vacuum hoses (Chapters 2A, 2B, and 4).

14 Engine lacks power

1 Obstructed exhaust system (Chapter 4).
2 Faulty or incorrectly-gapped spark plugs (Chapter 1).
3 Problem with the fuel injection system (Chapter 4).
4 Dirty air filter (Chapter 1).
5 Brakes binding (Chapter 9).
6 Automatic transaxle fluid level incorrect (Chapter 1).
7 Clutch slipping (Chapter 8).
8 Fuel filter clogged and/or impurities in the fuel system (Chapters 1 and 4).
9 Emission control system not functioning properly (Chapter 6).
10 Low or uneven cylinder compression pressures (Chapter 2A).

15 Engine backfires

1 Emission control system not functioning properly (Chapter 6).
2 Problem with the fuel injection system (Chapter 4).
3 Vacuum leak at fuel injector(s), intake manifold or vacuum hoses (Chapters 2A, 2B, and 4).
4 Valve clearances incorrectly set (four-cylinder engines) and/or valves sticking (Chapter 2A).

16 Pinging or knocking engine sounds during acceleration or uphill

1 Incorrect grade of fuel.
2 Fuel injection system faulty (Chapter 4).
3 Improper or damaged spark plugs or wires (Chapter 1).
4 Knock sensor defective (Chapter 6).
5 EGR valve not functioning (Chapter 6).
6 Vacuum leak (Chapters 2A, 2B, and 4).

17 Engine runs with oil pressure light on

1 Low oil level (Chapter 1).
2 Idle rpm below specification (Chapter 1).
3 Short in wiring circuit (Chapter 12).
4 Faulty oil pressure sender (Chapter 2A).
5 Worn engine bearings and/or oil pump (Chapter 2A).

18 Engine continues to run after switching off

Defective ignition switch (Chapter 12).

Engine electrical systems

19 Battery will not hold a charge

1 Drivebelt or tensioner defective (Chapter 1).
2 Battery electrolyte level low (Chapter 1).
3 Battery terminals loose or corroded (Chapter 1).
4 Alternator not charging properly (Chapter 5).
5 Loose, broken or faulty wiring in the charging circuit (Chapter 5).
6 Short in vehicle wiring (Chapter 12).
7 Internally defective battery (Chapters 1 and 5).

20 Alternator light fails to go out

1 Faulty alternator or charging circuit (Chapter 5).
2 Drivebelt or tensioner defective (Chapter 1).

21 Alternator light fails to come on when key is turned on

1 Instrument cluster defective (Chapter 12).
2 Fault in the Smart Junction Box or wiring harness (Chapter 12).

Fuel system

22 Excessive fuel consumption

1 Dirty air filter element (Chapter 1).
2 Emissions system not functioning properly (Chapter 6).
3 Fuel injection system not functioning properly (Chapter 4).
4 Low tire pressure or incorrect tire size (Chapter 1).

23 Fuel leakage and/or fuel odor

1 Leaking fuel line (Chapters 1 and 4).
2 Tank overfilled.
3 Evaporative emissions control system problem (Chapters 1 and 6).
4 Problem with the fuel injection system (Chapter 4).

Cooling system

24 Overheating

1 Insufficient coolant in system (Chapter 1).
2 Water pump drivebelt defective or out of adjustment (Chapter 1).
3 Radiator core blocked or grille restricted (Chapter 3).
4 Thermostat faulty (Chapter 3).
5 Electric coolant fan inoperative or blades broken (Chapter 3).
6 Expansion tank cap not maintaining proper pressure (Chapter 3).

25 Overcooling

1 Faulty thermostat (Chapter 3).
2 Inaccurate temperature gauge sending unit (Chapter 3).

26 External coolant leakage

1 Deteriorated/damaged hoses; loose clamps (Chapters 1 and 3).
2 Water pump defective (Chapter 3).
3 Leakage from radiator core or coolant reservoir (Chapter 3).
4 Engine drain or water jacket core plugs leaking (Chapter 2A).

27 Internal coolant leakage

1 Leaking cylinder head gasket (Chapter 2A).
2 Cracked cylinder bore or cylinder head (Chapter 2A).

28 Coolant loss

1 Too much coolant in reservoir (Chapter 1).
2 Coolant boiling away because of overheating (Chapter 3).
3 Internal or external leakage (Chapter 3).
4 Faulty radiator cap (Chapter 3).

29 Poor coolant circulation

1 Inoperative water pump (Chapter 3).
2 Restriction in cooling system (Chapter 3).
3 Drivebelt or tensioner defective (Chapter 1).
4 Thermostat sticking (Chapter 3).

Automatic transaxle

30 Fluid leakage

1 Automatic transaxle fluid is a deep red color. Fluid leaks should not be confused with engine oil, which can easily be blown onto the transaxle by air flow.
2 To pinpoint a leak, first remove all built-up dirt and grime from the transaxle housing with degreasing agents and/or steam cleaning. Then drive the vehicle at low speeds so air flow will not blow the leak far from its source. Raise the vehicle and determine where the leak is coming from. Common areas of leakage are:
 a) Dipstick tube (Chapters 1 and 7A).
 b) Transaxle oil lines (Chapter 7A).
 c) Speed sensor (Chapter 6).
 d) Driveaxle oil seals (Chapter 7A).

31 Transaxle fluid brown or has a burned smell

Transaxle fluid overheated (Chapter 1).

32 General shift mechanism problems

1 Chapter 7A deals with checking and adjusting the shift cable on automatic transaxles. Common problems which may be attributed to poorly adjusted linkage are:
 a) Engine starting in gears other than Park or Neutral.
 b) Indicator on shifter pointing to a gear other than the one actually being used.
 c) Vehicle moves when in Park.
2 Refer to Chapter 7A for the shift cable adjustment procedure.

33 Transaxle slips, shifts roughly, is noisy or has no drive in forward or reverse gears

There are many probable causes for the above problems, but the home mechanic should be concerned with only one possibility - fluid level. Before taking the vehicle to a repair shop, check the level and condition of the fluid as described in Chapter 1. Correct the fluid level as necessary or change the fluid and filter if needed. If the problem persists, have a professional diagnose the cause.

Driveaxles

34 Clicking noise in turns

Worn or damaged outboard CV joint (Chapter 8).

35 Shudder or vibration during acceleration

1 Excessive toe-in (Chapter 10).
2 Worn or damaged inboard or outboard CV joints (Chapter 8).
3 Sticking inboard CV joint assembly (Chapter 8).

36 Vibration at highway speeds

1 Out-of-balance front wheels and/or tires.
2 Out-of-round front tires.
3 Worn CV joint(s) (Chapter 8).

Brakes

37 Vehicle pulls to one side during braking

1 Incorrect tire pressures (Chapter 1).
2 Front end out of alignment (have the front end aligned).
3 Front, or rear, tire sizes not matched to one another.
4 Restricted brake lines or hoses (Chapter 9).
5 Malfunctioning caliper assembly (Chapter 9).
6 Loose suspension parts (Chapter 10).
7 Excessive wear of pad material or disc on one side (Chapter 9).
8 Contamination (grease or brake fluid) of brake pad material or disc on one side (Chapter 9).

38 Noise (high-pitched squeal when the brakes are applied)

Brake pads worn out. Replace pads with new ones immediately (Chapter 9).

39 Brake roughness or chatter (pedal pulsates)

1 Excessive lateral runout (Chapter 9).
2 Uneven pad wear (Chapter 9).
3 Defective disc (Chapter 9).

40 Excessive brake pedal effort required to stop vehicle

1 Malfunctioning power brake booster (Chapter 9).
2 Partial system failure (Chapter 9).
3 Excessively worn pads (Chapter 9).
4 Piston in caliper stuck or sluggish (Chapter 9).
5 Brake pads contaminated with oil or grease (Chapter 9).
6 Brake disc grooved and/or glazed (Chapter 9).

41 Excessive brake pedal travel

1 Partial brake system failure (Chapter 9).
2 Insufficient fluid in master cylinder (Chapters 1 and 9).
3 Air trapped in system (Chapter 9).

42 Dragging brakes

1 Incorrect adjustment of brake light switch (Chapter 9).
2 Master cylinder pistons not returning correctly (Chapter 9).
3 Caliper piston stuck (Chapter 9).
4 Restricted brakes lines or hoses (Chapter 9).
5 Incorrect parking brake adjustment (Chapter 9).

43 Grabbing or uneven braking action

1 Malfunction of proportioning valve (Chapter 9).
2 Binding brake pedal mechanism (Chapter 9).
3 Contaminated brake linings (Chapter 9).

44 Brake pedal feels spongy when depressed

1 Air in hydraulic lines (Chapter 9).
2 Master cylinder mounting bolts loose (Chapter 9).
3 Master cylinder defective (Chapter 9).

45 Brake pedal travels to the floor with little resistance

1 Little or no fluid in the master cylinder reservoir caused by leaking caliper piston(s) (Chapter 9).
2 Loose, damaged or disconnected brake lines (Chapter 9).

46 Parking brake does not hold

Parking brake improperly adjusted (Chapter 9).

Suspension and steering systems

47 Vehicle pulls to one side

1 Mismatched or uneven tires (Chapter 10).
2 Broken or sagging springs (Chapter 10).
3 Wheel alignment incorrect. Have the wheels professionally aligned.
4 Front brake dragging (Chapter 9).

48 Abnormal or excessive tire wear

1 Wheel alignment out-of-specification. Have the wheels professionally aligned.
2 Sagging or broken springs (Chapter 10).
3 Tire out-of-balance (Chapter 10).
4 Worn strut damper (Chapter 10).
5 Overloaded vehicle.
6 Tires not rotated regularly.

49 Wheel makes a thumping noise

1 Blister or bump on tire (Chapter 10).
2 Improper strut damper action (Chapter 10).

50 Shimmy, shake or vibration

1 Tire or wheel out-of-balance or out-of-round (Chapter 10).
2 Loose or worn wheel bearings (Chapter 10).
3 Worn tie-rod ends (Chapter 10).
4 Worn balljoints (Chapters 1 and 10).
5 Excessive wheel runout (Chapter 10).
6 Blister or bump on tire (Chapter 10).

51 Hard steering

1 Worn balljoints and/or tie-rod ends (Chapter 10).
2 Wheel alignment out-of-specifications. Have the wheels professionally aligned.
3 Low tire pressure(s) (Chapter 1).
4 Worn steering gear (Chapter 10).

52 Poor returnability of steering to center

1 Worn balljoints or tie-rod ends (Chapter 10).
2 Worn steering gear assembly (Chapter 10).
3 Wheel alignment out-of-specifications. Have the wheels professionally aligned.

53 Abnormal noise at the front end

1 Worn balljoints or tie-rod ends (Chapter 10).
2 Damaged shock absorber mounting (Chapter 10).
3 Worn control arm bushings or tie-rod ends (Chapter 10).
4 Loose stabilizer bar (Chapter 10).
5 Loose wheel nuts (Chapter 1).
6 Loose suspension bolts (Chapter 10).

54 Wander or poor steering stability

1 Mismatched or uneven tires (Chapter 10).
2 Worn balljoints and/or tie-rod ends (Chapters 1 and 10).
3 Worn shock absorber assemblies (Chapter 10).
4 Loose stabilizer bar (Chapter 10).
5 Broken or sagging springs (Chapter 10).
6 Wheels out of alignment. Have the wheels professionally aligned.

55 Erratic steering when braking

1 Wheel bearings worn (Chapter 10).
2 Broken or sagging springs (Chapter 10).
3 Leaking wheel cylinder or caliper (Chapter 10).
4 Excessive brake disc runout (Chapter 9).

56 Excessive pitching and/or rolling around corners or during braking

1 Loose stabilizer bar (Chapter 10).
2 Worn strut dampers or mountings (Chapter 10).
3 Broken or sagging springs (Chapter 10).
4 Overloaded vehicle.

57 Suspension bottoms

1 Overloaded vehicle.
2 Sagging springs (Chapter 10).

58 Cupped tires

1 Front wheel or rear wheel alignment out-of-specifications. Have the wheels professionally aligned.
2 Worn shock absorbers (Chapter 10).
3 Wheel bearings worn (Chapter 10).
4 Excessive tire or wheel runout (Chapter 10).
5 Worn balljoints (Chapter 10).

59 Excessive tire wear on outside edge

1 Inflation pressures incorrect (Chapter 1).
2 Excessive speed in turns.
3 Wheel alignment incorrect (excessive toe-in). Have professionally aligned.
4 Suspension arm bent or twisted (Chapter 10).

60 Excessive tire wear on inside edge

1 Inflation pressures incorrect (Chapter 1).
2 Wheel alignment incorrect (toe-out). Have professionally aligned.
3 Loose or damaged steering components (Chapter 10).

61 Tire tread worn in one place

1 Tires out-of-balance.
2 Damaged or buckled wheel. Inspect and replace if necessary.
3 Defective tire (Chapter 1).

62 Excessive play or looseness in steering system

1 Wheel bearing(s) worn (Chapter 10).
2 Tie-rod end loose (Chapter 10).
3 Steering gear loose (Chapter 10).
4 Worn or loose steering intermediate shaft U-joint (Chapter 10).

63 Rattling or clicking noise in steering gear

1 Steering gear loose (Chapter 10).
2 Steering gear defective.

Chapter 1
Tune-up and routine maintenance

Contents

	Section		Section
Air filter check and replacement	17	Fuel system check	15
Automatic transaxle fluid change	23	Introduction	2
Battery check, maintenance and charging	8	Maintenance schedule	1
Brake check	13	Seat belt check	11
Brake fluid change	21	Spark plug check and replacement	26
Cabin air filter replacement	18	Steering, suspension and driveaxle boot check	14
Cooling system check	9	Tire and tire pressure checks	5
Cooling system servicing	20	Tire rotation	10
Differential lubricant change (AWD models)	24	Transfer case lubricant change (AWD models)	25
Drivebelt check and replacement/tensioner replacement	22	Transfer case lubricant level check	16
Engine oil and filter change	6	Tune-up general information	3
Exhaust system check	19	Underhood hose check and replacement	12
Fluid level checks	4	Wiper blade inspection and replacement	7

Specifications

Recommended lubricants and fluids

Note: *Listed here are manufacturer recommendations at the time this manual was written. Manufacturers occasionally upgrade their fluid and lubricant specifications, so check with your local auto parts store for current recommendations.*

Engine oil	
Type	API "certified for gasoline engines"
Viscosity	
Four-cylinder engine	SAE 5W-30
V6 engines	
Non-turbocharged V6 engines	SAE 5W-20
Turbocharged V6 engines	SAE 5W-30
Fuel	Unleaded gasoline, 87 octane
Automatic transaxle fluid	MERCON LV Automatic Transmission Fluid

Caution: *Do not mix MERCON V and MERCON LV fluids. Transaxle damage could occur.*

Transfer case (AWD models)	SAE 75W-140 synthetic rear axle lubricant
Rear differential lubricant (AWD models)	SAE 80W-90 Premium Rear Axle Lubricant
Brake fluid	DOT 4 brake fluid
Engine coolant	50/50 mixture of Motorcraft Specialty Green Engine Coolant (green colored) or Motorcraft Orange Antifreeze/Coolant (orange colored) and distilled water

Caution: *Do not mix coolants of different colors. Doing so might damage the cooling system and/or the engine. The manufacturer specifies either a green colored coolant or an orange colored coolant to be used in these systems, depending on what was originally installed in the vehicle.*

Capacities*

Engine oil (including filter)
  Four-cylinder engine .... 5.7 quarts (5.4 liters)
  V6 engines .... 6.0 quarts (5.7 liters)
Coolant
  Four-cylinder engine .... 11.6 quarts (11.0 liters)
  V6 engines
    Without trailer tow option .... Up to 13.1 quarts (12.4 liters)
    With trailer tow option .... Up to 13.5 quarts (12.7 liters)
Automatic transaxle (dry fill)
  Four-cylinder models (6F35) .... Up to 9 quarts (8.5 liters)
  V6 models (6F50/6F55) .... Up to 11.1 quarts (10.5 liters)

Note: *Since this is a dry-fill specification, the amount required during a routine fluid change will be substantially less. The best way to determine the amount of fluid to add during a routine fluid change is to measure the amount drained. Begin the refill procedure by initially adding 1/3 of the amount drained. Then, with the engine running, add 1/2-pint at a time (cycling the shifter through each gear position between additions) until the level is correct on the dipstick. It is important to not overfill the transaxle. You will, however, need to purchase a few extra quarts, since the fluid replacement procedure involves flushing the torque converter (see Section 23).*

Transfer case (AWD models) .... Up to 18 ounces (0.53 liters)
Rear differential (AWD models) .... Up to 2.43 pints (1.15 liters)

* All capacities approximate. Add as necessary to bring up to appropriate level.

Ignition system

Spark plug type and gap
  Type
    Four-cylinder engine
      2.0L engine .... NGK T4025R or equivalent
      2.3L engine .... Motorcraft 12405 (SP-520) or equivalent
    V6 engines .... Motorcraft 12405 (SP-520) or equivalent
  Gap
    Four-cylinder engine .... 0.031 inch (0.8 mm)
    V6 engines
      3.5L Duratec engine .... 0.049 to 0.053 inch (1.25 to 1.35 mm)
      3.5L turbocharged engine .... 0.035 inch (0.889 mm)
Engine firing order
  Four-cylinder engine 1-3-4-2
  V6 engines 1-4-2-5-3-6

Cylinder location - four-cylinder engines

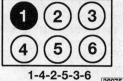

1-4-2-5-3-6

Cylinder location - V6 engines

Brakes

Disc brake pad lining thickness (minimum) .... 1/8 inch (3 mm)

Torque specifications

	Ft-lbs (unless otherwise indicated)	Nm

Note: *One foot-pound (ft-lb) of torque is equivalent to 12 inch-pounds (in-lbs) of torque. Torque values below approximately 15 ft-lbs are expressed in inch-pounds, because most foot-pound torque wrenches are not accurate at these smaller values.*

	Ft-lbs	Nm
Engine oil drain plug	20	27
Automatic transaxle drain plug	106 in-lbs	12
Automatic transaxle oil leveling plug (four-cylinder models - 6F35)	71 in-lbs	8
Rear differential cover bolts (AWD)	17	23
Rear differential check/fill plug (AWD)	21	29
Transfer case		
2016 and earlier models		
Drain plug	106 in-lbs	12
Check/fill plug		
Four-cylinder models	106 in-lbs	12
V6 models	177 in-lbs	20
2017 models (check/fill and drain plugs)	106 in-lbs	12
Spark plugs		
Four-cylinder engine	106 in-lbs	12
V6 engines	133 in-lbs	15
Drivebelt tensioner bolts		
Four-cylinder engines	18	25
V6 engines	97 in-lbs	11
Wheel lug nuts	100	135

Chapter 1 Tune-up and routine maintenance   1-3

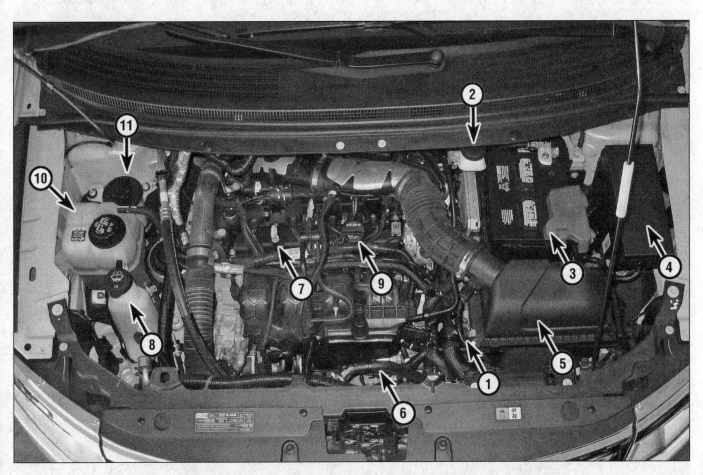

Typical engine compartment components (four-cylinder engine)

1 Automatic transaxle fluid filler tube
2 Brake fluid reservoir
3 Battery
4 Fuse/relay block
5 Air filter housing
6 Radiator hose
7 Engine oil dipstick
8 Windshield washer fluid reservoir
9 Engine oil filler cap
10 Coolant expansion tank
11 Power steering fluid reservoir

Chapter 1 Tune-up and routine maintenance

Engine compartment underside components (four-cylinder engine)

1. Engine oil filter
2. Driveaxle boot
3. Front disc brake caliper
4. Engine oil drain plug
5. Automatic transaxle drain plug
6. Exhaust system

Chapter 1 Tune-up and routine maintenance

Engine compartment components (non-turbocharged V6 engine)

1. Brake fluid reservoir
2. Battery
3. Fuse/relay block
4. Air filter housing
5. Oil dipstick
6. Oil filler cap
7. Windshield washer fluid reservoir
8. Coolant expansion tank

Engine compartment underside components (non-turbocharged V6 engine)

1. Engine oil filter
2. Front disc brake caliper
3. Outer driveaxle boots
4. Automatic transaxle drain plug
5. Exhaust system
6. Engine oil drain plug

Chapter 1 Tune-up and routine maintenance 1-7

Rear underside components (front-wheel drive model shown)

1 Rear disc brake caliper
2 Muffler
3 Exhaust system hanger
4 Coil springs
5 Parking brake cables
6 Stabilizer bar
7 Fuel tank

1 Maintenance schedule

The maintenance intervals in this manual are provided with the assumption that you, not the dealer, will be doing the work. These are the minimum maintenance intervals recommended by the factory for vehicles that are driven daily. If you wish to keep your vehicle in peak condition at all times, you may wish to perform some of these procedures even more often. Because frequent maintenance enhances the efficiency, performance and resale value of your car, we encourage you to do so. If you drive in dusty areas, tow a trailer, idle or drive at low speeds for extended periods or drive for short distances (less than four miles) in below freezing temperatures, shorter intervals are also recommended.

When your vehicle is new, it should be serviced by a factory authorized dealer service department to protect the factory warranty. In many cases, the initial maintenance check is done at no cost to the owner.

Every 250 miles (400 km) or weekly, whichever comes first

Check the engine oil level (Section 4)
Check the engine coolant level (Section 4)
Check the brake fluid level (Section 4)
Check the windshield washer fluid level (Section 4)
Check the tires and tire pressures (Section 5)

Every 3000 miles (4800 km) or 3 months

Change the engine oil and oil filter (Section 6)

Every 7500 miles (12,000 km) or 6 months, whichever comes first

All items listed above plus:
Change the engine oil and oil filter (Section 6)
Inspect (and replace, if necessary) the windshield wiper blades (Section 7)
Check and service the battery (Section 8)
Check the cooling system (Section 9)
Rotate the tires (Section 10)
Check the seat belts (Section 11)

Every 15,000 miles (24,000 km) or 12 months, whichever comes first

All items listed above plus:
Check the automatic transaxle fluid level (Section 4)
Check all underhood hoses (Section 12)
Inspect the brake system (Section 13)*
Inspect the suspension, steering and driveaxle boots (Section 14)*
Check the fuel system (Section 15)
Check the transfer case lubricant level (AWD models) (Section 16)
Check (and replace, if necessary) the air filter (Section 17)*
Replace the cabin air filter (Section 18)

Every 30,000 miles (48,000 km) or 24 months, whichever comes first

All items listed above plus:
Check the exhaust system (Section 19)
Service the cooling system (drain, flush and refill) (Section 20)
Change the brake fluid (Section 21)
Check/adjust the engine drivebelts (Section 22)

Every 60,000 miles (72,420 km) or 48 months, whichever comes first

Replace the automatic transaxle fluid (Section 23)**
Replace the rear differential lubricant (AWD) (Section 24)**
Check (and replace, if necessary) the spark plugs (conventional, non-platinum or iridium type) (Section 26)

Every 100,000 miles (160,000 km)

Replace the spark plugs (iridium or platinum-tipped type) (Section 26)

*This item is affected by "severe" operating conditions as described below. If your vehicle is operated under "severe" conditions, perform all maintenance indicated with an asterisk (*) at 3000 mile/3 month intervals. Severe conditions are indicated if you mainly operate your vehicle under one or more of the following conditions:

a) Operating in dusty areas
b) Towing a trailer
c) Idling for extended periods and/or low speed operation
d) Operating when outside temperatures remain below freezing and when most trips are less than 4 miles

**If operated under one or more of the following conditions, change the automatic transaxle fluid and differential lubricant every 30,000 miles:

a) Operating in dusty areas
b) In heavy city traffic where the outside temperature regularly reaches 90-degrees F (32-degrees C) or higher
c) In hilly or mountainous terrain
d) On flex-fuel models, using E85 fuel more than 50% of the time

2 Introduction

1 This Chapter is designed to help the home mechanic maintain the Ford Explorer with the goals of maximum performance, economy, safety and reliability in mind.
2 Included is a master maintenance schedule, followed by procedures dealing specifically with each item on the schedule. Visual checks, adjustments, component replacement and other helpful items are included. Refer to the accompanying illustrations of the engine compartment and the underside of the vehicle for the locations of various components.
3 Servicing the vehicle, in accordance with the mileage/time maintenance schedule and the step-by-step procedures will result in a planned maintenance program that should produce a long and reliable service life. Keep in mind that it is a comprehensive plan, so maintaining some items but not others at the specified intervals will not produce the same results.
4 As you service the vehicle, you will discover that many of the procedures can - and should - be grouped together because of the nature of the particular procedure you're performing or because of the close proximity of two otherwise unrelated components to one another.
5 For example, if the vehicle is raised for chassis lubrication, you should inspect the exhaust, suspension, steering and fuel systems while you're under the vehicle. When you're rotating the tires, it makes good sense to check the brakes since the wheels are already removed. Finally, let's suppose you have to borrow or rent a torque wrench. Even if you only need it to tighten the spark plugs, you might as well check the torque of as many critical fasteners as time allows.
6 The first step in this maintenance program is to prepare yourself before the actual work begins. Read through all the procedures you're planning to do, then gather up all the parts and tools needed. If it looks like you might run into problems during a particular job, seek advice from a mechanic or an experienced do-it-yourselfer.

Owner's Manual and VECI label information

7 Your vehicle owner's manual was written for your year and model and contains very specific information on component locations, specifications, fuse ratings, part numbers, etc. The owner's manual is an important resource for the do-it-yourselfer to have; if one was not supplied with your vehicle, it can generally be ordered from a dealer parts department.
8 Among other important information, the Vehicle Emissions Control Information (VECI) label contains specifications and procedures for applicable tune-up adjustments and, in some instances, spark plugs (see Chapter 6 for more information on the VECI label). The information on this label is the exact maintenance data recommended by the manufacturer. This data often varies by intended operating altitude, local emissions regulations, month of manufacture, etc.
9 This Chapter contains procedural details, safety information and more ambitious maintenance intervals than you might find in manufacturer's literature. However, you may also find procedures or specifications in your owner's manual or VECI label that differ with what's printed here. In these cases, the owner's manual or VECI label can be considered correct, since it is specific to your particular vehicle.

3 Tune-up general information

1 The term tune-up is used in this manual to represent a combination of individual operations rather than one specific procedure.
2 If, from the time the vehicle is new, the routine maintenance schedule is followed closely and frequent checks are made of fluid levels and high wear items, as suggested throughout this manual, the engine will be kept in relatively good running condition and the need for additional work will be minimized.
3 More likely than not, however, there will be times when the engine is running poorly due to lack of regular maintenance. This is even more likely if a used vehicle, which has not received regular and frequent maintenance checks, is purchased. In such cases, an engine tune-up will be needed outside of the regular routine maintenance intervals.
4 The first step in any tune-up or diagnostic procedure to help correct a poor running engine is a cylinder compression check. A compression check (see Chapter 2C) will help determine the condition of internal engine components and should be used as a guide for tune-up and repair procedures. If, for instance, a compression check indicates serious internal engine wear, a conventional tune-up will not improve the performance of the engine and would be a waste of time and money. Because of its importance, the compression check should be done by someone with the proper equipment and the knowledge to use it properly.
5 The following procedures are those most often needed to bring a generally poor running engine back into a proper state of tune.

Minor tune-up

Check all engine related fluids (Section 4)
Clean, inspect and test the battery (Section 8)
Check the cooling system (Section 9)
Check all underhood hoses (Section 12)
Check the fuel system (Section 15)
Check the air filter (Section 17)
Check the drivebelt (Section 22)

Major tune-up

All items listed under Minor tune-up, plus ...
Replace the air filter (Section 17)
Replace the PCV valve (See Chapter 6)
Replace the spark plugs (Section 26)

4 Fluid level checks (every 250 miles or weekly)

1 Fluids are an essential part of the lubrication, cooling, brake and windshield washer systems. Because the fluids gradually become depleted and/or contaminated during normal operation of the vehicle, they must be periodically replenished. See *Recommended lubricants and fluids* in this Chapter's Specifications before adding fluid to any of the following components.
Note: *The vehicle must be on level ground when fluid levels are checked.*

Engine oil

2 The oil level is checked with a dipstick, which is attached to the engine block (see illustrations). The dipstick extends through a metal tube down into the oil pan.
3 The oil level should be checked before

4.2a Engine oil dipstick (A) and oil filler cap (B) locations - 2.0L four-cylinder engine

4.2b Engine oil dipstick (A) an oil filler cap (B) locations - 3.5L V6 engine

1-10 Chapter 1 Tune-up and routine maintenance

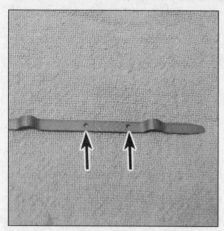

4.4 The oil level should be in the safe range - if it's below the MIN or ADD mark (lower hole), add enough oil to bring it up to or near the MAX or FULL mark (upper hole)

4.8 The cooling system expansion tank is located at the right side of the engine compartment

4.10 When the engine is cold, the coolant level should be in the COLD FILL range

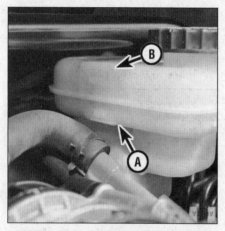

4.16 The brake fluid level should be kept between the MIN (A) and MAX (B) marks on the translucent plastic reservoir

the vehicle has been driven, or about 5 minutes after the engine has been shut off. If the oil is checked immediately after driving the vehicle, some of the oil will remain in the upper part of the engine, resulting in an inaccurate reading on the dipstick.

4 Pull the dipstick out of the tube and wipe all the oil from the end with a clean rag or paper towel. Insert the clean dipstick all the way back into the tube and pull it out again. Note the oil at the end of the dipstick. At its highest point, the level should be between the MIN and MAX marks on the dipstick (see illustration).

5 It takes one quart of oil to raise the level from the MIN mark to the MAX mark on the dipstick. Do not allow the level to drop below the MIN mark or oil starvation may cause engine damage. Conversely, overfilling the engine (adding oil above the MAX mark) may cause oil fouled spark plugs, oil leaks or oil seal failures. Maintaining the oil level above the MAX mark can cause excessive oil consumption.

6 To add oil, remove the filler cap from the valve cover (see illustrations 4.2a or 4.2b).

After adding oil, wait a few minutes to allow the level to stabilize, then pull out the dipstick and check the level again. Add more oil if required. Install the filler cap and tighten it by hand only.

7 Checking the oil level is an important preventive maintenance step. A consistently low oil level indicates oil leakage through damaged seals, defective gaskets or past worn rings or valve guides. If the oil looks milky in color or has water droplets in it, the cylinder head gasket(s) may be blown or the head(s) or block may be cracked. The engine should be checked immediately. The condition of the oil should also be checked. Whenever you check the oil level, slide your thumb and index finger up the dipstick before wiping off the oil. If you see small dirt or metal particles clinging to the dipstick, the oil should be changed (see Section 6).

Engine coolant

Warning: *Do not allow antifreeze to come in contact with your skin or painted surfaces of the vehicle. Flush contaminated areas immediately with plenty of water. Don't store new coolant or leave old coolant lying around where it's accessible to children or pets - they're attracted by its sweet smell. Ingestion of even a small amount of coolant can be fatal! Wipe up garage floor and drip pan spills immediately. Keep antifreeze containers covered and repair cooling system leaks as soon as they're noticed.*

8 All vehicles covered by this manual are equipped with a pressurized coolant recovery system. A plastic expansion tank located at the right front corner of the engine compartment is connected by hoses to the cooling system (see illustration). As the engine heats up during operation, the expanding coolant fills the tank.

9 The coolant level in the tank should be checked regularly.

Warning: *Do not remove the expansion tank cap when the engine is warm!*

10 The level in the tank varies with the temperature of the engine. When the engine is cold, the coolant level should be at the COLD FULL mark on the reservoir. If it isn't, remove the cap from the tank and add a 50/50 mixture of ethylene glycol-based antifreeze and water (see illustration).

11 Drive the vehicle, let the engine cool completely then recheck the coolant level. Don't use rust inhibitors or additives. If only a small amount of coolant is required to bring the system up to the proper level, water can be used. However, repeated additions of water will dilute the antifreeze and water solution. In order to maintain the proper ratio of antifreeze and water, always top up the coolant level with the correct mixture. An empty plastic milk jug or bleach bottle makes an excellent container for mixing coolant.

12 If the coolant level drops consistently, there may be a leak in the system. Inspect the radiator, hoses, filler cap, drain plugs and water pump (see Section 9). If no leaks are noted, have the expansion tank cap pressure tested by a service station.

13 If you have to remove the expansion tank cap wait until the engine has cooled completely, then wrap a thick cloth around the cap and unscrew it slowly, stopping if you hear a hissing noise. If coolant or steam escapes, let the engine cool down longer, then remove the cap.

14 Check the condition of the coolant as well. If it's brown or rust colored, the system should be drained, flushed and refilled. Even if the coolant appears to be normal, the corrosion inhibitors wear out, so it must be replaced at the specified intervals.

Brake fluid

15 The brake master cylinder is mounted on the front of the power booster unit in the engine compartment.

16 To check the fluid level, simply look at the MAX and MIN marks on the brake fluid reservoir (see illustration).

17 If the level is low, wipe the top of the reservoir cover with a clean rag to prevent con-

Chapter 1 Tune-up and routine maintenance   1-11

4.23 The windshield/rear window washer fluid reservoir is located in the right front corner of the engine compartment

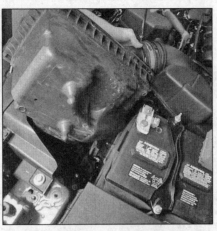

4.28 Reposition the air filter housing like this for the fluid level check

4.29 Location of the transaxle fluid filler cap/dipstick

tamination of the brake system before lifting the cover.

18 Add only the specified brake fluid to the reservoir (refer to *Recommended lubricants and fluids* in this Chapter's Specifications, or to your owner's manual). Mixing different types of brake fluid can damage the system. Fill the brake master cylinder reservoir only to the MAX line.

Warning: *Use caution when filling the reservoir - brake fluid can harm your eyes and damage painted surfaces. Do not use brake fluid that is more than one year old or has been left open. Brake fluid absorbs moisture from the air. Excess moisture can cause a dangerous loss of braking.*

19 While the reservoir cap is removed, inspect the master cylinder reservoir for contamination. If deposits, dirt particles or water droplets are present, the system should be drained and refilled.

20 After filling the reservoir to the proper level, make sure the lid is properly seated to prevent fluid leakage.

21 The fluid in the brake master cylinder will drop slightly as the brake pads at each wheel wear down during normal operation. If the master cylinder requires repeated replenishing to keep it at the proper level, this is an indication of leakage in the brake system, which should be corrected immediately. If the brake system shows an indication of leakage check all brake lines and connections, along with the calipers and booster (see Section 13 for more information).

22 If, upon checking the brake master cylinder fluid level, you discover the reservoir empty or nearly empty, the systems should be bled (see Chapter 9).

Windshield washer fluid

23 Fluid for the windshield washer system is stored in a plastic reservoir located at the right front of the engine compartment (see illustration).

24 In milder climates, plain water can be used in the reservoir, but it should be kept no more than 2/3 full to allow for expansion if the water freezes. In colder climates, use windshield washer system antifreeze, available at any auto parts store, to lower the freezing point of the fluid. Mix the antifreeze with water in accordance with the manufacturer's directions on the container.

Caution: *Do not use cooling system antifreeze - it will damage the vehicle's paint.*

Automatic transaxle fluid

Note: *It isn't necessary to check the transaxle fluid weekly; every 15,000 miles or 12 months is adequate (unless fluid leakage is suspected).*

25 The level of the automatic transaxle fluid should be carefully maintained. Low fluid level can lead to slipping or loss of drive, while overfilling can cause foaming, loss of fluid and transaxle damage.

26 The transaxle fluid level should only be checked when the transaxle is hot (at its normal operating temperature). If the vehicle has just been driven over 10 miles (15 miles in a frigid climate), and the fluid temperature is 160 to 175-degrees F, the transaxle is hot.

Caution: *If the vehicle has just been driven for a long time at high speed or in city traffic in hot weather, or if it has been pulling a trailer, an accurate fluid level reading cannot be obtained. Allow the fluid to cool down for about 30 minutes.*

27 If the vehicle has not just been driven, park the vehicle on level ground, set the parking brake and start the engine. While the engine is idling, depress the brake pedal and move the selector lever through all the gear ranges, beginning and ending in Park.

V6 models (6F50/6F55 transaxles)

Note: *To access the filler tube, the air filter housing must be removed to access the fluid filler cap/dipstick, then the intake duct must be reattached to the throttle body and the MAF sensor reconnected to run the engine for the fluid level check.*

28 Turn the engine off. Remove the air fil-

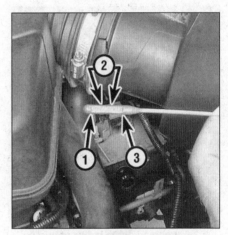

4.31 Transaxle fluid dipstick details (with fluid at normal operating temperature)

1 Low fluid level
2 Desired fluid level range
3 High fluid level

ter housing and reposition it sideways, then tighten the intake duct clamp (see illustration).

29 Restart the engine, move the shift lever through each range, then remove the filler cap/dipstick (see illustration).

30 Wipe the fluid from the dipstick with a clean rag and reinsert it back into the transaxle until the cap seats.

31 Pull the dipstick out again and note the fluid level (see illustration). The fluid level should be in the operating temperature range. If necessary, add the specified automatic transmission fluid with a funnel in small increments so as not to overfill the transaxle.

32 Add just enough of the recommended fluid to fill the transaxle to the proper level. It takes about one pint to raise the level from the low mark to the high mark when the fluid is hot, so add the fluid a little at a time and keep checking the level until it is correct.

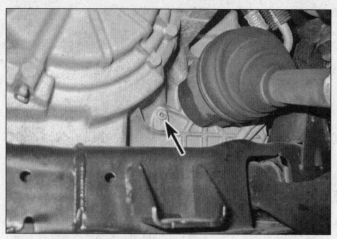

4.36 Location of the transaxle fluid leveling plug

4.37 Remove the cap from the transaxle fill tube to add transaxle fluid

5.2 A tire tread depth indicator should be used to monitor tire wear - they are available at auto parts stores and service stations and cost very little

33 The condition of the fluid should also be checked along with the level. If the fluid at the end of the dipstick is black or a dark reddish brown color, or if it emits a burned smell, the fluid should be changed (see Section 23). If you are in doubt about the condition of the fluid, purchase some new fluid and compare the two for color and smell.

Four-cylinder models (6F35 transaxle)

34 Remove the undervehicle splash shield.
35 Raise the vehicle on a hoist, keeping the vehicle in a level position, and place a drain pan under the transaxle. Alternatively, support the vehicle on jackstands in a level attitude.
36 Start the engine and allow it to idle. With your foot on the brake, move the shift lever through each range, returning it to Park. With the engine still idling, remove the oil leveling plug on the side of the transaxle (see illustration). Check the level of the fluid, it should be even with the bottom of the plug hole.
37 If the level is low, remove the filler tube cap (see illustration) and add the specified automatic transmission fluid through the filler tube with a funnel until the fluid starts to drip out of the oil leveling plug.
Note: *Allow all excess fluid to drip out of the plug hole.*
38 Once the fluid is even with the oil leveling plug hole, install the plug and tighten the plug to torque listed in this Chapter's Specifications.
39 The condition of the fluid should also be checked along with the level. If the fluid is black or a dark reddish brown color, or if it emits a burned smell, the fluid should be changed (see Section 23). If you are in doubt about the condition of the fluid, purchase some new fluid and compare the two for color and smell.

5   Tire and tire pressure checks (every 250 miles or weekly)

1 Periodic inspection of the tires may spare you the inconvenience of being stranded with a flat tire. It can also provide you with vital information regarding possible problems in the steering and suspension systems before major damage occurs.
2 The original tires on this vehicle are equipped with 1/2-inch wide bands that will appear when tread depth reaches 1/16-inch, at which point they can be considered worn out. Tread wear can be monitored with a simple, inexpensive device known as a tread depth indicator (see illustration).
3 Note any abnormal tread wear (see illustration). Tread pattern irregularities such as cupping, flat spots and more wear on one side than the other are indications of front end alignment and/or balance problems. If any of these conditions are noted, take the vehicle to a tire shop or service station to correct the problem.
4 Look closely for cuts, punctures and embedded nails or tacks. Sometimes a tire will hold air pressure for a short time or leak down very slowly after a nail has embedded itself in the tread. If a slow leak persists, check the valve stem core to make sure it is tight (see illustration). Examine the tread for an object that may have embedded itself in the tire or for a plug that may have begun to leak (radial tire punctures are repaired with a plug that is installed in a puncture). If a puncture is suspected, it can be easily verified by spraying a solution of soapy water onto the puncture area (see illustration). The soapy solution will bubble if there is a leak. Unless the puncture is unusually large, a tire shop or service station can usually repair the tire.
5 Carefully inspect the inner sidewall of each tire for evidence of brake fluid leakage. If you see any, inspect the brakes immediately.
6 Correct air pressure adds miles to the life span of the tires, improves mileage and enhances overall ride quality. Tire pressure cannot be accurately estimated by looking at a tire, especially if it's a radial. A tire pressure gauge is essential. Keep an accurate gauge in the glove compartment. The pressure gauges attached to the nozzles of air hoses at gas stations are often inaccurate.
7 Always check tire pressure when the tires are cold. Cold, in this case, means the vehicle has not been driven over a mile in the three hours preceding a tire pressure check. A pressure rise of four to eight pounds is not uncommon once the tires are warm.
8 Unscrew the valve cap protruding from the wheel or hubcap and push the gauge firmly onto the valve stem (see illustration). Note the reading on the gauge and compare the figure to the recommended tire pressure shown on the tire placard on the driver's side door. Be sure to reinstall the valve cap to keep dirt and moisture out of the valve stem mechanism. Check all four tires and, if necessary, add enough air to bring them up to the recommended pressure.
9 Don't forget to keep the spare tire inflated to the specified pressure (refer to the pressure molded into the tire sidewall).

Chapter 1 Tune-up and routine maintenance

UNDERINFLATION

CUPPING

Cupping may be caused by:
- Underinflation and/or mechanical irregularities such as out-of-balance condition of wheel and/or tire, and bent or damaged wheel.
- Loose or worn steering tie-rod or steering idler arm.
- Loose, damaged or worn front suspension parts.

OVERINFLATION

INCORRECT TOE-IN OR EXTREME CAMBER

FEATHERING DUE TO MISALIGNMENT

5.3 This chart will help you determine the condition of your tires, the probable cause(s) of abnormal wear and the corrective action necessary

5.4a If a tire loses air on a steady basis, check the valve core first to make sure it's snug (special inexpensive wrenches are commonly available at auto parts stores)

5.4b If the valve core is tight, raise the corner of the vehicle with the low tire and spray a soapy water solution onto the tread as the tire is turned slowly - slow leaks will cause small bubbles to appear

5.8 To extend the life of your tires, check the air pressure at least once a week with an accurate gauge (don't forget the spare!)

Chapter 1 Tune-up and routine maintenance

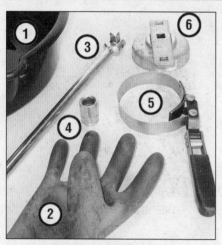

6.2 These tools are required when changing the engine oil and filter

1. **Drain pan** - It should be fairly shallow in depth, but wide to prevent spills
2. **Rubber gloves** - When removing the drain plug and filter, you will get oil on your hands (the gloves will prevent burns)
3. **Breaker bar** - Sometimes the oil drain plug is tight, and a long breaker bar is needed to loosen it
4. **Socket** – To be used with the breaker bar or a ratchet (must be the correct size to fit the drain plug - six-point preferred)
5. **Filter wrench** - This is a metal band-type wrench, which requires clearance around the filter to be effective
6. **Filter wrench** - This type fits on the bottom of the filter and can be turned with a ratchet or breaker bar (different-size wrenches are available for different types of filters)

6.6 Remove the under-vehicle air duct

6.7 Use a proper size box-end wrench or socket to remove the oil drain plug and avoid rounding it off

6 Engine oil and filter change (every 3000 miles or 3 months)

Note: Some models are equipped with an oil life indicator system that illuminates a light or message on the instrument panel when the system deems it necessary to change the oil. A number of factors are taken into consideration to determine when the oil should be considered worn out. Generally, this system will allow the vehicle to accumulate more miles between oil changes than the traditional 3000-mile interval, but we believe that frequent oil changes are cheap insurance and will prolong engine life. If you do decide not to change your oil every 3000 miles and rely on the oil life indicator instead, make sure you don't exceed 7,500 miles before the oil is changed, regardless of what the oil life indicator shows.

1   Frequent oil changes are the most important preventive maintenance procedures that can be done by the home mechanic. As engine oil ages, it becomes diluted and contaminated, which leads to premature engine wear.

2   Make sure that you have all the necessary tools before you begin this procedure (see illustration). You should also have plenty of rags or newspapers handy for mopping up oil spills.

3   Access to the oil drain plug and filter will be improved if the vehicle can be lifted on a hoist, driven onto ramps or supported by jackstands.

Warning: *Do not work under a vehicle supported only by a jack - always use jackstands!*

4   If you haven't changed the oil on this vehicle before, get under it and locate the oil drain plug and the oil filter. The exhaust components will be warm as you work, so note how they are routed to avoid touching them when you are under the vehicle.

5   Start the engine and allow it to reach normal operating temperature - oil and sludge will flow out more easily when warm. If new oil, a filter or tools are needed, use the vehicle to go get them and warm up the engine/oil at the same time. Park on a level surface and shut off the engine when it's warmed up. Remove the oil filler cap from the valve cover.

6   Raise the vehicle and support it on jackstands. Remove the under-vehicle air duct (see illustration).

7   Being careful not to touch the hot exhaust components, position a drain pan under the plug in the bottom of the engine, then remove the plug (see illustration). It's a good idea to wear a rubber glove while unscrewing the plug the final few turns to avoid being scalded by hot oil.

8   It may be necessary to move the drain pan slightly as oil flow slows to a trickle. Inspect the old oil for the presence of metal particles.

9   After all the oil has drained, wipe off the drain plug with a clean rag. Any small metal particles clinging to the plug would immediately contaminate the new oil.

10   Clean the area around the drain plug opening, reinstall the plug and tighten it to the torque listed in this Chapter's Specifications.

11   Move the drain pan into position under the oil filter.

12   Loosen the oil filter by turning it counter-clockwise with a filter wrench (see illustration). Any standard filter wrench will work.

13   Once the filter is loose, use your hands to unscrew it from the block. Just as the filter is detached from the block, immediately tilt the open end up to prevent the oil inside the filter from spilling out.

14   Using a clean rag, wipe off the mounting surface on the block. Also, make sure that none of the old gasket remains stuck to the mounting surface. It can be removed with a scraper if necessary.

15   Compare the old filter with the new one to make sure they are the same type. Smear some engine oil on the rubber gasket or O-ring seal of the new filter and screw it into place (see illustration). On typical filters, over-tightening the filter will damage the gasket, so don't use a filter wrench. Most filter manufac-

6.12 Use an oil filter wrench to remove the filter (four-cylinder engine shown)

Chapter 1 Tune-up and routine maintenance

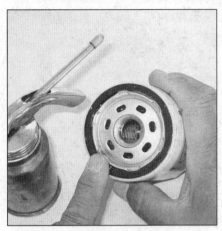

6.15 Lubricate the oil filter gasket with clean engine oil before installing the filter on the engine

7.4a To release the blade holder, squeeze the release tabs...

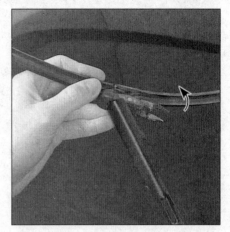

7.4b ... then rotate the wiper blade in the direction of the arrow to separate it from the arm

turers recommend tightening the filter by hand only. Normally they should be tightened 3/4-turn after the gasket contacts the block, but be sure to follow the directions on the filter or container. On cartridge type filters, install the new oil filter element and use the oil filter wrench to tighten the filter cover securely. Do not overtighten the oil filter cover.

16 Remove all tools and materials from under the vehicle, being careful not to spill the oil in the drain pan, then lower the vehicle.

17 Add new oil to the engine through the oil filler cap. Use a funnel to prevent oil from spilling onto the top of the engine. Pour four quarts of fresh oil into the engine. Wait a few minutes to allow the oil to drain into the pan, then check the level on the dipstick (see Section 4 if necessary). If the oil level is in the OK range, install the filler cap.

18 Start the engine and run it for about a minute. While the engine is running, look under the vehicle and check for leaks at the oil pan drain plug and around the oil filter. If either one is leaking, stop the engine and tighten the plug or filter slightly.

19 Wait a few minutes, then recheck the level on the dipstick. Add oil as necessary to bring the level into the OK range.

20 During the first few trips after an oil change, make it a point to check frequently for leaks and proper oil level.

21 The old oil drained from the engine cannot be reused in its present state and should be disposed of. Check with your local auto parts store, disposal facility or environmental agency to see if they will accept the oil for recycling. After the oil has cooled it can be drained into a container (capped plastic jugs, topped bottles, milk cartons, etc.) for transport to one of these disposal sites. Don't dispose of the oil by pouring it on the ground or down a drain!

Oil life monitor resetting

Note: *Not all models are equipped with an oil life monitor.*

22 Press the SETUP button to display "OIL LIFE XXX% HOLD RESET=NEW."

23 Press and hold the RESET button for two seconds, then release it. The indicator should now read "OIL LIFE SET TO 100%," which is approximately a 7,500-mile interval. This interval can be shortened by pressing and releasing the RESET button; each push of the button reduces the interval by 10-percent.

7 Wiper blade inspection and replacement (every 7500 miles or 6 months)

1 The windshield and rear wiper and blade assemblies should be inspected periodically for damage, loose components and cracked or worn blade elements.

2 Road film can build up on the wiper blades and affect their efficiency, so they should be washed regularly with a mild detergent solution.

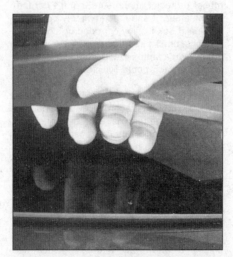

7.6 Pry the wiper blade straight out of the arm

3 If the wiper blade elements are cracked, worn or warped, or no longer clean adequately, they should be replaced with new ones.

Windshield wiper blades

4 Lift the arm assembly away from the glass for clearance, squeeze the locking tabs, then rotate the blade away from the arm to detach it (see illustrations).

5 Attach the new wiper to the arm. Connection can be confirmed by an audible click.

Rear wiper blade

6 Lift the wiper arm away from the glass, then use a plastic trim tool to pry the blade from the arm (see illustration).

7 To install, align the slot in the wiper blade with the pin in the wiper arm, then push until it clicks into place (see illustration).

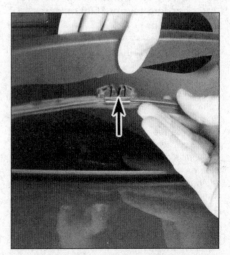

7.7 Align the opening in the blade with the pin in the arm, then push it into place

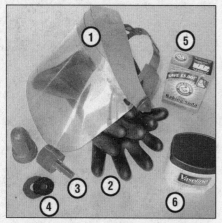

8.1 Tools and materials required for battery maintenance

1. **Face shield/safety goggles** - When removing corrosion with a brush, the acidic particles can easily fly up into your eyes
2. **Rubber gloves** - Another safety item to consider when servicing the battery; remember that's acid inside the battery!
3. **Battery terminal/cable cleaner** - This wire brush cleaning tool will remove all traces of corrosion from the battery and cable
4. **Treated felt washers** - Placing one of these on each terminal, directly under the cable end, will help prevent corrosion (be sure to get the correct type for side terminal batteries)
5. **Baking soda** - A solution of baking soda and water can be used to neutralize corrosion
6. **Petroleum jelly** - A layer of this on the battery terminal bolts will help prevent corrosion

8 Battery check, maintenance and charging (every 7500 miles or 6 months)

Warning: *Certain precautions must be followed when checking and servicing the battery. Hydrogen gas, which is highly flammable, is always present in the battery cells, so keep lighted tobacco and all other open flames and sparks away from the battery. The electrolyte inside the battery is actually diluted sulfuric acid, which will cause injury if splashed on your skin or in your eyes. It will also ruin clothes and painted surfaces. When removing the battery cables, always detach the negative cable first and hook it up last!*

1   A routine preventive maintenance program for the battery in your vehicle is the only way to ensure quick and reliable starts. But before performing any battery maintenance, make sure that you have the proper equipment necessary to work safely around the battery (see illustration).

8.6a Battery terminal corrosion usually appears as light, fluffy powder

2   There are also several precautions that should be taken whenever battery maintenance is performed. Before servicing the battery, always turn the engine and all accessories off and disconnect the cables from the negative terminal of the battery (see Chapter 5).
3   The battery produces hydrogen gas, which is both flammable and explosive. Never create a spark, smoke or light a match around the battery. Always charge the battery in a ventilated area.
4   Electrolyte contains poisonous and corrosive sulfuric acid. Do not allow it to get in your eyes, on your skin on your clothes. Never ingest it. Wear protective safety glasses when working near the battery. Keep children away from the battery.
5   Note the external condition of the battery. If the positive terminal and cable clamp on your vehicle's battery is equipped with a rubber protector, make sure that it's not torn or damaged. It should completely cover the terminal. Look for any corroded or loose connections, cracks in the case or cover or loose hold-down clamps. Also check the entire length of each cable for cracks and frayed conductors.
6   If corrosion, which looks like white, fluffy deposits (see illustration) is evident, particularly around the terminals, the battery should be removed for cleaning. Loosen the cable clamp bolts with a wrench, being careful to remove the ground cable first, and slide them off the terminals (see illustration). Then disconnect the hold-down clamp bolt and nut, remove the clamp and lift the battery from the engine compartment.
7   Clean the cable clamps thoroughly with a battery brush or a terminal cleaner and a solution of warm water and baking soda (see illustration). Wash the terminals and the top of the battery case with the same solution but make sure that the solution doesn't get into

8.6b Removing a cable from the battery post with a wrench - sometimes a pair of special battery pliers are required for this procedure if corrosion has caused deterioration of the nut hex. Always remove the ground (-) cable first and hook it up last!

the battery. When cleaning the cables, terminals and battery top, wear safety goggles and rubber gloves to prevent any solution from coming in contact with your eyes or hands. Wear old clothes too - even diluted, sulfuric acid splashed onto clothes will burn holes in them. If the terminals have been extensively corroded, clean them up with a terminal cleaner (see illustration). Thoroughly wash all cleaned areas with plain water.
8   Make sure that the battery tray is in good condition and the hold-down clamp fasteners are tight (see illustration). If the battery is removed from the tray, make sure no parts remain in the bottom of the tray when the battery is reinstalled. When reinstalling the hold-down clamp bolts, do not overtighten them.
9   Information on removing and installing the battery can be found in Chapter 5. If you disconnected the cable(s) from the negative and/or positive battery terminals, the powertrain control module (PCM) must relearn its idle and fuel trim strategy for optimum driveability and performance (see Chapter 5 for this procedure). Information on jump starting can be found at the front of this manual. For more detailed battery checking procedures, refer to the *Haynes Automotive Electrical Manual*.

Cleaning

10   Corrosion on the hold-down components, battery case and surrounding areas can be removed with a solution of water and baking soda. Thoroughly rinse all cleaned areas with plain water.
11   Any metal parts of the vehicle damaged by corrosion should be covered with a zinc-based primer, then painted.

Charging

Warning: *When batteries are being charged, hydrogen gas, which is very explosive and flammable, is produced. Do not smoke or al-*

Chapter 1 Tune-up and routine maintenance

8.7a When cleaning the cable clamps, all corrosion must be removed

8.7b Regardless of the type of tool used to clean the battery posts, a clean, shiny surface should be the result

8.8 Make sure the battery hold-down fasteners are tight

low open flames near a charging or a recently charged battery. Wear eye protection when near the battery during charging. Also, make sure the charger is unplugged before connecting or disconnecting the battery from the charger.

12 Slow-rate charging is the best way to restore a battery that's discharged to the point where it will not start the engine. It's also a good way to maintain the battery charge in a vehicle that's only driven a few miles between starts. Maintaining the battery charge is particularly important in the winter when the battery must work harder to start the engine and electrical accessories that drain the battery are in greater use.

13 It's best to use a one or two-amp battery charger (sometimes called a trickle charger). They are the safest and put the least strain on the battery. They are also the least expensive. For a faster charge, you can use a higher amperage charger, but don't use one rated more than 1/10th the amp/hour rating of the battery. Rapid boost charges that claim to restore the power of the battery in one to two hours are hardest on the battery and can damage batteries not in good condition. This type of charging should only be used in emergency situations.

14 The average time necessary to charge a battery should be listed in the instructions that come with the charger. As a general rule, a trickle charger will charge a battery in 12 to 16 hours.

9 Cooling system check (every 7,500 miles or 6 months)

1 Many major engine failures can be caused by a faulty cooling system.
2 The engine must be cold for the cooling system check, so perform the following procedure before the vehicle is driven for the day or after it has been shut off for at least three hours.

3 Remove the pressure cap from the expansion tank at the right side of the engine compartment. Clean the cap thoroughly, inside and out, with clean water. The presence of rust or corrosion in the expansion tank means the coolant should be changed (see Section 20). The coolant inside the expansion tank should be relatively clean and transparent. If it's rust colored, drain the system and refill it with new coolant.
4 Carefully check the radiator hoses and the smaller diameter heater hoses (see illustrations in Chapter 3). Inspect each coolant hose along its entire length, replacing any hose which is cracked, swollen or deteriorated (see illustration). Cracks will show up better if the hose is squeezed. Pay close attention to hose clamps that secure the hoses to cooling system components. Hose clamps can pinch and puncture hoses, resulting in coolant leaks.
5 Make sure that all hose connections are tight. A leak in the cooling system will usually show up as white or rust colored deposits on the area adjoining the leak. If wire-type clamps are used on the hoses, it may be a good idea to replace them with screw-type clamps.
6 Clean the front of the radiator and air conditioning condenser with compressed air, if available, or a soft brush. Remove all bugs, leaves, etc. embedded in the radiator fins. Be extremely careful not to damage the cooling fins or cut your fingers on them.
7 If the coolant level has been dropping consistently and no leaks are detectable, have the expansion tank cap and cooling system pressure checked at a service station.

10 Tire rotation (every 7,500 miles or 6 months)

1 The tires should be rotated at the specified intervals and whenever uneven wear is noticed. Since the vehicle will be raised and the tires removed anyway, check the brakes also (see Section 13).

Check for a chafed area that could fail prematurely.

Check for a soft area indicating the hose has deteriorated inside.

Overtightening the clamp on a hardened hose will damage the hose and cause a leak.

Check each hose for swelling and oil-soaked ends. Cracks and breaks can be located by squeezing the hose

9.4 Hoses, like drivebelts, have a habit of failing at the worst possible time - to prevent the inconvenience of a blown radiator or heater hose, inspect them carefully as shown here

2 Radial tires must be rotated in a specific pattern (see illustration). Don't include the spare tire in the rotation pattern.
3 Refer to the information in *Jacking and towing* at the front of this manual for the proper procedure to follow when raising the vehicle and changing a tire. If the brakes must be checked, don't apply the parking brake as stated.
4 The vehicle must be raised on a hoist or supported on jackstands to get all four wheels off the ground. Make sure the vehicle is safely supported!
5 After the rotation procedure is finished, check and adjust the tire pressures as necessary and be sure to check the lug nut tightness.

11 Seat belt check (every 7,500 miles or 6 months)

1 Check seat belts, buckles, latch plates and guide loops for obvious damage and signs of wear.
2 See if the seat belt reminder light comes on when the key is turned to the Run or Start position. A chime should also sound.
3 The seat belts are designed to lock up during a sudden stop or impact, yet allow free movement during normal driving. Make sure the retractors return the belt against your chest while driving and rewind the belt fully when the buckle is unlatched.
4 If any of the above checks reveal problems with the seat belt system, replace parts as necessary.

12 Underhood hose check and replacement (every 15,000 miles or 12 months)

Warning: *Replacement of air conditioning hoses must be left to a dealer service department or air conditioning shop that has the equipment to depressurize the system safely. Never remove air conditioning components or hoses until the system has been depressurized.*

General

1 High temperatures under the hood can cause deterioration of the rubber and plastic hoses used for engine, accessory and emission systems operation. Periodic inspection should be made for cracks, loose clamps, material hardening and leaks.
2 Information specific to the cooling system hoses can be found in Section 9.
3 Most (but not all) hoses are secured to the fittings with clamps. Where clamps are used, check to be sure they haven't lost their tension, allowing the hose to leak. If clamps aren't used, make sure the hose has not expanded and/or hardened where it slips over the fitting, allowing it to leak.

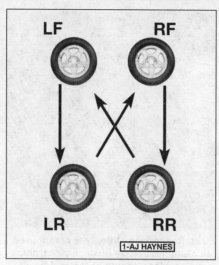

10.2 The recommended tire rotation pattern for these vehicles (FWD and AWD)

PCV system hose

4 To reduce hydrocarbon emissions, crankcase blow-by gas is vented through the PCV valve in the valve cover to the intake manifold via a rubber hose on most models. The blow-by gases mix with incoming air in the intake manifold before being burned in the combustion chambers.
5 Check the PCV hose for cracks, leaks and other damage. Remove the intake manifold (see Chapter 2A) and disconnect it from the valve cover then check the inside for obstructions. If it's clogged, clean it out with solvent.

Vacuum hoses

6 It's quite common for vacuum hoses, especially those in the emissions system, to be color coded or identified by colored stripes molded into them. Various systems require hoses with different wall thickness, collapse resistance and temperature resistance. When replacing hoses, be sure the new ones are made of the same material.
7 Often the only effective way to check a hose is to remove it completely from the vehicle. If more than one hose is removed, be sure to label the hoses and fittings to ensure correct installation.
8 When checking vacuum hoses, be sure to include any plastic T-fittings in the check. Inspect the fittings for cracks and the hose where it fits over each fitting for distortion, which could cause leakage.
9 A small piece of vacuum hose (1/4-inch inside diameter) can be used as a stethoscope to detect vacuum leaks. Hold one end of the hose to your ear and probe around vacuum hoses and fittings, listening for the hissing sound characteristic of a vacuum leak.
Warning: *When probing with the vacuum hose stethoscope, be careful not to come into contact with moving engine components such as drivebelts, the cooling fan, etc.*

Fuel hose

Warning: *Gasoline is flammable, so take extra precautions when you work on any part of the fuel system. Don't smoke or allow open flames or bare light bulbs near the work area, and don't work in a garage where a gas-type appliance (such as a water heater or clothes dryer) is present. Since fuel is carcinogenic, wear fuel-resistant gloves when there's a possibility of being exposed to fuel, and, if you spill any fuel on your skin, rinse it off immediately with soap and water. Mop up any spills immediately and do not store fuel-soaked rags where they could ignite. The fuel system is under constant pressure, so, if any fuel lines are to be disconnected, the fuel pressure in the system must be relieved first (see Chapter 4 for more information). When you perform any kind of work on the fuel system, wear safety glasses and have a Class B type fire extinguisher on hand.*

10 The fuel lines are usually under pressure, so if any fuel lines are to be disconnected be prepared to catch spilled fuel.
Warning: *Your vehicle is equipped with fuel injection and you must relieve the fuel system pressure before servicing the fuel lines. Refer to Chapter 4 for the fuel system pressure relief procedure.*
11 Check all flexible fuel lines for deterioration and chafing. Check especially for cracks in areas where the hose bends and just before fittings, such as where the fuel line attaches to the fuel rail.
12 When replacing a hose, use only hose that is specifically designed for your fuel injection system.
13 Some fuel lines use quick-connect fittings, which require a special tool to disconnect. See Chapter 4 for more information on these types of fittings.

Metal lines

14 Sections of metal line are often used for fuel line that runs underneath the vehicle. Check carefully to make sure the line isn't bent, crimped or cracked.
15 If a section of metal fuel line must be replaced, use seamless steel tubing only, since copper and aluminum tubing do not have the strength necessary to withstand vibration caused by the engine.
16 Check the metal brake lines where they enter the master cylinder and brake proportioning unit (if used) for cracks in the lines and loose fittings. Any sign of brake fluid leakage calls for an immediate thorough inspection of the brake system.

13 Brake check (every 15,000 miles or 12 months)

Warning: *Dust created by the brake system is harmful to your health. Never blow it out with compressed air and don't inhale any of it. An approved filtering mask should be worn when working on brakes. Do not, under any circumstances, use petroleum-based solvents to clean brake parts. Use brake system cleaner only!*

Chapter 1 Tune-up and routine maintenance

13.5a You will find an inspection hole like this in each caliper through which you can view the thickness of remaining friction material for the inner pad

13.5b Use a small ruler to check and measure the thickness of the pad materials, too

1   The brakes should be inspected every time the wheels are removed or whenever a defect is suspected. Indications of a potential brake system problem include the vehicle pulling to one side when the brake pedal is depressed, noises coming from the brakes when they are applied, excessive brake pedal travel, a pulsating pedal and leakage of fluid, usually seen on the inside of the tire or wheel.
Note: *It is normal for a vehicle equipped with an Anti-lock Brake System (ABS) to exhibit brake pedal pulsations during severe braking conditions.*

Disc brakes

2   Disc brakes can be visually checked without removing any parts except the wheels. Remove the hub caps (if applicable) and loosen the wheel lug nuts a quarter turn each.
3   Raise the vehicle and place it securely on jackstands.
Warning: *Never work under a vehicle that is supported only by a jack!*
4   Remove the wheels. Now visible is the disc brake caliper which contains the pads. There is an outer brake pad and an inner pad. Both must be checked for wear.
Note: *Usually the inner pad wears faster than the outer pad.*
5   Measure the thickness of the outer pad at each end of the caliper and the inner pad through the inspection hole in the caliper body (see illustrations). Compare the measurement with the limit given in this Chapter's Specifications; if any brake pad thickness is less than specified, then all brake pads must be replaced (see Chapter 9).
6   If you're in doubt as to the exact pad thickness or quality, remove them for measurement and further inspection (see Chapter 9).
7   Check the disc for score marks, wear and burned spots. If any of these conditions exist, the disc should be removed for servic-

ing or replacement (see Chapter 9).
8   Before installing the wheels, check all the brake lines and hoses for damage, wear, deformation, cracks, corrosion, leakage, bends and twists, particularly in the vicinity of the rubber hoses and calipers.
9   Install the wheels, lower the vehicle and tighten the wheel lug nuts to the torque given in this Chapter's Specifications.

Parking brake

10   Park the vehicle on a steep hill with the engine running (so you can apply the brakes if necessary) with the parking brake set and the transaxle in Neutral. If the parking brake cannot prevent the vehicle from rolling, it needs adjustment (see Chapter 9).

---

14   Steering, suspension and driveaxle boot check (every 15,000 miles or 12 months)

Note: *For detailed illustrations of the steering and suspension components, refer to Chapter 10.*

Shock absorber check

1   Park the vehicle on level ground, turn the engine off and set the parking brake. Check the tire pressures.
2   Push down at one corner of the vehicle, then release it while noting the movement of the body. It should stop moving and come to rest in a level position within one or two bounces.
3   If the vehicle continues to move up-and-down or if it fails to return to its original position, a worn or weak shock absorber is probably the reason.
4   Repeat the above check at each of the three remaining corners of the vehicle.
5   Raise the vehicle and support it securely on jackstands.

14.6 Check the shocks for leakage at the indicated area

6   Check the shock absorbers for evidence of fluid leakage (see illustration). A light film of fluid is no cause for concern. Make sure that any fluid noted is from the shocks and not from some other source. If leakage is noted, replace the shocks as a set.
7   Check the shocks to be sure that they are securely mounted and undamaged. Check the upper mounts for damage and wear. If damage or wear is noted, replace the shocks as a set (front or rear).
8   If the shocks must be replaced, refer to Chapter 10 for the procedure.

Steering and suspension check

9   Check the tires for irregular wear patterns and proper inflation. See Section 5 in this Chapter for information regarding tire wear and Chapter 10 for information on wheel bearing replacement.
10   Inspect the universal joint between the steering shaft and the steering gear housing. Check the steering gear housing for lubricant

14.11 To check a balljoint for wear, try to pry the control arm up and down to make sure there is no play in the balljoint (if there is, replace it)

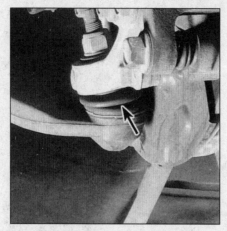

14.12 Check the balljoint boots for damage

14.15 Flex the driveaxle boots by hand to check for cracks and/or leaking grease

leakage. Make sure that the dust boots are not damaged and that the boot clamps are not loose. Check the tie-rod ends for excessive play. Look for loose bolts, broken or disconnected parts and deteriorated rubber bushings on all suspension and steering components. While an assistant turns the steering wheel from side to side, check the steering components for free movement, chafing and binding. If the steering components do not seem to be reacting with the movement of the steering wheel, try to determine where the slack is located.

11   Check the balljoints for wear by trying to move each control arm up and down with a prybar (see illustration) to ensure that its balljoint has no play. If any balljoint does have play, it's worn out. See Chapter 10 for the control arm replacement procedure (the balljoints aren't replaceable separately).

12   Inspect the balljoint boots for damage and leaking grease (see illustration).

13   At the rear of the vehicle, inspect the suspension arm bushings for deterioration.

Additional information on suspension components can be found in Chapter 10.

Driveaxle boot check

14   The driveaxle boots are very important because they prevent dirt, water and foreign material from entering and damaging the constant velocity (CV) joints. Because it constantly pivots back and forth following the steering action of the front hub, the outer CV boot wears out sooner and should be inspected regularly.

15   Inspect the boots for tears and cracks as well as loose clamps (see illustration). If there is any evidence of cracks or leaking lubricant, they must be replaced as described in Chapter 8.

15   Fuel system check (every 15,000 miles or 12 months)

Warning: *Gasoline is flammable, so take extra precautions when you work on any part of the fuel system. Don't smoke or allow open flames or bare light bulbs near the work area, and don't work in a garage where a gas-type appliance (such as a water heater or clothes dryer) is present. Since fuel is carcinogenic, wear fuel-resistant gloves when there's a possibility of being exposed to fuel, and, if you spill any fuel on your skin, rinse it off immediately with soap and water. Mop up any spills immediately and do not store fuel-soaked rags where they could ignite. When you perform any kind of work on the fuel system, wear safety glasses and have a Class B type fire extinguisher on hand. The fuel system is under constant pressure, so, before any lines are disconnected, the fuel system pressure must be relieved (see Chapter 4).*

1   If you smell gasoline while driving or after the vehicle has been sitting in the sun, inspect the fuel system immediately.

2   Remove the fuel filler cap and inspect it for damage and corrosion. The gasket should have an unbroken sealing imprint. If the gasket is damaged or corroded, install a new cap.

3   Inspect the fuel feed line for cracks. Make sure that the connections between the fuel lines and the fuel injection system, and between the fuel lines and the fuel tank (inspect from below) are tight and dry.

Warning: *Your vehicle is fuel injected, so you must relieve the fuel system pressure before servicing fuel system components. The fuel system pressure relief procedure is outlined in Chapter 4.*

4   Since some components of the fuel system - the fuel tank and part of the fuel feed line, for example - are underneath the vehicle, they can be inspected more easily with the vehicle raised on a hoist. If that's not possible, raise the vehicle and support it on jackstands.

5   With the vehicle raised and safely supported, inspect the gas tank and filler neck for punctures, cracks and other damage. The connection between the filler neck and the tank is particularly critical. Sometimes a rubber filler neck will leak because of loose clamps or deteriorated rubber. Inspect all fuel tank mounting brackets and straps to be sure that the tank is securely attached to the vehicle.

Warning: *Do not, under any circumstances, try to repair a fuel tank (except rubber components). A welding torch or any open flame can easily cause fuel vapors inside the tank to explode.*

6   Carefully check all rubber hoses and metal lines leading away from the fuel tank. Check for loose connections, deteriorated hoses, crimped lines and other damage. Repair or replace damaged sections as necessary (see Chapter 4).

16   Transfer case lubricant level check (AWD models)

Note: *Do not service the transfer case fluid unless the vehicle is used in extreme off-road conditions or in sand.*

1   Raise the vehicle and support it securely on jackstands. Remove the under-vehicle splash shield, if equipped.

2   Unscrew the check/fill plug from the right side of the transfer case.

3   Using your little finger, reach inside the housing to feel the lubricant level. The level should be at or near the bottom of the plug hole. If it isn't, add the recommended lubricant through the plug hole with a syringe or squeeze bottle.

4   Install and tighten the plug. Check for leaks after the first few miles of driving.

Note: *The manufacturer states that if the transfer case is ever submerged in water, the entire unit must be replaced (see Chapter 7B). However, it may be worth trying to gain more service life from the unit by draining the lubricant and refilling the unit with fresh lubricant (see Section 25).*

Chapter 1 Tune-up and routine maintenance 1-21

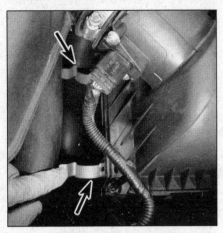

17.1a Unlatch these clips . . .

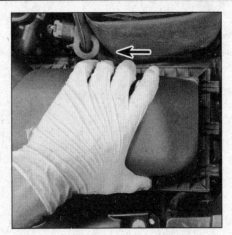

17.1b ... slide the cover to release the locking tabs from the housing ...

17.1c ... then pull the cover out of the way and lift the element out

17 Air filter check and replacement (every 15,000 miles or 12 months)

1 The air filter is located inside a housing at the left (driver's) side of the engine compartment. To remove the air filter, release the clamps that secure the two halves of the air filter housing together, then slide the cover from the housing to separate the cover halves and remove the air filter element (see illustrations).

2 Inspect the outer surface of the filter element. If it is dirty, replace it. If it is only moderately dusty, it can be reused by blowing it clean from the back to the front surface with compressed air. Because it is a pleated paper type filter, it cannot be washed or oiled. If it cannot be cleaned satisfactorily with compressed air, discard and replace it. While the cover is off, be careful not to drop anything down into the housing.

Caution: *Never drive the vehicle with the air filter removed. Excessive engine wear could result and backfiring could even cause a fire under the hood.*

3 Wipe out the inside of the air filter housing.
4 Place the new filter into the filter housing, making sure it seats properly.
5 Make sure the cover is seated properly, then secure it with the clamps.

18 Cabin air filter replacement (every 15,000 miles or 12 months)

Note: *Not all models were equipped from the factory with a cabin air filter. If a filter was not installed, a screen was installed in its place. Check with your local auto parts store or dealer parts department for a retrofit filter available for these models.*

1 Unlatch the glove box door stop from the top, and lower the door.
2 Pull back on the retaining tabs of the filter door and remove it, then pull the filter from the housing (see illustrations).
3 Installation is the reverse of removal.

19 Exhaust system check (every 30,000 miles or 24 months)

1 With the engine cold (at least three hours after the vehicle has been driven), check the complete exhaust system from the engine to the end of the tailpipe. Ideally, the inspection should be done with the vehicle on a hoist to permit unrestricted access. If a hoist isn't available, raise the vehicle and support it securely on jackstands.

2 Check the exhaust pipes and connections for evidence of leaks, severe corrosion and damage. Make sure that all brackets and hangers are in good condition and tight (see illustration).

3 At the same time, inspect the underside of the body for holes, corrosion, open seams, etc. which may allow exhaust gases to enter the passenger compartment. Seal all body openings with silicone or body putty.

4 Rattles and other noises can often be traced to the exhaust system, especially the mounts and hangers. Try to move the pipes,

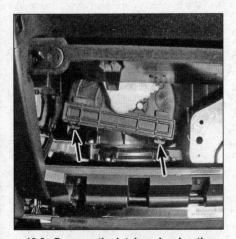

18.2a Depress the latch and swing the access panel open . . .

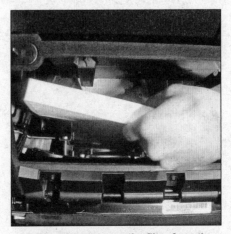

18.2b ... then remove the filter from the housing. When installing, make note of any airflow directional marks

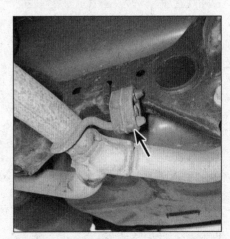

19.2 Check each exhaust system rubber hanger for damage

20.4 The radiator drain fitting is located at the lower right corner of the radiator - before opening the valve, it is helpful to push a short length of rubber hose onto the plastic fitting to prevent the coolant from splashing

muffler and catalytic converter. If the components can come in contact with the body or suspension parts, secure the exhaust system with new mounts.

5 Check the running condition of the engine by inspecting inside the end of the tailpipe. The exhaust deposits here are an indication of engine state-of-tune. If the pipe is black and sooty or coated with white deposits, the engine may need a tune-up, including a thorough fuel system inspection and adjustment.

20 Cooling system servicing (draining, flushing and refilling) (every 30,000 miles or 24 months)

Warning: *Do not allow antifreeze to come in contact with your skin or painted surfaces of the vehicle. Rinse off spills immediately with plenty of water. Antifreeze is highly toxic if ingested. Never leave antifreeze lying around in an open container or in puddles on the floor; children and pets are attracted by its sweet smell and may drink it. Check with local authorities about disposing of used antifreeze. Many communities have collection centers which will see that antifreeze is disposed of safely. Never dump used antifreeze on the ground or pour it into drains.*
Caution: *Do not mix coolants of different colors. Doing so might damage the cooling system and/or the engine. The manufacturer specifies either a green colored coolant or an orange colored coolant to be used in these systems. Read the warning label in the engine compartment for additional information.*
Note: *Non-toxic antifreeze is now manufactured and available at local auto parts stores, but even this type must be disposed of properly.*

1 Periodically, the cooling system should be drained, flushed and refilled to replenish the antifreeze mixture and prevent formation of rust and corrosion, which can impair the performance of the cooling system and cause engine damage. When the cooling system is serviced, all hoses and the expansion tank cap should be checked and replaced if necessary.

Draining

Warning: *The engine must be completely cool before beginning this procedure.*
2 Apply the parking brake and block the rear wheels, then raise the front of the vehicle and support it securely on jackstands. If the vehicle has just been driven, wait several hours to allow the engine to cool down before beginning this procedure.
3 Once the engine is completely cool, remove the expansion tank cap.
4 Move a large container under the radiator drain to catch the coolant, then open the drain fitting (a pair of pliers may be required to turn it) (see illustration).
5 While the coolant is draining, check the condition of the radiator hoses, heater hoses and clamps (refer to Section 9 if necessary). Replace any damaged clamps or hoses. When the system is finished draining, close the drain valve.

Flushing

6 Fill the cooling system with clean water, following the *Refilling* procedure (see Step 13).
7 Start the engine and allow it to reach normal operating temperature, then rev up the engine a few times.
8 Turn the engine off and allow it to cool completely, then drain the system as described earlier.
9 Repeat Steps 6 through 8 until the water being drained is free of contaminants.
10 In severe cases of contamination or clogging of the radiator, remove the radiator (see Chapter 3) and have a radiator repair facility clean and repair it if necessary.
11 Many deposits can be removed by the chemical action of a cleaner available at auto parts stores. Follow the procedure outlined in the manufacturer's instructions.
Note: *When the coolant is regularly drained and the system refilled with the correct antifreeze/water mixture, there should be no need to use chemical cleaners or descalers.*

Refilling

12 Close and tighten the radiator drain.
13 Place the heater temperature control in the maximum heat position.
14 Slowly, add coolant to the expansion tank until the level is at the MAX fill mark on the expansion tank.
15 Install the expansion tank cap and run the engine at 2500 rpm for ten minutes.
Caution: *If at any time the engine begins to overheat, or the coolant level falls below the MIN fill line on the expansion tank, turn off the engine, allow it to cool completely, then add* coolant to the expansion tank to the MAX fill line.
16 Turn the engine off and let it cool. Add more coolant mixture to bring the level to the MAX fill mark on the expansion tank.
17 Repeat Steps 14 through 16 if necessary.
18 Start the engine, allow it to reach normal operating temperature and check for leaks. Also, set the heater and blower controls to the maximum setting and check to see that the heater output from the air ducts is warm. This is a good indication that all air has been purged from the cooling system.

21 Brake fluid change (every 30,000 miles or 24 months)

Warning: *Brake fluid can harm your eyes and damage painted surfaces, so use extreme caution when handling or pouring it. Do not use brake fluid that has been standing open or is more than one year old. Brake fluid absorbs moisture from the air. Excess moisture can cause a dangerous loss of braking effectiveness.*

1 At the specified intervals, the brake fluid should be drained and replaced. Since the brake fluid may drip or splash when pouring it, place plenty of rags around the master cylinder to protect any surrounding painted surfaces.
2 Before beginning work, purchase the specified brake fluid (see *Recommended lubricants and fluids* in this Chapter's Specifications).
3 Remove the cap from the master cylinder reservoir.
4 Using a hand suction pump or similar device, withdraw the fluid from the master cylinder reservoir.
5 Add new fluid to the master cylinder until it rises to the base of the filler neck.
6 Bleed the brake system (see Chapter 9) at all four brakes until new and uncontaminated fluid is expelled from the bleeder screw. Be sure to maintain the fluid level in the master cylinder as you perform the bleeding process. If you allow the master cylinder to run dry, air will enter the system.
7 Refill the master cylinder with fluid and check the operation of the brakes. The pedal should feel solid when depressed, with no sponginess.
Warning: *Do not operate the vehicle if you are in doubt about the effectiveness of the brake system.*

22 Drivebelt check and replacement/ tensioner replacement (every 30,000 miles or 24 months)

Drivebelt

1 A single serpentine drivebelt is located at the front of the engine and plays an important role in the overall operation of the engine

Chapter 1 Tune-up and routine maintenance

1-23

22.2a Remove the fasteners from the front area of the inner fender liner...

22.2b ... then remove the remaining fasteners and remove the splash shield

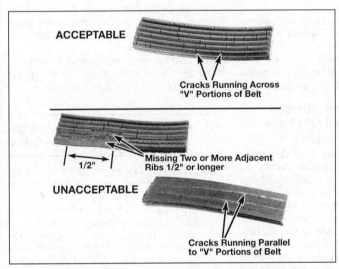

22.4 Small cracks in the underside of a V-ribbed belt are acceptable - lengthwise cracks, or missing pieces that cause the belt to make noise, are cause for replacement

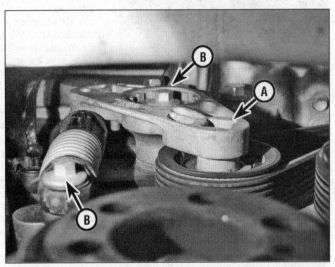

22.7a Insert a ratchet or breaker bar into the square hole (A) and rotate the tensioner arm counterclockwise to relieve belt tension. (B) indicates the tensioner mounting bolts (early build four-cylinder model shown)

and its components. Due to its function and material make up, the belt is prone to wear and should be periodically inspected. The serpentine belt drives the alternator, water pump (four-cylinder models only) and air conditioning compressor. Although the belt should be inspected at the recommended intervals, replacement may not be necessary for more than 100,000 miles.

Check

2 Since the drivebelt is located very close to the right-hand side of the engine compartment, it is possible to gain better access by raising the front of the vehicle and removing the right-hand wheel, then removing the splash shield in the right fenderwell (see illustrations). Be sure to support the front of the vehicle securely on jackstands.

3 With the engine stopped, inspect the full length of the drivebelt for cracks and separation of the belt plies. It will be necessary to turn the engine (using a wrench or socket and bar on the crankshaft pulley bolt, working clockwise only) in order to move the belt from the pulleys so that the belt can be inspected thoroughly. Twist the belt between the pulleys so that both sides can be viewed. Also check for fraying, and glazing which gives the belt a shiny appearance. Check the pulleys for nicks, cracks, distortion and corrosion.

4 Note that it is not unusual for a ribbed belt to exhibit small cracks in the edges of the belt ribs, and unless these are extensive or very deep, belt replacement is not essential (see illustration).

Replacement - main drivebelt

5 Disconnect the cable from the negative terminal of the battery (see Chapter 5). Loosen the right front wheel lug nuts, then raise the front of the vehicle and support it securely on jackstands. Remove the right front wheel and the splash shield (see illustration 22.2).

6 On late-build 2013 and later four-cylinder models, remove the air conditioning compressor drivebelt (see Step 13).

7 Note how the drivebelt is routed, then remove the belt from the pulleys. On all V6 and 2011 through early-build 2013 four-cylinder models, insert a ratchet or breaker bar into the tensioner hole and pull the handle (counterclockwise on four-cylinder models, clockwise on V6 models) to release the drivebelt tension (see illustrations). Once tension

Chapter 1 Tune-up and routine maintenance

22.7b On V6 models, rotate the tensioner clockwise to release tension

22.17 Install the belt installation tool and rotate the crankshaft pulley until the belt is lifted onto the pulley (typical shown)

22.21 Drivebelt tensioner mounting bolts - V6 models

23.6 Location of the transaxle drain plug - V6 models (6F50/6F55 transaxle)

has been released, remove the belt from the pulleys.

8 On 2011 through early-build 2013 four-cylinder models, loosen the drivebelt tensioner mounting bolts and pull the tensioner away from the engine far enough to allow the belt to slide past the tensioner pulley.

9 On late-build 2013 and later four-cylinder models, remove the engine cover by pulling straight up on the cover to disengage the grommets from the cover studs. Working from above, note how the drivebelt is routed, then rotate the tensioner counterclockwise to release the drivebelt tension. Once tension has been released, remove the belt from the pulleys.

Caution: *The engine cover must be pulled straight up - do not pull the cover forward or sideways to remove it. The cover mounting points and studs are easily damaged if the cover is not pulled straight upward from the underside of the cover.*

10 Install the new drivebelt onto the crankshaft, alternator, and air conditioning compressor pulleys, then turn the tensioner back and locate the drivebelt on the pulley. Make sure that the drivebelt is correctly seated in all of the pulley grooves, then release the tensioner.

11 On late-build 2013 and later four-cylinder models, install the air conditioning compressor drivebelt (see Step 15).

12 Install the splash shield and wheel, then lower the vehicle. Tighten the lug nuts to the torque listed in this Chapter's Specifications.

Replacement - air conditioning compressor drivebelt (2013 and later four-cylinder models)

13 Disconnect the cable from the negative terminal of the battery (see Chapter 5). Remove the under-vehicle splash shield and the inner fender splash shield (see Step 2).

14 Using a pair of diagonal cutting pliers, cut the air conditioning belt to remove it.

Note: *A nylon strap wrapped around the belt, while rotating the crankshaft and pulling on the belt, can be used to remove the belt without cutting it. Some tool manufacturers make belt removal tools that can remove the belt without cutting it off as well. We only recommend using this method if the belt if relatively new. Always check the belt for damage after the removal process.*

15 Place the new belt on the crankshaft pulley, making sure the belt is fully seated in the pulley grooves.

16 Place the installation tool on the compressor pulley, then loop the belt over the tool.

17 Rotate the crankshaft pulley and allow the tool to turn, which will lift the belt onto the pulley (see illustration).

18 The remainder of installation is the reverse of removal.

Tensioner replacement

19 On early-build 2013 and earlier four-cylinder models, remove the two tensioner mounting bolts (see illustration 22.7a) and remove the tensioner.

20 On late-build 2013 and later four-cylinder models, remove the two bolts from the side of the tensioner body, then remove the tensioner.

21 On V6 models, remove the three bolts securing the tensioner to the timing chain cover, then detach the tensioner from the cover (see illustration).

22 Installation is the reverse of removal. Tighten the tensioner bolt(s) to the torque listed in this Chapter's Specifications.

23 Automatic transaxle fluid change (every 60,000 miles or 48 months)

Note: *For transaxle identification, refer to the Vehicle Identification Numbers information at the front of this manual.*

1 The automatic transaxle fluid should be changed at the recommended intervals.

2 Before beginning work, purchase the specified transmission fluid (see *Recommended lubricants and fluids* in this Chapter's Specifications).

3 Other tools necessary for this job include jackstands to support the vehicle in a raised position, wrenches, a drain pan, newspapers and clean rags.

4 The fluid should be drained immediately after the vehicle has been driven. Hot fluid is more effective than cold fluid at removing built-up sediment.

Warning: *Fluid temperature can exceed 350-degrees F in a hot transaxle. Wear protective gloves.*

5 After the vehicle has been driven to warm up the fluid, raise the front of the vehicle and support it securely on jackstands.

Warning: *Never work under a vehicle that is supported only by a jack!*

V6 models (6F50/6F55 transaxle)

6 Place the drain pan under the drain plug in the transaxle and remove the drain plug. Once the fluid has drained, reinstall the drain plug and tighten it to the torque listed in this Chapter's Specifications (see illustration).

Chapter 1 Tune-up and routine maintenance

7   Lower the vehicle and unscrew fluid filler cap/dipstick from the transaxle (see illustration 4.29).
8   Add five quarts of new fluid to the transaxle through the fill plug hole (see *Recommended lubricants and fluids* for the recommended fluid type), then install the fluid filler cap/dipstick.
9   Start the engine and, while depressing the brake pedal, cycle the shifter through each gear position and return it to Park. With the engine running, add fluid through the dipstick tube, 1/2-pint at a time as necessary (cycling the shifter through each gear position between additions) until the level is up to the bottom of the desired range on the dipstick (see illustration 4.31). Reinstall the filler cap/dipstick.
10 Drive the vehicle a few miles until the fluid is up to normal operating temperature, then recheck the fluid level (see Section 4); it should be in the desired range indicated on the dipstick. If not, add fluid a little at a time (cycling the shifter through each gear position between additions) until the level is correct. Tighten the filler cap/dipstick securely.
11 If you wish to flush the torque converter of old fluid, repeat Steps 6 through 10 one or two more times.

Four-cylinder models (6F35 transaxle)

12 Place the drain pan under the drain plug in the transaxle and remove the drain plug (see illustration). Once the fluid has drained, reinstall the drain plug and tighten it to the torque listed in this Chapter's Specifications.
13 Lower the vehicle.
14 Measure the amount of fluid drained and record this figure for reference when refilling.
15 With the engine off, add new fluid to the transaxle through the fill tube (see illustration 4.37). Begin the refill procedure by initially adding 1/3 of the amount drained. Then, with the engine running, add 1/2-pint at a time (cycling the shifter through each gear position between additions) until the level is correct on the dipstick.
16 Repeat Steps 12 through 15 two more times to flush any contaminated fluid from the torque converter.
17 Raise the vehicle on a hoist keeping the vehicle in a level position and place a drain pan under the transaxle.
18 With the engine still idling and in the park position, remove the oil leveling plug on the side of the transaxle (see illustration 4.36). Check the level of the fluid, it should be even with the bottom of the plug hole.
19 If the level is low, remove the filler tube cap and add the specified automatic transmission fluid through the filler tube with a funnel until the fluid starts to drip out of the oil leveling plug.
Note: *Allow all excess fluid to drip out of the plug hole.*
20 Once the fluid is even with the oil leveling plug hole install the plug and tighten the plug to torque listed in this Chapter's Specifications.

23.12 Location of the transaxle drain plug (6F35 transaxle)

21 Drive the vehicle a few miles until the fluid is up to normal operating temperature, then recheck the fluid level (see Section 4); it should be in the cross-hatched range (or in the range between the hot operation lines). If not, add fluid a little at a time (cycling the shifter through each gear position between additions) until the level is correct.

Fluid disposal

22 The old fluid drained from the transaxle cannot be reused in its present state and should be disposed of. Check with your local auto parts store, disposal facility or environmental agency to see if they will accept the fluid for recycling. After the fluid has cooled it can be drained into a container (capped plastic jugs, topped bottles, milk cartons, etc.) for transport to one of these disposal sites. Don't dispose of the fluid by pouring it on the ground or down a drain!

24 Differential lubricant change (AWD models) (every 60,000 miles or 48 months)

Drain

1   This procedure should be performed after the vehicle has been driven so the lubricant will be warm and therefore flow out of the differential more easily.
2   Raise the vehicle and support it securely on jackstands.
3   The easiest way to drain the differential is to remove the lubricant through the filler plug hole with a suction pump. If the differential cover gasket is leaking, it will be necessary to remove the cover to drain the lubricant (which will also allow you to inspect the differential).

Changing the lubricant with a suction pump

4   Remove the filler plug from the differential (see illustration).
5   Insert the flexible hose.

24.4 Rear differential check/fill plug

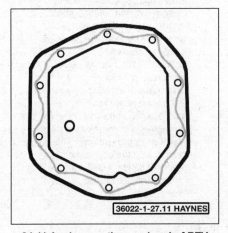

24.11 Apply a continuous bead of RTV sealant to the cover

6   Work the hose down to the bottom of the differential housing and pump the lubricant out.

Changing the lubricant by removing the cover

7   Move a drain pan, rags, newspapers and wrenches under the vehicle.
8   Remove the bolts on the lower half of the cover. Loosen the bolts on the upper half and use them to loosely retain the cover. Allow the oil to drain into the pan, then completely remove the cover.
9   Using a lint-free rag, clean the inside of the cover and the accessible areas of the differential housing. As this is done, check for chipped gears and metal particles in the lubricant, indicating that the differential should be more thoroughly inspected and/or repaired.
10 Thoroughly clean the gasket mating surfaces of the differential housing and the cover plate. Use a gasket scraper or putty knife to remove all traces of the old gasket.
11 Apply a bead of RTV sealant to the cover flange (see illustration). Make sure the bolt holes align properly, then install the cover and tighten the fasteners to the torque listed in this Chapter's Specifications.

1-26   Chapter 1 Tune-up and routine maintenance

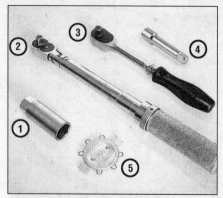

26.2 Tools required for changing spark plugs

1. **Spark plug socket** - This will have special padding inside to protect the spark plug's porcelain insulator
2. **Torque wrench** - Although not mandatory, using this tool is the best way to ensure the plugs are tightened properly
3. **Ratchet** - Standard hand tool to fit the spark plug socket
4. **Extension** - Depending on model and accessories, you may need special extensions and universal joints to reach one or more of the plugs
5. **Spark plug gap gauge** - This gauge for checking the gap comes in a variety of styles. Make sure the gap for your engine is included

26.5 When checking the spark plug gap, the wire should slide between the electrodes with a slight drag

Refill

12 Use a hand pump, syringe or funnel to fill the differential housing with the specified lubricant until it's within 1/8 to 3/16-inch (3 to 5 mm) from the bottom of the filler plug hole.
13 Install the fill plug and tighten it to the torque listed in this Chapter's Specifications.
14 The old lubricant drained from the differential cannot be reused in its present state and should be disposed of. Check with your local auto parts store, disposal facility or environmental agency to see if they will accept the lubricant for recycling. After the lubricant has cooled, it can be drained into a container (capped plastic jugs, topped bottles, milk cartons, etc.) for transport to one of these disposal sites. Don't dispose of the lubricant by pouring it on the ground or down a drain!

25 Transfer case lubricant change (AWD models)

Note: *Technically, the transfer case is filled for life; routine fluid changes aren't necessary unless the fluid somehow becomes contaminated.*

Drain

1 This procedure should be performed after the vehicle has been driven so the lubricant will be warm and therefore flow out of the transfer case more easily.
2 Raise the vehicle and support it securely on jackstands.
3 On models equipped with a drain plug, remove the drain plug and allow the fluid to drain.
4 On models without a drain plug, remove the filler plug and insert the flexible hose. Work the hose down to the bottom of the unit and pump the lubricant out.

Refill

5 Use a hand pump, syringe or funnel to fill the housing with the specified lubricant until it's even with the bottom of the filler plug hole.
6 Install the fill plug and tighten it to the torque listed in this Chapter's Specifications.
7 The old lubricant cannot be reused in its present state and should be disposed of. Check with your local auto parts store, disposal facility or environmental agency to see if they will accept the lubricant for recycling. After the lubricant has cooled, it can be drained into a container (capped plastic jugs, topped bottles, milk cartons, etc.) for transport to one of these disposal sites. Don't dispose of the lubricant by pouring it on the ground or down a drain!

26 Spark plug check and replacement (see Maintenance schedule for service intervals)

1 The spark plugs are located in the center of the valve cover(s). Remove the engine cover by pulling straight up on the cover to disengage the grommets from the cover studs.
Caution: *On 2.3L models, the engine cover must be pulled straight up, do not pull the cover forward or sideways to remove it. The cover mounting points and studs are easily damaged if the cover is not pulled straight upward from the underside of the cover.*
Note: *Access to the rear cylinder bank spark plugs on V6 models requires removal of the upper intake manifold (see Chapter 2B).*
2 In most cases, the tools necessary for spark plug replacement include a spark plug socket which fits onto a ratchet (spark plug sockets are padded inside to prevent damage to the porcelain insulators on the new plugs), various extensions and a gap gauge to check and adjust the gaps on the new plugs (see illustration). A torque wrench should be used to tighten the new plugs.
3 The best approach when replacing the spark plugs is to purchase the new ones in advance, adjust them to the proper gap and replace the plugs one at a time. When buying the new spark plugs, be sure to obtain the correct plug type for your particular engine. This information can be found in this Chapter's Specifications or in the factory owner's manual.
4 Allow the engine to cool completely before attempting to remove any of the plugs. These engines are equipped with aluminum cylinder heads, which can be damaged if the spark plugs are removed when the engine is hot. While you are waiting for the engine to cool, check the new plugs for defects and adjust the gaps.
5 The gap is checked by inserting the proper-thickness gauge between the electrodes at the tip of the plug (see illustration). The gap between the electrodes should be the same as the one specified in this Chapter's Specifications. The gauge should just slide between the electrodes with a slight amount of drag. If the gap is incorrect, use the adjuster on the gauge body to bend the curved side electrode slightly until the proper gap is obtained. If the side electrode is not exactly over the center electrode, bend it with the adjuster until it is. Check for cracks in the porcelain insulator (if any are found, the plug should not be used).
Note: *We recommend using a wire-type thickness gauge when checking platinum- or iridium-type spark plugs. Other types of gauges may scrape the thin coating from the electrodes, thus dramatically shortening the life of the plugs.*
6 All models are equipped with individual ignition coils which must be removed first to access the spark plugs (see illustrations).
7 If compressed air is available, use it to blow any dirt or foreign material away from the

Chapter 1 Tune-up and routine maintenance

26.6a All models are equipped with individual coils which must be removed to access the spark plugs (four-cylinder shown, V6 similar)

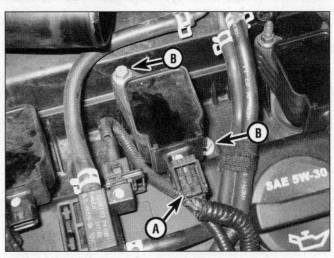

26.6b Depress the tab (A) and disconnect the electrical connector, remove the coil retaining bolts (B), then pull the coil straight up to remove it

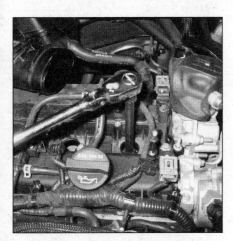

26.8 Use a ratchet and extension to remove the spark plugs

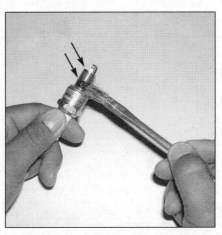

26.10a Apply a thin coat of anti-seize compound to the spark plug threads

26.10b A length of snug-fitting rubber hose will save time and prevent damaged threads when installing the spark plugs

spark plug hole. The idea here is to eliminate the possibility of debris falling into the cylinder as the spark plug is removed.

8 Place the spark plug socket over the plug and remove it from the engine by turning it in a counterclockwise direction (see illustration).

9 Compare each old spark plug to those shown on the inside of the back cover to get an indication of the general running condition of the engine.

10 Apply a small amount of anti-seize compound to the spark plug threads (see illustration). Install one of the new plugs into the hole until you can no longer turn it with your fingers, then tighten it with a torque wrench (if available) or the ratchet. It is a good idea to slip a short length of rubber hose over the end of the plug to use as a tool to thread it into place (see illustration). The hose will grip the plug well enough to turn it, but will start to slip if the plug begins to cross-thread in the hole - this will prevent damaged threads and the accompanying repair costs.

11 Repeat the procedure for the remaining spark plugs.

Notes

Chapter 2 Part A
Four-cylinder engines

Contents

	Section		Section
Camshafts and lifters - removal, inspection and installation	10	Oil pump - removal and installation	13
Crankshaft front oil seal - replacement	9	Rear main oil seal - replacement	15
Crankshaft pulley - removal and installation	7	Repair operations possible with the engine in the vehicle	2
Cylinder head - removal and installation	11	Timing chain cover, timing chain and tensioner - removal and installation	8
Driveplate - removal, inspection and installation	14	Top Dead Center (TDC) for number 1 piston - locating	3
Engine mount - check and replacement	16	Valve clearances - check and adjustment	5
General information	1	Valve cover - removal and installation	4
Intake manifold - removal and installation	6		
Oil pan - removal and installation	12		

Specifications

General

Engine type... Four-cylinder, in-line, DOHC
Displacement
  2.0L engine... 122 cubic inches (1999 cc)
  2.3L engine... 140 cubic inches (2294 cc)
Engine VIN code
  2.0L engine... 9
  2.3L engine... H
Firing order.. 1-3-4-2
Bore... 3.4449 inches (87.5 mm)
Stroke
  2.0L engines... 3.2717 inches (83.1 mm)
  2.3L engines... 3.7008 inches (94 mm)
Compression ratio
  2.0L engine... 9.3:1
  2.3L engine... 10:1
Compression pressure... See Chapter 2C
Oil pressure... See Chapter 2C

Camshafts

Lobe height
  Intake
    2.0L engines... 0.326 inch (8.3 mm)
    2.3L engines... 0.353 inch (8.9 mm)
  Exhaust
    2.0L engines... 0.291 inch (7.4 mm)
    2.3L engines... 0.307 inch (7.8 mm)
Bearing journal diameter... 0.9827 to 0.9835 inch (24.96 to 24.98 mm)
Journal-to-bore clearance.. 0.0014 to 0.0031 inch (0.35 to 0.08 mm)
Runout... 0.0012 inch (0.03 mm)

Valve clearances (cold)
Ideal clearance
   Intake .... 0.0098 inch (0.25 mm)
   Exhaust .... 0.0142 inch (0.36 mm)
Acceptable clearance
   Intake .... 0.007 to 0.012 inch (0.19 to 0.31 mm)
   Exhaust .... 0.012 to 0.017 inch (0.30 to 0.42 mm)

Warpage limits
Cylinder head gasket surfaces (head and block) .... 0.0020 inch (0.05 mm)
Exhaust manifold .... 0.030 inch (0.76 mm)

Torque specifications   Ft-lbs (unless otherwise indicated)   Nm

Note: *One foot-pound (ft-lb) of torque is equivalent to 12 inch-pounds (in-lbs) of torque. Torque values below approximately 15 foot-pounds are expressed in inch-pounds, because most foot-pound torque wrenches are not accurate at these smaller values.*

Item	Ft-lbs (unless otherwise indicated)	Nm
Camshaft bearing cap bolts (in sequence - see illustration 10.26)		
Step 1, All but front bearing cap	62 in-lbs	7
Step 2, All but front bearing cap	142 in-lbs	16
Step 3, Front bearing cap (3-bolt)	62 in-lbs	7
Step 4, Front bearing cap (3-bolt)	142 in-lbs	16
Camshaft phaser and sprocket bolts		
Step 1	30	40
Step 2	Tighten an additional 60-degrees	
Crankshaft pulley bolt		
Step 1	74	100
Step 2	Tighten an additional 90-degrees	
Cylinder head bolts (in sequence - see illustration 11.30a)		
Step 1	62 in-lbs	7
Step 2	133 in-lbs	15
Step 3	41	55
Step 4	Tighten an additional 90-degrees	
Step 5	Tighten an additional 90-degrees	
Variable Camshaft Timing (VCT) solenoid bolt	89 in-lbs	10
Valve cover bolts (in sequence - see illustration 4.13)	89 in-lbs	10
Turbocharger mounting nuts	37	50
Engine mount nuts\bolts	66	90
Driveplate bolts		
Step 1	37	50
Step 2	59	80
Step 3	89	120
Intake manifold bolts	177 in-lbs	20
Crankshaft rear oil seal and retainer	89 in-lbs	10
Oil pan-to-bellhousing bolts	35	48
Timing chain cover-to-oil pan bolts	89 in-lbs	10
Oil pan-to-engine block bolts	177 in-lbs	20
Oil pick-up pipe bolts	89 in-lbs	10
Oil pump-to-cylinder block bolts		
Step 1	89 in-lbs	10
Step 2	177 in-lbs	20
Sprocket bolt	18	25
Oil pump chain tensioner and guide bolts	89 in-lbs	10
Timing chain cover (in sequence - see illustrations 8.11a or 8.11b)		
8 mm bolts	89 in-lbs	10
13 mm bolts	35	48
Timing chain guide bolts	89 in-lbs	10
Timing chain tensioner bolts	89 in-lbs	10
Transale-to-engine bolts	35	48

Chapter 2 Part A Four-cylinder engines

1 General information

How to use this Chapter

1 This Part of Chapter 2A is devoted to repair procedures for the four-cylinder engine possible while the engine is still installed in the vehicle. Since these procedures are based on the assumption that the engine is installed in the vehicle, if the engine has been removed from the vehicle and mounted on a stand, some of the preliminary dismantling steps outlined will not apply.

2 Information concerning engine/transaxle removal and replacement and engine overhaul can be found in Part C of this Chapter.

Engine description

3 These engines are sixteen-valve, double overhead camshaft (DOHC), four-cylinder, in-line type, mounted transversely at the front of the vehicle, with the transaxle on the left-hand end. They incorporate an aluminum cylinder head, an aluminum cylinder block, direct injection and a turbocharger.

4 The two camshafts are driven by a timing chain, each operating eight valves via conventional lifters. Each camshaft rotates in five bearings that are line-bored directly in the cylinder head and the (bolted-on) bearing caps. This means that the bearing caps are not available separately from the cylinder head, and must not be interchanged with caps from another engine. The exhaust manifold is integral with the cylinder head, and the turbocharger bolts directly to the cylinder head.

5 These engines incorporate an aluminum timing chain cover and oil pan, and the crankshaft main caps are part of a one-piece lower block support. When working on these engines, note that Torx-type (both male and female heads) and hexagon socket (Allen head) fasteners are widely used. A good selection of sockets, with the necessary adapters, will be required, so that these can be unscrewed without damage and, on reassembly, tightened to the specified torque settings.

Lubrication system

6 The oil pump is driven via a chain from the front of the crankshaft. The pump forces oil through an externally mounted full-flow cartridge-type filter. From the filter, the oil is pumped into a main gallery in the cylinder block/crankcase, from where it is distributed to the crankshaft (main bearings) and cylinder head.

7 The connecting rod bearings are supplied with oil via internal drillings in the crankshaft. Each piston crown and connecting rod is cooled by a spray of oil.

8 The cylinder head is provided with two oil galleries, one on the intake side and one on the exhaust side, to ensure constant oil supply to the camshaft bearings and lifters. A retaining valve (inserted into the cylinder head's top surface, in the middle, on the intake side) prevents these galleries from being drained when the engine is switched off. The valve incorporates a ventilation hole in its upper end, to allow air bubbles to escape from the system when the engine is restarted.

2 Repair operations possible with the engine in the vehicle

1 Many major repair operations can be accomplished without removing the engine from the vehicle.

2 Clean the engine compartment and the exterior of the engine with some type of degreaser before any work is done. It will make the job easier and help keep dirt out of the internal areas of the engine.

3 Depending on the components involved, it may be helpful to remove the hood to improve access to the engine as repairs are performed (refer to Chapter 11 if necessary). Cover the fenders to prevent damage to the paint. Special pads are available, but an old bedspread or blanket will also work.

4 If vacuum, exhaust, oil or coolant leaks develop, indicating a need for gasket or seal replacement, the repairs can generally be made with the engine in the vehicle. The intake and exhaust manifold gaskets, oil pan gasket, crankshaft oil seals and cylinder head gasket are all accessible with the engine in place.

5 Exterior engine components, such as the intake and exhaust manifolds, the oil pan, the oil pump, the water pump, the starter motor, the alternator and the fuel system components can be removed for repair with the engine in place.

6 Since the camshaft(s) and cylinder head can be removed without pulling the engine, valve component servicing can also be accomplished with the engine in the vehicle. Replacement of the timing chain and sprockets is also possible with the engine in the vehicle.

7 In extreme cases caused by a lack of necessary equipment, repair or replacement of piston rings, pistons, connecting rods and rod bearings is possible with the engine in the vehicle. However, this practice is not recommended because of the cleaning and preparation work that must be done to the components involved.

3 Top Dead Center (TDC) for number 1 piston - locating

Note: *You will need two special tools for this procedure: the camshaft positioning tool (303-1565, or equivalent) and the timing pin (303-507).*

1 Relieve the fuel system pressure (see Chapter 4), then disconnect the battery negative cable (see Chapter 5).

2 Remove the passenger's side inner fender splash shield (see Chapter 11).

3 Remove the engine splash shield fasteners and splash shield.

4 Remove the drivebelt (see Chapter 1).

5 Remove the brake vacuum pump (see Chapter 9) and the high-pressure fuel pump (see Chapter 4).

6 Remove the valve cover (see Section 4), and remove the rear intake camshaft bearing cap (see Section 10).

7 Prevent the camshaft from turning by using a wrench on the flats of the intake camshaft, then loosen and remove the brake vacuum pump adapter from the end of the camshaft.

8 Rotate the crankshaft clockwise, until the No.1 piston is 45-degrees BTDC using the guide holes on the engine front cover and the crankshaft pulley.

9 Remove the right-side driveaxle and support bracket (see Chapter 8).

10 The TDC timing hole is located near the lower right front corner of the engine block (on the firewall side), hidden behind the driveaxle support bracket; it provides a means of accurately positioning the no. 1 cylinder at TDC. When you locate this hole, remove the timing pin plug bolt (see illustration).

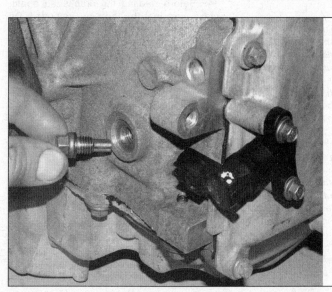

3.10 Remove the timing hole plug . . .

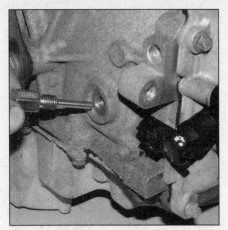

3.11 ... and insert the timing pin tool

3.13a Remove the rear intake camshaft cap

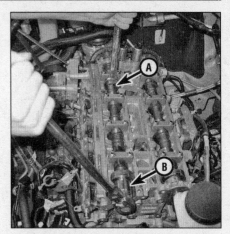

3.13b Prevent the camshaft from turning by holding it with a wrench on the hex surface (A), then unscrew the vacuum pump adapter (B)

3.13c When the engine is at TDC for cylinder no. 1, the slots in the ends of the camshafts will be horizontal (and the offset portions will be even with the cylinder head surface). Manufacturer tool no. 303-1565 can then be inserted into the slots

11 Screw in the timing pin (see illustration).
Caution: *We don't recommend trying to fabricate a timing pin with a bolt because while you would be able to determine the correct bolt diameter and thread pitch, it is impossible to determine what the length of the bolt should be. These pins come in several lengths, depending on the engine family. There is no way to determine the correct pin length without comparing it to a factory or aftermarket tool designed to be used with this engine. Using a bolt of the wrong length could damage the engine. Also, never use the timing pin as a means to stop the engine from rotating - tool breakage and/or engine damage can result.*
Note: *With the timing pin installed the engine can still be rotated counterclockwise.*
12 Turn the crankshaft slowly clockwise until the crankshaft counterweight comes into contact with the timing pin and the hole on the crankshaft pulley is in a center line with the hole on the timing cover (6 o'clock position) - in this position, the engine is set to TDC on no. 1 cylinder.
13 The camshafts each have a machined slot at the transaxle end of the engine. Both slots will be completely horizontal, and at the same height as the cylinder head machined surface, when the engine is at TDC on the Number 1 cylinder. Manufacturer service tool 303-1565 is used to check this position, and to positively locate the camshafts in position. The rear intake camshaft cap and the vacuum pump drive adapter will have to be removed from the intake camshaft (see illustrations).
Note: *The timing slots in the camshafts are offset. If the service tool cannot be installed, remove the timing pin and carefully rotate the crankshaft three fourths of a turn clockwise and repeat Steps 12 and 13 of this procedure.*
Caution: *Never use the camshaft alignment tool as a means to stop the engine from rotating - engine damage can result.*
14 Before rotating the crankshaft again, make sure that the tools are removed. Do not forget to install the brake vacuum pump adapter and tighten it securely.
15 Once no. 1 cylinder has been positioned at TDC on the compression stroke, TDC for any of the other cylinders can then be located by rotating the crankshaft clockwise 180-degrees at a time and following the firing order (see this Chapter's Specifications).

4 **Valve cover - removal and installation**

Removal

1 Disconnect the cable from the negative battery terminal (see Chapter 5).
2 Remove the engine cover by pulling straight up on the cover to disengage the grommets from the cover studs. Remove the individual ignition coil assemblies from the spark plugs (see Chapter 5).
Caution: *The engine cover must be pulled straight up - do not pull the cover forward or sideways to remove it. The cover mounting points and studs are easily damaged if the cover is not pulled straight upward from the underside of the cover.*
3 Remove the Charge Air Cooler (CAC) inlet and outlet tubes (see Chapter 4), then remove the CAC bracket fasteners and bracket.
4 Disconnect the electrical connectors for the Variable Camshaft Timing (VCT) solenoids, EVAP canister purge, Fuel Rail Pressure (FRP) sensor and the Manifold Absolute Pressure (MAP) sensor.
5 Disconnect the vacuum lines and retainers.
6 Remove the bracket for the wiring harness on the valve cover stud, then set the harness aside.
7 Remove the engine oil dipstick.
8 Working progressively, unscrew the valve cover retaining fasteners, noting the captive spacer sleeve and rubber seal, then detach the cover.
9 Discard the cover gasket. This must be replaced whenever it is disturbed. Check that the sealing faces are undamaged and that the rubber seal at each bolt hole is serviceable. Replace any worn or damaged spark plug tube seals or Variable Camshaft Timing (VCT) solenoid seals.

Installation

10 On installation, clean the cover and cylinder head gasket faces carefully. Install a new gasket onto the valve cover, ensuring that it is located correctly by the rubber seals and spacer sleeves.
11 At the top of the timing chain cover, apply a small bead of RTV sealant to the joints where the timing chain cover meets the cylinder head (see illustration).
12 Install the cover to the cylinder head,

Chapter 2 Part A Four-cylinder engines

4.11 Apply a bead of RTV sealant to the joints where the timing chain cover meets the cylinder head

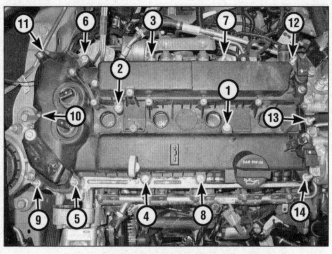

4.13 Valve cover bolt tightening sequence

ensuring that the gasket remains seated as the cover is tightened.
13 Tighten the cover bolts, a little at a time in sequence (see illustration), to the torque listed in this Chapter Specifications.
14 Reconnect the battery.
15 Run the engine and check for signs of oil leakage.

5 Valve clearances - check and adjustment

1 Disconnect the cable from the negative battery terminal (see Chapter 5).
2 Remove the spark plugs (see Chapter 1), then remove the valve cover (see Section 4).
3 Loosen the right front wheel lug nuts, raise the front of the vehicle and support it securely on jackstands, then remove the wheel. Remove the fender splash shield.
4 Using a wrench or socket on the crankshaft pulley bolt, rotate the crankshaft clockwise and check each lifter when its camshaft lobe is straight up, ensuring that the measurement is between the base circle of the camshaft lobe and the top of the lifter. Use feeler gauges to measure the clearances (see illustration).
5 The clearance must be checked between each camshaft lobe and the lifter it operates. Keep careful notes of the measurements recorded for each lifter.
6 If some measurements fall outside the recommended clearances in this Chapter's Specifications, the camshafts must be removed and the out-of-spec lifters removed (see Section 10). New lifters are available with various thicknesses to correct the valve clearances (see illustration). Each lifter is marked with a thickness number. Only refer to the numbers after the decimal point. A "0.650" marking refers to an actual thickness of 3.650 mm.

7 To arrive at the desired thickness for new lifters: add the thickness of the original lifter (such as 0.650 mm) to the clearance you measured. Subtract the midrange figure for ideal clearance (see the Specifications) from that number and you have the proper lifter thickness to order. Every thickness is not available, so choose the closest to your requirement.
8 Refer to Section 10 for installation of the camshafts. Once the camshafts and timing chain have been installed, recheck the valve clearances.
9 The remainder of installation is the reverse of removal. Run the engine and check for oil leaks.

6 Intake manifold - removal and installation

Removal

1 Remove the engine cover by pulling straight up on the cover to disengage the grommets from the cover studs.
Caution: *The engine cover must be pulled straight up, do not pull the cover forward or sideways to remove it. The cover mounting points and studs are easily damaged if the cover is not pulled straight upward from the underside of the cover.*
2 Disconnect the cable from the negative battery terminal (see Chapter 5).
3 On 2.3L engines, remove the drivebelt and drivebelt tensioner (see Chapter 1).
4 Raise the vehicle and support it securely on jackstands.
5 Remove the engine splash shield fasteners and shield.
6 Disconnect the EVAP tube from the Charge Air Cooler (CAC) outlet tube, then loosen the CAC outlet tube clamp and pull the tube off of the throttle body.
7 Disconnect the electrical connectors

5.4 Check the valve clearances with a feeler gauge of the specified thickness. If the clearance is correct, you should feel a slight drag as the feeler gauge is slid between the lifter and the camshaft

5.6 Use a micrometer to measure the thickness of the lifter head

6.10 Push in on the lock ring while pulling the brake booster vacuum hose from the intake manifold

6.12 Intake manifold fasteners - 2.0L shown, 2.3L engine similar

from the throttle body, Fuel Rail Pressure (FRP) sensor and the Manifold Absolute Pressure (MAP) sensor.

8 Disengage the EVAP line and heater hose retainers underneath of the manifold.

9 Remove the EVAP canister purge valve (see Chapter 6).

10 Release the power brake booster vacuum hose from the intake manifold by depressing the red quick-connect lock ring and pulling the hose outward at the same time (see illustration).

11 Release the harness clips and set aside the wiring harness at the top of the intake manifold.

12 Remove the mounting bolts and pull the intake manifold away from the engine enough to access and disconnect the crankcase vent hose from the oil separator (see illustration). Squeeze the two clips on the vent hose to release it then on 2.3L engines, disconnect the knock sensor harness from the manifold.

Installation

13 There are individual gaskets for each of the four ports of the intake manifold. Using new manifold gaskets, install the intake manifold. Tighten the bolts a little at a time, working from the center out, to the torque listed in this Chapter's Specifications.

14 Installation is otherwise the reverse of removal.

7 Crankshaft pulley - removal and installation

Removal

Caution: *Once the crankshaft pulley is loosened, the crankshaft (timing) sprocket will be loosened as well. The engine is considered out-of-time at this point. The installation procedure in this Section must be followed exactly to re-time the engine properly. Severe engine damage may occur otherwise.*

Note: *You will need two special tools for this procedure: the camshaft positioning tool (303-1565, or equivalent) and the timing pin (303-507). Using these tools to prevent engine rotation can result in engine damage.*

1 Remove the right inner fender splash shield, (see Chapter 11), then remove the drivebelt (see Chapter 1).

2 Remove the brake vacuum pump (see Chapter 9).

3 Remove the fasteners from the right driveaxle intermediate shaft bracket.

4 Set the engine to TDC using the camshaft and crankshaft positioning tools (see Section 3).

5 The crankshaft must be held to prevent its rotation while the pulley bolt is unscrewed. Use a strap wrench around the crankshaft pulley to hold it. The manufacturer recommends not using an 1/2-inch drive air or electric impact gun to remove the pulley bolt.

Caution: *Failure to hold the crankshaft pulley securely while removing the pulley bolt could result in engine damage. NEVER use the timing pin or the camshaft alignment tool as a means of locking the crankshaft - they are designed for calibration only. Engine damage could occur by using these tools for anything other than their intended purpose.*

6 Unscrew the pulley bolt.

7 Remove the pulley and the diamond washer behind it. Obtain a new pulley bolt and diamond washer.

Installation

8 Install a new diamond washer onto the nose of the crankshaft.

9 Lightly coat the crankshaft front seal with clean engine oil, then install the crankshaft pulley.

Note: *If the seal shows signs of leakage, you may want to replace it before installing the crankshaft pulley (see Section 9).*

10 Install the crankshaft pulley with the small hole in the pulley at the 6 o'clock position.

11 Install the Crankshaft Position (CKP) sensor alignment tool (#303-1521) onto the sensor. Install a new crankshaft pulley bolt and washer, hand tight only, and remove the tool.

12 Insert a M6 bolt through the hole in the crankshaft pulley, hand tighten the bolt, ONLY. This will correctly align the crankshaft pulley.

13 Using the strap wrench to hold the pulley, tighten the crankshaft pulley bolt in stages, to the torque listed in this Chapter's Specifications.

Caution: *Do not use an impact gun to tighten the crankshaft pulley bolt.*

14 Remove the bolt from the pulley then remove the timing pin from the cylinder block.

15 Remove the camshaft alignment tool.

16 Remove the spark plugs and rotate the engine clockwise two complete revolutions by turning the crankshaft pulley bolt with a wrench or large socket.

Caution: *If you feel resistance at any point, stop and find out why. If the valve timing is incorrect, the valves may be contacting the pistons.*

17 Rotate the engine again to achieve TDC (see Section 3).

Note: *Rotate the engine in the clockwise direction only.*

18 Install the timing pin into the cylinder block.

19 With the tooth on the crankshaft pulley in a center line with the CKP sensor, install the camshaft alignment tool and check the position of the camshafts. If the tool cannot be installed, the engine timing must be corrected by repeating the TDC procedure (see Section 3).

20 The correct engine timing is achieved when the camshaft alignment tool is in place, the timing pin is inserted, and the crankshaft pulley TDC tooth is in a center line with the CKP sensor, simultaneously.

21 Once correct engine timing is achieved, remove all the alignment tools and bolts and install the timing pin plug.

22 Reinstall all components removed previously.

Chapter 2 Part A Four-cylinder engines

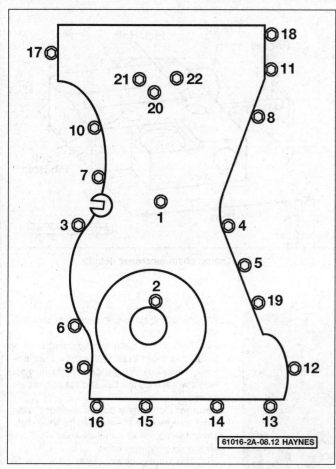

8.11a Timing chain cover bolt tightening sequence - 2.0L engines

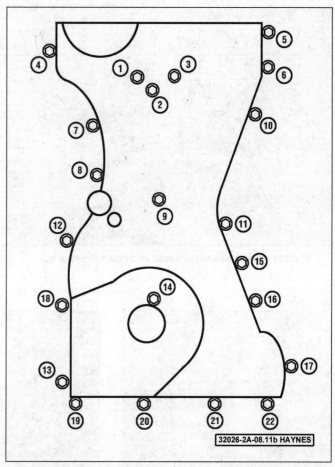

8.11b Timing chain cover bolt tightening sequence - 2.3L engines

8 Timing chain cover, timing chain and tensioner - removal and installation

Timing chain cover

Removal

1 Remove the engine cover by pulling straight up on the cover to disengage the grommets from the cover studs then relieve the fuel system pressure (see Chapter 4), then disconnect the negative battery cable (see Chapter 5).

Caution: *The engine cover must be pulled straight up, do not pull the cover forward or sideways to remove it. The cover mounting points and studs are easily damaged if the cover is not pulled straight upward from the underside of the cover.*

2 Loosen the water pump pulley bolts, remove the drivebelt and drivebelt idler pulley (see Chapter 1), then remove the water pump pulley (see Chapter 3).

3 Remove the fuel injection high pressure pump and housing (see Chapter 4).

4 Remove the crankshaft pulley (see Section 7). After this Step, the engine must remain at TDC with the valve cover removed.

5 Disconnect the Crankshaft Position (CKP) sensor electrical connector, then remove the sensor (see Chapter 6).

Note: *Unfortunately with this engine, any time this sensor is removed, a new sensor must be installed. The new sensor is packaged with a special sensor alignment tool that is critical for installation.*

6 Remove the power steering hose bracket fastener and bracket from the bottom of cover.

7 Support the engine from above with an engine support fixture (see Chapter 7A, illustration 5.11). Raise the engine enough to remove the right engine mount (see Section 16).

8 Unbolt and set aside the power steering pump without disconnecting the hoses (see Chapter 10), then remove the pump bracket bolts and bracket.

9 Remove the crankshaft front oil seal (see Section 9).

10 Remove the bolts, stud bolts and the timing chain cover (see illustrations 8.11a or 8.11b).

Installation

11 Installation is the reverse of removal, noting the following:
  a) Clean the mating surfaces of all sealant.
  b) Install the engine cover within ten minutes of applying a 2.5 mm bead of RTV sealant.
  c) Tighten the bolts a little at a time, in sequence (see illustrations), to the torque listed in this Chapter's Specifications.
  d) Install a new crankshaft front oil seal.
  e) Verify proper crankshaft/camshaft timing (see Section 7).
  f) Install a new CKP sensor, which includes the necessary alignment tool (see Chapter 6). Do not tighten the sensor mounting bolts until the installation tool is in place.

Timing chain and tensioner

Removal

12 Remove the timing chain cover (see Steps 1 through 10).

13 Compress the timing chain tensioner and place a pin (a drill bit or paper clip will work)

2A-8 Chapter 2 Part A Four-cylinder engines

8.13a Compress the tensioner and insert the lock pin

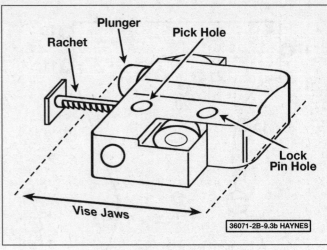

8.13b Timing chain tensioner details

8.18 Use a wrench on the hex portion to hold the camshaft while removing/installing the sprocket bolt

8.22 Compress the tensioner to release the lock pin

23 Install the camshaft alignment tool 303-1565.
24 Tighten the camshaft sprocket bolts to the torque listed in this Chapter's Specifications; while holding the camshafts in place with a wrench on the hex surface of the camshafts.
Caution: *Do not rely on the camshaft alignment tool to hold the camshafts while tightening the camshaft sprocket bolts. Tool and engine damage may occur.*
25 The timing chain cover can now be installed (see Step 11).
26 Install the fuel injection pump housing, making sure the camshaft lobe is at its lowest point on the lobe (see Chapter 4).
27 The remainder of installation is the reverse of removal.

9 Crankshaft front oil seal - replacement

1 Remove the crankshaft pulley (see Section 7).
Caution: *Once the crankshaft pulley is loosened, the crankshaft (timing) sprocket will be loosened as well. The engine is considered out-of-time at this point. The installation procedure in this Section must be followed exactly to re-time the engine properly. Severe engine damage may occur otherwise.*
2 Use a screwdriver or hook tool to carefully pry out the seal.
Note: *Be careful not to damage the timing chain cover bore where the seal is seated or the nose and sealing surface of the crankshaft.*
3 Another procedure for removing the seal is to drill a small hole on each side of the seal and place a self-tapping screw in each hole. Use these screws as a means of pulling the seal out without having to pry on the seal itself.
4 Wipe the sealing surfaces in the engine cover and on the crankshaft. Clean and coat

into the hole to hold it in the compressed position (see illustration).
Note: *The tensioner contacts the right-hand chain guide.*
Caution: *Compress only the round plunger on the tensioner and not the ratchet mechanism. The ratchet is next to the plunger and has square sides. If the ratchet needs to be reset, perform the following (see illustration):*
   a) Remove the tensioner and place it lightly in a vise, with the jaws contacting the plunger and tensioner housing.
   b) Place a pick-type tool in the hole closest to the ratchet to relieve the tension on the ratchet mechanism.
   c) While holding the pick tool in place, move the ratchet back into the tensioner, then install a pin into the other hole to keep the plunger and ratchet compressed.
   d) Remove the tensioner from the vise.
14 Remove the two tensioner mounting bolts, then remove the tensioner.
15 Remove the loose timing chain guide

(right). Remove the timing chain.
16 The left chain guide and camshaft sprockets can now be removed if necessary.

Installation
17 Remove the camshaft alignment tool (if installed).
18 Loosen both camshaft phaser and sprocket bolts but don't remove them. Use a wrench on the hexagonal area of the camshaft to hold it while turning the camshaft sprocket bolt (see illustration).
Caution: *Damage to the valves or pistons may occur if the camshafts are rotated during this procedure.*
19 Install the left chain guide (if removed).
20 Install the timing chain.
21 Install the right chain guide.
22 Install the timing chain tensioner and tighten the fasteners to the torque listed in this Chapter's Specifications. Remove the pin to release the tensioner to engage the chain guide (see illustration).

Chapter 2 Part A Four-cylinder engines

9.5 Make certain that the oil seal is kept square as it is placed in the bore

9.6 A socket of the correct size can be used to install the new seal

10.3 Remove the Variable Camshaft Timing (VCT) solenoid retaining bolts - 2.0L engine shown, 2.3L engine similar

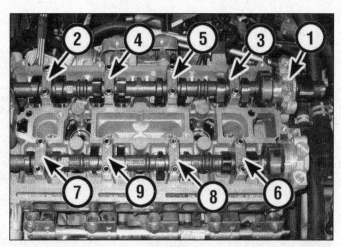

10.6 Camshaft bearing cap loosening sequence - loosen each pair of bolts on the designated bearing cap in sequence

them with clean engine oil.

5 Start installing the new seal by pressing it into the timing chain cover (see illustration).

6 Once started, use a seal driver or a suitable socket of the correct size to carefully drive the seal squarely into place (see illustration).

7 The seal should be flush with the timing chain cover and remain square when installed.

8 Coat the lip of the seal (where it contacts the crankshaft) with clean engine oil.

9 Install the crankshaft pulley (see Section 7).

10 Camshafts and lifters - removal, inspection and installation

Caution: *Damage to the valves or pistons may occur if the crankshaft or camshafts are rotated during this procedure.*

Note: *Whenever the camshafts are to be removed for a procedure, it's a good idea to check the valve clearances before disassembly (see Section 5), so any required new lifters can be ordered from a dealership.*

Removal

1 Remove the timing chain (see Section 8).

Note: *Before removing the timing chain, note the positions of the no. 1 cylinder cam lobes and the slots in the ends of the camshafts (for the alignment tool 303-1565). When installing the camshafts, the lobes and the slots in the ends of the camshafts must be in the same positions.*

2 Remove the camshaft phaser and sprockets.

Note: *The camshaft phaser and sprockets should be marked with indelible ink so that it can be reinstalled in the same position. When loosening the camshaft sprocket bolts, place a wrench on the hexagonal area of the camshaft to prevent it from turning (see illustration 8.18).*

3 Remove the Variable Camshaft Timing (VCT) solenoid fasteners and pull the solenoids out of the front camshaft bearing cap (see illustration).

4 Remove the front camshaft bearing cap bolts and lift the cap from both camshafts.

5 Mark the bearing caps for each position and the direction it faces - the bearing caps must be reinstalled in their original positions.

Caution: *The camshaft bearing caps must be loosened in order or the camshafts can be damaged.*

6 Working in the sequence shown, loosen the camshaft bearing cap bolts progressively by half a turn at a time (see illustration). Work only as described, to release gradually and evenly the pressure of the valve springs on the caps.

7 Withdraw the caps, noting their markings and the presence of the locating dowels, then remove the camshafts. The exhaust camshaft

2A-10 Chapter 2 Part A Four-cylinder engines

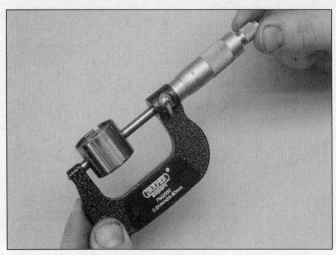

10.10 Measure the lifter outside diameter at several points

10.13 Check that the camshaft bearing oil ways are not blocked with debris

can be identified by the reference lobe for the high pressure fuel pump on the end (and it is longer, as well); therefore, there is no need to mark the camshafts.

8   Obtain sixteen small, clean containers, and number them 1 to 16. Using a rubber suction tool (such as a valve-lapping tool), withdraw each lifter in turn and place them in the containers. Do not interchange the lifters.

Inspection

9   With the camshafts and lifters removed, check each for signs of obvious wear (scoring, pitting, etc.) and for roundness and replace if necessary.

10   Measure the outside diameter of each lifter - take measurements at the top and bottom of each lifter, then a second set at right-angles to the first; if any measurement is significantly different from the others, the lifter is tapered or oval (as applicable) and must be replaced (see illustration). If the necessary equipment is available, measure the inside diameter of the corresponding cylinder head bore. No manufacturer's specifications were available at the time of writing; if the lifters or the cylinder head bores are excessively worn, new lifters and/or a new cylinder head may be required.

11   If the engine's valve components have sounded noisy, it may be just that the valve clearances need adjusting (see Section 5).

12   Visually examine the camshaft lobes for score marks, pitting, galling (wear due to rubbing) and evidence of overheating (blue, discolored areas). Look for flaking away of the hardened surface layer of each lobe. If any such signs are evident, replace the component concerned.

13   Examine the camshaft bearing journals and the cylinder head bearing surfaces for signs of obvious wear or pitting. If any such signs are evident, consult an automotive machine shop for advice. Also check that the bearing oil ways in the cylinder head are clear (see illustration).

14   Using a micrometer, measure the diameter of each journal at several points. If the diameter of any one journal is less than the specified value, replace the camshaft.

Caution: *If any new parts are being installed the valve clearance must be checked and properly adjusted (see Step 19), or serious engine damage may occur. If no new parts are being installed it is not necessary to check the valve clearance.*

15   To check the bearing journal running clearance, remove the lifters, use a suitable solvent and a clean lint-free rag to carefully clean all bearing surfaces, then install the camshafts and bearing caps with a strand of Plastigage across each journal. Tighten the bearing cap bolts in the proper sequence (see illustration 10.26) to the specified torque setting (do not rotate the camshafts), then remove the bearing caps and use the scale provided to measure the width of the compressed strands. Scrape off the Plastigage with your fingernail or the edge of a credit card - don't scratch or nick the journals or bearing caps.

16   If the running clearance of any bearing is found to be worn to beyond the specified service limits, install a new camshaft and repeat the check; if the clearance is still excessive, the cylinder head must be replaced.

17   To check camshaft endplay, remove the lifters, clean the bearing surfaces carefully and install the camshafts and bearing caps. Tighten the bearing cap bolts to the specified torque wrench setting, then measure the endplay using a dial indicator mounted on the cylinder head so that its tip bears on the camshaft right-hand end.

18   Tap the camshaft fully towards the gauge, zero the gauge, then tap the camshaft fully away from the gauge and note the gauge reading. If the endplay measured is found to be at or beyond the specified service limit, install a new camshaft and repeat the check; if the clearance is still excessive, the cylinder head must be replaced.

Installation

Note: *If no new parts are being installed, skip Step 19.*

19   If new parts were installed, follow these steps;

a) *Install the crankshaft pulley bolt and washer then paint or mark a line on the bolt washer at the 12 o'clock position.*

b) *Remove the special service tool peg (303-507) from the side of the engine block.*

c) *Rotate the engine clockwise 270-degrees, using the crankshaft bolt until the painted line is at the 9 o'clock position. Rotating the crankshaft will position all of the pistons below the top of the cylinder block and allow the camshafts to be installed and the valve clearance checked without the possibility of damage to the valves or pistons.*

d) *Install the camshafts, then see Section 5 and check the valve adjustment.*

e) *Once the adjustment is properly completed, remove the camshafts.*

f) *Install the special service tool peg (303-507) into the side of the engine block. Rotate the engine 90-degrees clockwise to the original TDC position, then proceed to Step 20.*

20   Confirm that the crankshaft is still positioned at TDC and that the timing pin is in place.

21   Liberally oil the cylinder head lifter bores and the lifters. Carefully install the lifters to the cylinder head, ensuring that each lifter is replaced to its original bore.

22   Liberally oil the camshaft bearing surfaces in the cylinder head, taking care not to get any on the camshaft cap mating surface.

23   Ensuring that each camshaft is in its original location, install the camshafts, locating each so that lobes for cylinder no. 1 are in the same position as noted in Step 1 and the slot in its left-hand end is parallel to, and just above, the cylinder head mating surface.

Check that, as each camshaft is laid in position, the TDC setting tool will fit into the slot.

Caution: *When the camshaft bearing caps are tightened, it is imperative that the camshafts do not rotate from their TDC positions.*

24 Check that all mating surfaces are completely clean, unmarked and free from oil. Apply a 0.039-inch (1 mm) bead of RTV sealant to the exhaust camshaft rear bearing cap, making sure to keep the sealant out of the oil galleys.

25 Apply a little oil to the camshaft journals and lobes, then install each of the camshaft bearing caps to its previously-noted position, except the front bearing cap, so that its numbered side faces outwards, to the front (exhaust) or to the rear (intake).

26 Ensuring that each cap is kept square to the cylinder head as it is tightened down and working in the sequence shown, tighten the camshaft bearing cap bolts slowly and by one turn at a time, until each cap touches the cylinder head (see illustration). This is the Step 1 torque.

27 Next, using the same sequence, tighten the bearing cap bolts to the Step 2 torque listed in this Chapter's Specifications.

28 Install the front bearing cap and tighten the cap in tighten the camshaft bearing cap bolts slowly and by one turn at a time, until the cap touches the cylinder head. This is the Step 3 torque then tighten the front bearing cap bolts to the Step 4 torque listed in this Chapter's Specifications.

29 Install the variable camshaft timing (VCT) solenoids into the front camshaft bearing cap and tighten the fasteners to the torque listed in this Chapter's Specifications.

30 Install the phasers and sprockets to the camshafts, tightening the retaining bolts loosely.

31 The remainder of the reassembly procedure, including replacement of the timing chain and setting the valve timing, is as described in Section 8.

32 Before installing the valve cover, check the valve clearances (see Section 5).

11 Cylinder head - removal and installation

Warning: *Wait until the engine is completely cool before beginning this procedure.*

Removal

1 Remove the engine cover by pulling straight up on the cover to disengage the grommets from the cover studs. Relieve the fuel pressure (see Chapter 4).

Caution: *The engine cover must be pulled straight up, do not pull the cover forward or sideways to remove it. The cover mounting points and studs are easily damaged if the cover is not pulled straight upward from the underside of the cover.*

2 Disconnect the cable from the negative battery terminal (see Chapter 5).

10.26 Camshaft bearing cap tightening sequence

3 Remove the air filter housing and charger air cooler inlet and outlet pipes (see Chapter 4).

4 Remove the power steering pump and bracket (see Chapter 10), and the alternator (see Chapter 5).

5 Drain the cooling system (see Chapter 1).

6 Remove the turbocharger (see Chapter 4).

7 Remove the coolant expansion tank (see Chapter 3).

8 Remove the timing chain (see Section 8).

Note: *Whenever the camshafts are to be removed for a procedure, it's a good idea to check the valve clearances before disassembly, so any required new lifters can be ordered from a dealership.*

9 Disconnect the Camshaft Position (CMP) sensor electrical connector (see Chapter 6).

10 Unbolt and remove the VCT solenoid valves from the front camshaft bearing cap (see Section 10).

11 Remove the camshafts and lifters (see Section 10), keeping the lifters in order.

12 Remove the fuel rail and injectors (see Chapter 4).

13 Disconnect the hoses from the coolant outlet housing. Remove the coolant tube mounting bolts and remove the tubes. Always replace the coolant tube O-rings.

14 Remove the intake manifold (see Section 6).

15 If an engine support fixture or hoist is being used to hold the engine up and it interferes with removal of the cylinder head, use a floor jack and block of wood to support the engine from below.

16 Loosen the cylinder head bolts progressively and by half a turn at a time, working in the reverse order of the tightening sequence (see illustration 11.30a).

Caution: *The head bolts are torque-to-yield bolts that must be replaced with new ones on installation.*

17 Lift the cylinder head from the engine compartment.

18 If the head is stuck, be careful how you choose to free it. Remember that the cylinder head is made of aluminum alloy, which is easily damaged. Striking the head with tools carries the risk of damage, and the head is located on two dowels, so its movement will be limited. Do not, under any circumstances, pry the head between the mating surfaces, as this will certainly damage the sealing surfaces for the gasket, leading to leaks. Try rocking the head free, to break the seal, taking care not to damage any of the surrounding components.

19 Once the head has been removed, remove and discard the gasket. Check for the presence of locating dowels in the cylinder block and cylinder head. If dowels are present, make sure they are returned to their original locations after cleaning the components.

Inspection

20 The mating faces of the cylinder head and cylinder block must be perfectly clean before replacing the head. Use spray-on gasket remover and a hard plastic or wood scraper to remove all traces of gasket and carbon.

21 Take particular care during the cleaning operations, as aluminum alloy is easily damaged. Also, make sure that the carbon is not allowed to enter the oil and water passages - this is particularly important for the lubrication system, as carbon could block the oil supply to the engine's components.

22 To prevent carbon entering the gap between the pistons and bores, smear a little grease in the gap. After cleaning each piston, use a small brush to remove all traces of grease and carbon from the gap, then wipe away the remainder with a clean rag.

23 Check the mating surfaces of the cylinder block and the cylinder head for nicks, deep scratches and other damage. Also check the cylinder head gasket surface and the cylinder block gasket surface with a precision straight-edge and feeler gauges. If either surface exceeds the warpage limit listed in this Chapter's Specifications, the manufacturer states that the component out of specification must be replaced. If the gasket mating surface of your cylinder head or block is out of specification or is severely nicked or scratched, you may want to consult with an automotive machine shop for advice.

2A-12   Chapter 2 Part A   Four-cylinder engines

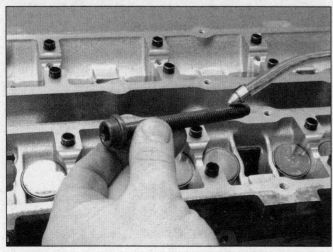

11.29 Apply a light coat of oil to the cylinder head bolt threads

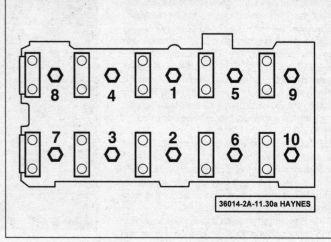

11.30a Cylinder head bolt tightening sequence - 2.0L engine shown, 2.3L engine identical

11.30b You can use a torque angle gauge, or you can carefully note the starting and stopping points of the wrench handle

Installation

24 Wipe clean the mating surfaces of the cylinder head and cylinder block. If equipped, install the alignment dowels into their original locations.

25 The cylinder head bolt holes must be free from oil or water. This is most important, because a hydraulic lock in a cylinder head bolt hole can cause a fracture of the block casting when the bolt is tightened. Note the location of the cylinder head alignment dowels in the block.

26 Position a new gasket on the cylinder block surface, so that the "TOP" mark is facing up.

27 As the cylinder head is such a heavy and awkward assembly to install, it is helpful to make up a pair of guide studs from two 10 mm (thread size) studs approximately 90 mm long, with a screwdriver slot cut in one end - you can use two of the old cylinder head bolts with their heads cut off. Screw these guide studs, screwdriver slot upwards to permit removal, into the bolt holes at diagonally-opposite corners of the cylinder block surface; ensure that approximately 70 mm of stud protrudes above the gasket.

28 Install the cylinder head, sliding it down the guide studs (if used) and locating it on the dowels. Unscrew the guide studs (if used) when the head is in place.

29 Coat the threads with engine oil - do not apply more than a light film of oil (see illustration). Install the NEW cylinder head bolts and screw them in by hand only until finger-tight.

Note: *New cylinder head bolts must be used.*

30 Working progressively and in the sequence shown, first tighten all the bolts to the specified Step 1 torque setting listed in this Chapter's Specifications (see illustration). On these engines there are five tightening stages, the final two using the angle torque method (see illustration).

31 Replacement of the other components removed is a reversal of removal.

32 Change the engine oil and filter and refill the cooling system (see Chapter 1).

12 Oil pan - removal and installation

Removal

1 Raise the vehicle and support it securely on jackstands.

2 On 2.3L models, remove the engine cover by pulling straight up on the cover to disengage the grommets from the cover studs.

Caution: *The engine cover must be pulled straight up, do not pull the cover forward or sideways to remove it. The cover mounting points and studs are easily damaged if the cover is not pulled straight upward from the underside of the cover.*

3 Drain the engine oil (see Chapter 1), then clean and install the engine oil drain plug, tightening it to the torque listed in the Chapter 1 Specifications. Remove and discard the oil filter, so that it can be replaced with the oil.

4 Remove the engine lower splash shield.

5 Remove the air filter housing (see Chapter 4).

6 The transaxle must be moved slightly back from the engine. Support the engine from above with a support fixture (see Chapter 2C) connected to the left end of the engine, near the transaxle, and support the transaxle with a floor jack. Loosen the upper engine-to-transaxle mounting bolts and back them off about 3/16-inch (5 mm), loosen the left-side engine-to-bellhousing bolts, then loosen the right-side bolts.

Caution: *The transaxle must not be moved back more than 0.19-inch (5 mm) or the transaxle can be damaged.*

7 Remove the transaxle roll restrictor mount through bolts and remove the mount (see Chapter 7A).

8 Remove the catalytic converter (see Chapter 4).

9 Remove the three bellhousing-to-pan bolts, and one oil pan-to-bellhousing bolt. Also remove the timing chain cover-to-oil pan fasteners.

10 Remove the a/c compressor and tie the compressor out of the way (see Chapter 3), without disconnecting the compressor lines.

11 Use a screwdriver to pry between the engine and transaxle until the bellhousing has moved away from the block to the limit of the loosened bolts (about 0.20-inch [5 mm]).

12 Progressively unscrew the oil pan retaining bolts evenly until all the bolts are removed. Use a rubber mallet to loosen the oil pan seal, then lower the oil pan, turning it as necessary to clear the exhaust system. Unfortunately, the use of sealant can make removal of the oil pan more difficult. Be careful when prying between the mating surfaces, otherwise they will be damaged, resulting in leaks when finished. With care, a putty knife can be used to cut through the sealant.

Chapter 2 Part A Four-cylinder engines

13.6 The oil pump pick-up tube is held by two mounting bolts (typical)

13.8 Using a holding tool on the oil pump drive sprocket to remove the retaining bolt

13.9 Remove the four mounting bolts for the oil pump

Installation

13   Thoroughly clean and degrease the mating surfaces of the lower engine block/crankcase and oil pan, removing all traces of sealant, then use a clean rag to wipe out the oil pan.
Caution: *Do not use wire brush or power abrasive discs to clean the mounting surfaces or damage to the pan may occur.*
14   Apply a 1/8-inch wide bead of sealant to the oil pan flange so that the bead is approximately 3/16-inch from the outside edge of the flange. Make sure the bead is around the inside edge of the bolt holes. Also apply sealant to the front flange of the oil pan where it meets the timing chain cover.
Note: *The oil pan must be installed within 10 minutes of applying the sealant.*
15   Install the oil pan bolts, only tightening them finger tight at this time.
16   Install the timing chain cover-to-oil pan fasteners and tighten them to the torque listed in this Chapter's Specifications.
17   Tighten the oil pan-to engine block bolts, a little at a time, working from the center outwards in a circular pattern, to the torque listed in this Chapter's Specifications.
18   Tighten the oil pan-to-bellhousing bolts and the transaxle to engine bolts, a little at a time to draw them together evenly, to the torque listed in this Chapter's Specifications.
19   Lower the vehicle to the ground. Before refilling the engine with oil, wait at least 1 hour for the sealant to cure, or whatever time is indicated by the sealant manufacturer. Trim off the excess sealant with a sharp knife. Install a new oil filter (see Chapter 1).

13   Oil pump - removal and installation

Note: *The oil pump is serviced as a complete unit without any sub-assembly or internal inspection.*

1   Drain the engine oil and remove the oil filter (see Chapter 1).
2   Support the engine from above with a support fixture (see Chapter 2C).
3   Remove the timing chain cover (see Section 8).
4   Remove the a/c compressor fasteners and tie the compressor out of the way without disconnecting the compressor lines.
5   Remove the oil pan (see Section 12).
6   Remove the oil pick-up tube (see illustration) and remove the O-ring from the sealing surface of the tube.
7   Use a screwdriver to pry the end of the oil pump drive chain tensioner's spring from under the shouldered bolt. Remove the two bolts and the tensioner.
8   Remove the chain from the oil pump sprocket. While holding the oil pump drive sprocket with a suitable tool, remove the sprocket bolt from the oil pump, then remove the sprocket (see illustration).
9   Remove the oil pump mounting bolts, then remove the pump (see illustration).
10   Installation is the reverse of removal, noting the following points:

   a) *Replace all gaskets with new ones.*
   b) *Tighten the oil pump mounting bolts to the torque listed in this Chapter's Specifications in a criss-cross pattern.*
   c) *After installing the oil pan, install a new oil filter and refill the crankcase with oil (see Chapter 1).*
   d) *Be certain to check for any oil warning lights in the instrument panel after the vehicle has been started and idling.*

14   Driveplate - removal, inspection and installation

Removal

1   Remove the transaxle (see Chapter 7A). Now is a good time to check components such as oil seals and replace them if necessary.
2   Use a center-punch or paint to make alignment marks on the driveplate and crankshaft to make replacement easier - the bolt holes are slightly offset, and will only line up one way, but making a mark eliminates the guesswork.
3   Hold the driveplate stationary and unscrew the bolts. To prevent the driveplate from turning, insert one of the transaxle mounting bolts into the cylinder block and have an assistant engage a wide-bladed screwdriver with the starter ring gear teeth while the driveplate bolts are loosened.
4   Loosen and remove each bolt in turn and ensure that new replacements are obtained for reassembly. These bolts are subjected to severe stresses and so must be replaced, regardless of their apparent condition, whenever they are removed.
5   Remove the driveplate - do not drop it.

Inspection

6   Clean the driveplate to remove grease and oil. Check for cracked and broken ring gear teeth.
7   Clean and inspect the mating surfaces of the flywheel and the crankshaft. If the oil seal is leaking, replace it (see Section 15) before replacing the driveplate. If the engine has high mileage, it may be worth installing a new seal as a matter of course, given the amount of work needed to access it.

Installation

8   On installation, ensure that the engine/transaxle adapter plate is in place (where necessary), then install the driveplate on the crankshaft so that all bolt holes align - it will fit only one way - check this using the marks made on removal. Install the new bolts, tightening them by hand.
9   Lock the driveplate by the method used on disassembly. Working in a diagonal sequence to tighten them evenly and increasing to the final amount in two or three stages, tighten the new bolts to the torque listed in this Chapter's Specifications.
10   The remainder of reassembly is the reverse of the removal procedure

15 Rear main oil seal - replacement

1 The one-piece rear main oil seal is pressed into the rear main oil seal carrier mounted at the rear of the block. Remove the transaxle (see Chapter 7A) and the driveplate (see Section 14).
2 Remove the oil pan (see Section 12).
3 Unbolt the oil seal and carrier.
4 Clean the mating surface for the oil seal carrier on the cylinder block and the crankshaft. Carefully remove and polish any burrs or raised edges on the crankshaft that may have caused the seal to fail.
5 Apply a 3 mm bead of silicone gasket RTV sealant to the oil pan surface and a 5 mm bead at the T-joints.
Caution: *If the crankshaft rear seal is not installed within 10 minutes of sealant application, the sealant must be removed and the surface area cleaned. Failure to install the seal within the allotted time can cause an oil leakage*
6 Lightly coat the inside lip of the new seal with clean engine oil. Use a thin (but durable) two-inch wide plastic strip (or a two-liter plastic beverage bottle cut to size) around the inside circumference of the seal to act as a liner for installation. The manufacturer tool (#303-328) for this purpose also works well.
7 With the plastic seal liner or tool in place, carefully move the new carrier (with seal factory-installed) into position by sliding it onto the contact surface of the crankshaft.
Note: *Start install the seal housing at a slight angle to prevent scraping the RTV sealant off the oil pan surface.*
8 Install the oil seal carrier bolts and finger tighten them while holding the carrier in place. Align the bottom of the seal carrier precisely with the bottom edge of the engine block to ensure that the surfaces are flush before tightening the carrier mounting bolts.
Caution: *The oil pan may leak if the two surfaces are not perfectly flush.*
9 Carefully remove the plastic liner or tool so that the new seal contacts the crankshaft mating surface correctly.
10 Tighten the oil seal carrier to the torque listed in this Chapter's Specifications using the proper sequence (see illustration).
11 The remainder of installation is the reverse of removal.

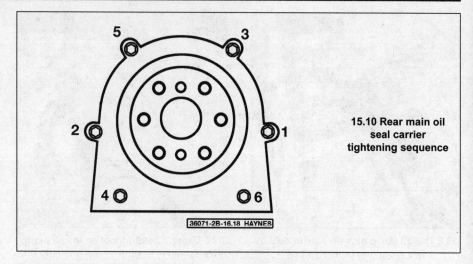

15.10 Rear main oil seal carrier tightening sequence

16 Engine mount - check and replacement

1 Engine mounts seldom require attention, but broken or deteriorated mounts should be replaced immediately or the added strain placed on the driveline components may cause damage or wear.

Check

2 During the check, the engine must be raised slightly to remove the weight from the mounts.
3 Raise the vehicle and support it securely on jackstands, then position a jack under the engine oil pan. Place a large wood block between the jack head and the oil pan to prevent oil pan damage, then carefully raise the engine just enough to take the weight off the mounts.
Warning: *DO NOT place any part of your body under the engine when it's supported only by a jack!*
4 Check the mount to see if the rubber is cracked, hardened or separated from the bushing in the center of the mount.
5 Check for relative movement between the mount and the engine or chassis. Use a large screwdriver or prybar to attempt to move the mounts. If movement is noted, lower the engine and tighten the mount fasteners.

Replacement

Warning: *Wait until the engine is completely cool before beginning this procedure.*
Note: *Refer to Chapter 7A for information on the transaxle mount.*
6 Disconnect the cable from the negative terminal of the battery (see Chapter 5).
7 Remove the engine splash shield fasteners and shield.
8 Remove the coolant expansion tank (see Chapter 3).
9 Place a floor jack under the engine with a wood block between the jack head and oil pan, then raise the engine slightly to relieve the weight from the mount.
10 Remove the power steering fluid reservoir nuts and place the reservoir out of the way.
11 Disengage the wiring harness retainer from the valve cover stud and move the harness out of the way.
12 Remove the fasteners and detach the mount from the frame and engine.
Caution: *Do not disconnect more than one mount at a time, except during engine removal.*
13 Installation is the reverse of removal. Use thread-locking compound on the mount bolts and be sure to tighten them securely.

Chapter 2 Part B
V6 engines

Contents

	Section		Section
Camshafts and tappets/roller followers and lash adjusters - removal, inspection and installation	11	Oil pan - removal and installation	12
		Oil pump - removal and installation	13
Crankshaft pulley and crankshaft front oil seal - replacement	7	Rear main oil seal - replacement	15
Cylinder heads - removal and installation	8	Repair operations possible with the engine in the vehicle	2
Driveplate - removal and installation	14	Timing chain cover - removal and installation	9
Engine mounts - check and replacement	16	Timing chains and sprockets - removal and installation	10
Exhaust manifolds - removal and installation	6	Valve clearance - check and adjustment	3
General information	1	Valve covers - removal and installation	4
Intake manifold(s) - removal and installation	5		

Specifications

General

Displacement	214 cubic inches
Engine VIN code	
3.5L Duratec engine	8
3.5L EcoBoost engine	T
Bore	3.641 inches (92.5 mm)
Stroke	3.413 inches (86.7 mm)
Cylinder numbers (front-to-rear)	
Front cylinder bank	4-5-6
Rear cylinder bank	1-2-3
Firing order	1-4-2-5-3-6
Compression ratio	
3.5L Duratec engine	10.8: 1
3.5L EcoBoost engine	10.0: 1
Cylinder head warpage limit	
Lengthwise	0.003 inch
Widthwise	0.002 inch
Compression pressure	See Chapter 2C
Oil pressure	See Chapter 2C

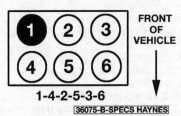

Cylinder locations

Camshaft

Lobe lift
  3.5L Duratec engine
    Intake .... 0.394 inch (10 mm)
    Exhaust .... 0.380 inch (9.68 mm)
  3.5L EcoBoost engine
    Intake .... 0.354 inch (9.0 mm)
    Exhaust .... 0.358 inch (9.1 mm)
Lobe wear limit .... 0.0024 inch
Endplay
  Standard .... 0.0011 to 0.0067 inch (0.27 to 0.17 mm)
  Service limit .... 0.0075 inch (0.19 mm)
Camshaft journal bore inside diameter
  First journal .... 1.537 to 1.538 inches (39.0375 to 39.0625 mm)
  Intermediate journals .... 1.023 to 1.024 inches (25.9875 to 26.0125 mm)
Camshaft bearing outside diameter
  First journal .... 1.535 to 1.536 inches (38.99 to 39.01 mm)
  Intermediate journals .... 1.021 to 1.022 inches (25.937 to 25.963 mm)
Journal-to-bearing (oil) clearance
  First journal .... 0.0029 inch (.0725 mm)
  Intermediate journals .... 0.0030 inch (.0755 mm)
Journal runout limit .... 0.0016 inch (0.04 mm)
Valve clearance - room temperature, measured with camshaft at base circle
  Intake .... 0.006 to 0.010 in (0.15 to 0.25 mm)
  Exhaust .... 0.010 to 0.020 in (0.36 to 0.46 mm)

Torque specifications* Ft-lbs (unless otherwise indicated)

Note: *One foot-pound (ft-lb) of torque is equivalent to 12 inch-pounds (in-lbs) of torque. Torque values below approximately 15 ft-lbs are expressed in inch-pounds, since most foot-pound torque wrenches are not accurate at these smaller values.*

Variable Camshaft Timing (VCT) unit (camshaft sprocket) bolt (new)
  3.5L EcoBoost engine
    Step 1 30
    Step 2 .... Loosen one full turn
    Step 3 .... 89 in-lbs
    Step 4 .... Tighten an additional 90-degrees
  3.5L Duratec engine
    Step 1 30
    Step 2 .... Loosen one full turn
    Step 3 18
    Step 4 .... Tighten an additional 180-degrees
Camshaft cap bolts (see illustration 11.25c, 11.25d or 11.39)
  Step 1 .... 71 in-lbs
  Step 2 .... Tighten an additional 45 degrees
  Step 3 .... Loosen bolts 8 through 11
  Step 4 .... Repeat Steps 1 and 2 on bolts 8 through 11
Crankshaft pulley bolt
  3.5L Duratec engines and 2015 and earlier 3.5L EcoBoost engines
    Step 1 .... 89 in-lbs
    Step 2 .... Loosen one turn
    Step 3 37
    Step 4 .... Tighten an additional 90-degrees
  2016 and later 3.5L EcoBoost engine
    Step 1 89
    Step 2 .... Loosen one turn
    Step 3 37
    Step 4 66
Crankshaft rear seal retainer .... 89 in-lbs
  Crankshaft rear seal retainer-to-engine block bolts .... 89 in-lbs
  Crankshaft rear seal retainer-to-oil pan bolts
    2015 and earlier models
      Step 1 .... 89 in-lbs
      Step 2 .... Tighten an additional 45-degrees
    2016 and later models .... 89 in-lbs
Crossmember bolts 66

Chapter 2B V6 engines

Torque specifications*   Ft-lbs (unless otherwise indicated)

Note: *One foot-pound (ft-lb) of torque is equivalent to 12 inch-pounds (in-lbs) of torque. Torque values below approximately 15 ft-lbs are expressed in inch-pounds, since most foot-pound torque wrenches are not accurate at these smaller values.*

Specification	Value
Cylinder head bolts (new, oiled) (in sequence - see illustration 8.22a or 8.22b)	
Step 1	177 in-lbs
Step 2	26
Step 3	Tighten an additional 90-degrees
Step 4	Tighten an additional 90-degrees
Step 5	Tighten an additional 45-degrees
M6 bolt at front of cylinder head	89 in-lbs
Drivebelt tensioner pulley mounting bolt(s)	18
Exhaust manifold nuts	
Step 1	168 in-lbs
Step 2	18
Exhaust manifold studs	106 in-lbs
Driveplate bolts	59
Intake manifold bolts	
3.5L turbocharged engine (see illustration 5.43)	
Step 1	89 in-lbs
Step 2	Tighten an additional 45-degrees
3.5L non-turbochrged engine	
Lower intake manifold-to-cylinder heads see illustration 5.22)	89 in-lbs
Upper intake manifold-to-lower intake manifold (see illustration 5.28)	
Step 1	89 in-lbs
Step 2	Tighten an additional 45-degrees
Oil pan-to-block bolts (see illustration 12.17)	89 in-lbs
Oil pump screen nuts-to-engine block nuts	89 in-lbs
Oil pump mounting bolts	89 in-lbs
Oil filter adapter bolt	
Step 1	89 in-lbs
Step 2	Tighten an additional 45 degrees
Engine front cover-to-block bolts (see illustration 9.25)	
Step 1, bolts 1 through 21	89 in-lbs
Step 2, bolts 1 through 20	18
Step 3, bolt 21	177 in-lbs
Step 4, bolt 21	Tighten an additional 90-degrees
Step 5, bolt 22	89 in-lbs
Step 6, bolt 22	Tighten an additional 45 degrees
Oil pan-to-timing chain cover bolts	89 in-lbs
Engine mount fasteners	
Through-bolts	258
Mount bracket-to-engine block bolts	57
Mount-to-frame bolts	129
Mount nuts	111
Timing chain tensioner bolts	96 in-lbs
Turbocharger-to-exhaust manifold bolts	24
Turbocharger bracket-to-block bolts	89 in-lbs
Turbocharger bracket-to-turbocharger bolt	159 in-lbs
Valve cover bolts	89 in-lbs
Variable Camshaft Timing oil control solenoid bolts	
Step 1	71 in-lbs
Step 2	Tighten an additional 20 degrees

*Note: *Refer to Part C for additional specifications.*

3.5 Measure the clearance for each valve with a feeler gauge of the specified thickness - if the clearance is correct, you should feel a slight drag on the gauge as you pull it out

3.8 Use a micrometer to measure the thickness of the head of the tappet

1 General information

1 Chapter 2B is devoted to in-vehicle repair procedures for the 3.5L turbocharged EcoBoost V6 and the 3.5L Duratec non-turbocharged V6. These engines utilize a cast-aluminum block, all have six cylinders arranged in a V-shape at a 60-degree angle between the two banks. The cylinder heads on all models are cast-aluminum and the valve actuation is via two overhead camshafts in each cylinder head with four valves per cylinder (one for each intake valve one for each exhaust valve). 3.5L turbocharged engines are equipped with direct injection of fuel into the cylinders and use two turbochargers, while the 3.5L non-turbocharged engine does not have direct injection or turbochargers. Information concerning engine removal and installation and engine overhaul can be found in Chapter 2C. The following repair procedures are based on the assumption that the engine is installed in the vehicle. If the engine has been removed from the vehicle and mounted on a stand, many of the steps outlined in this Chapter will not apply.

2 Repair operations possible with the engine in the vehicle

1 Many major repair operations can be accomplished without removing the engine from the vehicle.
2 Clean the engine compartment and the exterior of the engine with some type of pressure washer before any work is done. It will make the job easier and help keep dirt out of the internal areas of the engine.
3 If vacuum, exhaust, oil or coolant leaks develop, indicating a need for gasket or seal replacement, the repairs can generally be made with the engine in the vehicle. The intake and exhaust manifold gaskets, timing chain cover gasket, oil pan gasket and crankshaft oil seals are all accessible with the engine in place.
4 Exterior engine components, such as the intake and exhaust manifolds, the oil pan (and the oil pump), the water pump, the starter motor, the alternator, and the fuel system components can be removed for repair with the engine in place.
5 Replacement of the timing chain and sprockets is also possible with the engine in the vehicle. Camshaft and cylinder hear removal and installation requires removal of the engine.
6 In extreme cases caused by a lack of necessary equipment, repair or replacement of piston rings, pistons, connecting rods and rod bearings is possible with the engine in the vehicle. However, this practice is not recommended because of the cleaning and preparation work that must be done to the components involved.

3 Valve clearance - check and adjustment

Note: *The engine must be room temperature before checking the valve clearances.*
Note: *Checking and, if necessary, adjusting the valve clearance is only necessary after replacement of the camshaft(s) or other valve-related parts, or if the valvetrain is making excessive noise.*

1 Disconnect the cable from the negative terminal of the battery (see Chapter 5).
2 Remove the spark plugs (see Chapter 1).
3 Remove the valve covers (see Section 4).
4 Using a socket and a breaker bar on the crankshaft pulley center bolt, rotate the engine until the cam lobes on the cylinder to be checked are pointing away from the tappets.
5 Measure the clearance of the indicated valves with a feeler gauge (see illustration). Record each measurement and compare your measurements with the desired valve clearance found in this Chapter's Specifications. Note which are out of specification, as this data will be used later to determine the required replacement tappets.
6 Repeat Steps 4 and 5 until the clearances for all valves have been measured.
7 If a clearance is out of specification, the tappet must be replaced with a new tappet that has a different thickness head to correct the clearance. Refer to Section 11 and remove the camshafts to access the tappets.
8 Mark the lifters that are to be replaced, and record which valve they came from. Use a micrometer to measure the thickness of the head of the lifter, making sure the measurement is precise and on the center projection on the underside of the lifter (see illustration).
9 To calculate the correct thickness for a replacement lifter that will place the valve clearance within the specified value, use the following formula:

$N = R + (M1 - M2)$
N = thickness of new tappet
R = measured valve clearance
$M1$ = thickness of old tappet
$M2$ = standard valve clearance

10 Lifters are marked on the underside as to their size. A marking of 3.310 on the underside of the lifter indicates a thickness value of 3.31 mm, or 0.13 inch.
11 Mark the new lifters as to their destination, lubricate them with engine assembly lube and install them. After replacing the lifters, refer to Section 11 to reinstall the camshafts and Section 10 to reinstall the timing chains, then re-check the valve clearances.

4 Valve covers - removal and installation

Removal

1 If you're working on a turbocharged model, relieve the fuel system pressure (see Chapter 4).
2 Disconnect the cable from the negative battery terminal (see Chapter 5, Section 3).
3 Remove the engine cover, if equipped.

Front cylinder bank

Non-turbocharged engine

4 Disconnect the vent tube from the valve cover (see illustration).
5 Remove the dipstick.
6 Remove the ground wire fasteners from the strut tower and move the wires out of the way.

Turbocharged engine

7 Disconnect the electrical connector to the turbocharger boost pressure sensor/charge air cooler, then loosen the clamps and remove the charge air cooler pipe.
8 Remove the dipstick.
9 Disconnect the quick-connectors for the PCV hose then detach the PCV hose from the valve cover and manifold.

Chapter 2B V6 engines

4.4 Detach the vent tube from the valve cover - non-turbocharged model shown

4.17 Disconnect the electrical connectors from the VCT oil control solenoids in each valve cover. It's a good idea to mark one of the electrical connectors and solenoids to prevent mixing up the connectors on reassembly

10 Remove the sound insulator shield for the high-pressure fuel injection pump.
11 Remove the intake manifold (see Section 5).
12 Remove the nut securing the high-pressure fuel tube to the valve cover, then remove the high-pressure fuel tube flare nuts from the fuel injection pump and fuel rail (see Chapter 4).
13 Remove the high-pressure fuel pump (see Chapter 4).

Rear cylinder bank

Non-turbocharged engine

14 Remove the upper intake manifold (see Section 5).
15 Detach the PCV hose from the valve cover.

Both valve covers

16 Remove the ignition coils from the cover to be removed (see Chapter 5).
17 Disconnect the two electrical connectors at the valve cover for the VCT solenoids (being careful not to twist the solenoids) (see illustration) and the electrical connectors from the camshaft position sensors (CMP).
18 Remove the valve cover fasteners, then separate the wiring/clips and remove the valve cover from the cylinder head (see illustration). Wiggle the valve cover by hand to detach it from the cylinder head if it's stuck.
Caution: *Do not use a hammer to knock the cover loose. Try to slip a flexible putty knife between the cylinder head and cover to break the gasket seal. Don't pry at the cover-*

to-cylinder head joint or damage to the sealing surfaces may occur (leading to oil leaks in the future).

Installation

19 The mating surfaces of each cylinder head and valve cover must be perfectly clean when the covers are installed. Remove all traces of sealant and old gasket material, then clean the mating surfaces with brake cleaner. If there's sealant or oil on the mating surfaces when the cover is installed, oil leaks may develop.
20 Inspect the spark plug seals, the seals for the VCT solenoids and the valve cover gasket (see illustration). Do not replace them unless necessary.

4.18 Valve cover fasteners - non-turbocharged model shown

4.20 Check the spark plug tube seals, VCT solenoid seals and valve cover gasket. If they are in good condition they can be reused

Chapter 2B V6 engines

4.21 Check the bolt seals - replace them if they're not in good condition

4.22 Apply RTV sealant to the seams where the timing chain cover meets the cylinder head

21 Make sure the bolt seals are in good condition, too (see illustration).
22 Apply a bead of RTV sealant to the joints where the timing chain cover meets the cylinder head and install the valve cover within four minutes (see illustration).
23 Carefully position the cover on the cylinder head and install the bolts.
24 Tighten the bolts a little at a time, in the proper sequence (see illustrations), to the torque listed in this Chapter's Specifications.
25 The remaining installation steps are the reverse of removal.
26 Start the engine and check carefully for oil leaks as the engine warms up.

5 Intake manifold(s) - removal and installation

Warning: *If you're working on a 3.5L non-turbocharged engine and are going to be removing the lower intake manifold, relieve the fuel system pressure (see Chapter 4).*

1 Disconnect the cable from the negative battery terminal (see Chapter 5, Section 3).
2 Remove the engine cover.
3 Remove the air filter housing and intake duct (see Chapter 4).

Non-turbocharged engine
Removal
4 If you're going to be removing the lower intake manifold, relieve the fuel system pressure (if not already done) (see Chapter 4).

Upper intake manifold
5 Remove the PCV tube from the manifold.
6 Disconnect the electrical connector from the throttle body (see Chapter 6).
7 Disconnect the EVAP tube and the EVAP canister purge valve electrical connector (see Chapter 6).
8 Detach the coolant tube retainers and move the coolant tube out of the way.
9 Disconnect the vacuum hose connector and the brake booster vacuum hose from the upper intake manifold.
10 Disconnect the PCV valve hoses from the valve covers and remove them from the engine (see Chapter 6).
11 Remove the upper intake support bracket bolt, located under the throttle body.
12 Remove the mounting bolts from the upper intake manifold, then lift the upper manifold off of the lower manifold.
13 Cover the air intake passages with a shop towel to prevent tools or debris from falling inside the engine.

Lower intake manifold
Warning: *Wait until the engine is completely cool before beginning this procedure.*
14 Remove the upper intake manifold.
15 Drain the cooling system (see Chapter 1).
16 Cover the drivebelt with plastic wrap, then remove the thermostat housing (see Chapter 3). Disconnect the heater hose from the manifold.
17 Detach the fuel line from the fuel rail (see Chapter 4) then remove the fuel rail insulators.

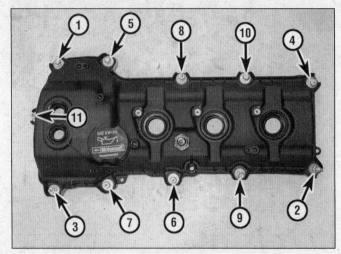

4.24a Valve cover bolt tightening sequence - non-turbocharged model (front valve cover shown, rear cover similar)

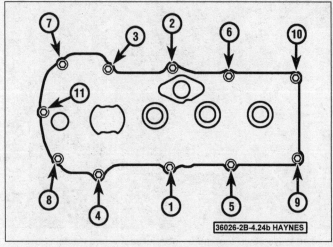

4.24b Valve cover bolt tightening sequence - turbocharged model (front valve cover shown, rear cover similar)

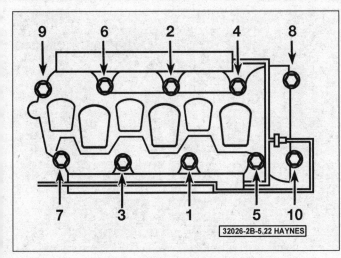

5.22 Lower intake manifold bolt tightening sequence - non-turbocharged engine

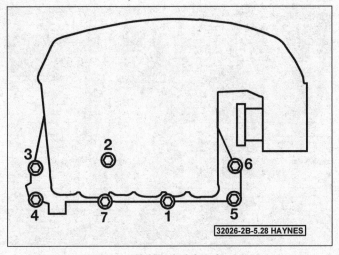

5.28 Upper intake manifold bolt tightening sequence - non-turbocharged engine

18 Disconnect the electrical connectors from the injectors.
19 Loosen the intake manifold mounting bolts in 1/4-turn increments in the reverse of the tightening sequence (see illustration 5.22) until they can be removed by hand.

Installation

Lower intake manifold

20 Remove the gaskets from the grooves in the manifold. Clean the manifold and cylinder head mating surfaces.
21 Install new gaskets into the grooves in the manifold, then carefully set the manifold in place.
22 Install the bolts and tighten them to the torque listed in this Chapter's Specifications, following the recommended sequence (see illustration).
23 The remaining installation steps are the reverse of removal.
24 Refill the cooling system (see Chapter 1).
25 Start the engine and check carefully for oil or coolant leaks at the intake manifold joints.

Upper intake manifold

26 Remove the gaskets from the grooves in the manifold. Clean the mating surfaces of the upper and lower manifolds.
27 Install new gaskets into the grooves in the manifold, then carefully set the manifold in place.
28 Install the bolts and tighten them to the torque listed in this Chapter's Specifications, following the recommended sequence (see illustration).
29 The remaining installation steps are the reverse of removal.

Turbocharged engine

Removal

30 Drain the cooling system (see Chapter 1).
31 Disconnect the electrical connector to the turbocharger boost pressure sensor/charge air cooler, then loosen the clamps and

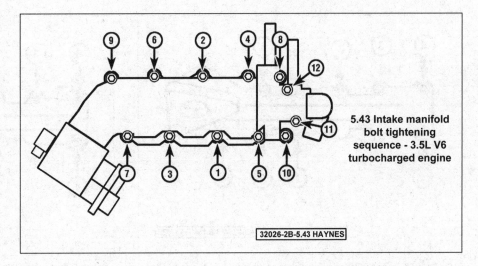

5.43 Intake manifold bolt tightening sequence - 3.5L V6 turbocharged engine

remove the charge air cooler pipe (see Chapter 4).
32 Disconnect both turbocharger bypass valve electrical connectors and turbocharger bypass valve hoses.
33 Remove the left and right turbocharger wastegate regulating valve hoses from the charge air cooler tubes and turbocharger wastegate regulating valves.
Note: *It's a good idea to make a index mark on the hose connections so they can be installed in their original location.*
34 Loosen the large hose clamp and detach the air intake duct from the throttle body.
35 Disconnect the electrical connector at the throttle body.
36 Disconnect the PCV electrical connector.
37 Disconnect the brake booster vacuum hose from the intake manifold, then disconnect the MAP sensor and the EVAP canister purge valve connectors.
38 Disconnect the turbocharger vacuum regulator hose from the intake manifold, then release the pushpin securing the turbocharger vacuum regulator to the intake manifold.

39 Disconnect the heater hoses to the rear of the manifold then disconnect the turbocharger coolant hoses located below the heater hose fittings from the manifold.
40 Disconnect the upper radiator hose then remove the fuel tube mounting bolt, once the mounting bolt is removed the coolant tube fitting can be disconnected and the tube removed.
41 Remove the manifold mounting bolts and detach the manifold from the engine.
Note: *Do not use a hammer, and once the manifold is off, cover the intake ports of the engine with clean shop rags.*

Installation

42 Remove the old gaskets from the manifold. Clean the manifold and cylinder head mating surfaces.
43 Install the new manifold gaskets to the grooves in the manifold. Install the manifold and bolts. Tighten the bolts in the indicated sequence to the torque listed in this Chapter's Specifications (see illustration).
44 The remainder of the installation is the reverse of removal.

6.3 Underbody air duct fastener locations

6.7 Apply penetrating oil to the exhaust Y-pipe-to-manifold studs

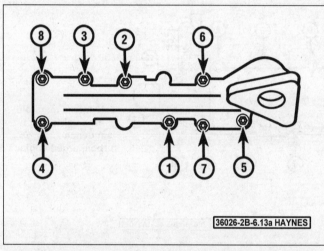

6.13a 3.5L Rear cylinder bank exhaust manifold bolt tightening sequence (turbocharged engine)

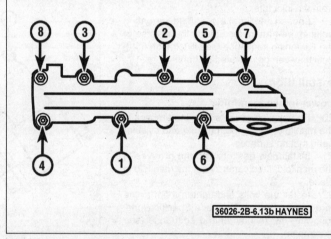

6.13b 3.5L Front cylinder bank exhaust manifold bolt tightening sequence (turbocharged engine)

6 Exhaust manifolds - removal and installation

Removal

Note: *On non turbocharged engines, the exhaust manifold is integrated into the catalytic converter and is not serviceable separately from the converter.*

1 Disconnect the cable from the negative battery terminal (see Chapter 5).
2 Raise the vehicle and support it securely on jackstands.
3 Twist the lock retainers securing the underbody air duct (see illustration) and remove the air duct.
4 Remove the upper and lower exhaust manifold heat shields.
5 On turbocharged models, remove the turbocharger from the manifold you're working on (see Chapter 4).
6 Disconnect the oxygen sensor electrical connectors (see Chapter 6).
7 On non-turbocharged models, working under the vehicle, apply penetrating oil to the exhaust Y-pipe-to-manifold studs (see illustrations) and nuts (they're usually rusty).
8 On 3.5L non-turbocharged engines, remove the nuts holding the catalytic converter-to-exhaust manifolds.
9 Remove the manifold mounting nuts and detach the manifold from the cylinder head.
10 Remove the exhaust gaskets and clean the gasket surfaces on the manifold and cylinder heads.

Installation

Note: *The manufacturer recommends removing the exhaust mounting studs from the cylinder heads and installing new studs.*

11 Check the exhaust manifold for cracks and make sure all the stud threads are clean and undamaged. The exhaust manifold and cylinder head mating surfaces must be clean before the manifolds are reinstalled.
12 Position the exhaust manifold and new gasket over the studs on the cylinder head and install the mounting nuts.
13 Tighten the nuts in sequence (see illustrations) to the torque listed in this Chapter's Specifications is reached.
14 The remaining installation steps are the reverse of removal.
15 Start the engine and check for exhaust leaks.

7 Crankshaft pulley and crankshaft front oil seal - replacement

Replacement with timing chain cover in place

1 Disconnect the cable from the negative battery terminal (see Chapter 5).
2 Remove the drivebelt(s) (see Chapter 1).
3 Secure the crankshaft from rotating with a strap or chain wrench (wrap a length of rag or old drivebelt around the pulley to protect it). Remove the bolt from the front of the crankshaft, then use a three-jaw puller to detach the crankshaft pulley. Clean the crankshaft nose and the seal contact surface on the pulley with RTV remover. Leave the Woodruff key in place in the crankshaft keyway.
Caution: *Be sure to use an adapter between the puller screw and the nose of the crankshaft so as not to damage the threads in the crankshaft. Also, don't use a puller with jaws that grip the outer edge of the damper. The puller must be the type that applies force to the damper hub only.*
4 Carefully remove the seal from the timing chain cover with a screwdriver or seal removal tool. Be careful not to damage the cover or scratch the wall of the seal bore. If the engine has accumulated a lot of miles, apply penetrating oil to the seal-to-cover joint and allow it to soak in before attempting to remove the seal.
5 Check the seal bore and crankshaft, as well as the seal contact surface on the pulley for nicks and burrs. Position the new seal in the bore with the open end of the seal facing IN. A film of engine oil applied to the outer edge of the new seal will make installation easier.
6 Drive the seal into the bore with a seal driver or a large socket and hammer until it's completely seated. If you're using a socket, select one that's the same outside diameter as the seal.
7 Apply clean engine oil to the seal contact surface of the crankshaft pulley and coat the keyway (groove) with a thin layer of RTV sealant.
8 Install the pulley on the end of the crankshaft. The keyway in the pulley bore must be aligned with the Woodruff key in the crankshaft nose. If the pulley can't be seated by hand, tap it into place with a soft-face hammer or slip a large washer over the bolt, install the bolt and tighten it to press the pulley into place. Remove the large washer, then install the bolt and tighten it in Steps to the torque listed in this Chapter's Specifications.
9 Install the remaining parts removed for access to the seal.
10 Start the engine and check for leaks.

Replacement with engine front cover removed

11 Remove the engine front cover (see Section 9).

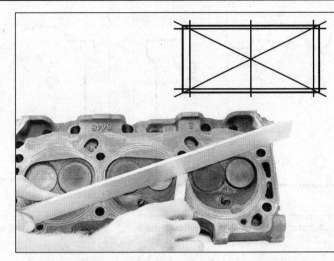

8.17 Check the cylinder head and block surfaces for flatness with a straightedge

12 Use a punch or screwdriver and hammer to drive the seal out of the cover from the back side. Support the cover as close to the seal bore as possible, using blocks of wood. Be careful not to distort the cover or scratch the wall of the seal bore. If the engine has accumulated a lot of miles, apply penetrating oil to the seal-to-cover joint on each side and allow it to soak in before attempting to drive the seal out.
13 Clean the bore to remove any old seal material and corrosion. Support the cover on blocks of wood and position the new seal in the bore with the open end of the seal facing IN. A film of oil applied to the outer edge of the new seal will make installation easier.
14 Drive the seal into the bore with a seal driver or a large socket and hammer until it's completely seated. If you're using a socket, select one that's the same outside diameter as the seal.
15 Reinstall the engine front cover (see Section 9) and crankshaft pulley (see Step 8).
16 Start the engine and check for leaks.

8 Cylinder heads - removal and installation

Warning: *Wait until the engine is completely cool before beginning this procedure.*
Note: *The engine must be removed to remove the cylinder head(s) (see Chapter 2C).*

Removal

1 Relieve the fuel system pressure (see Chapter 4), then disconnect the cable from the negative battery terminal (see Chapter 5, Section 3).
2 Remove the engine and transaxle assembly (see Chapter 2C).
3 Remove the drivebelt and the drivebelt tensioner (see Chapter 1).
4 Remove the intake manifold(s) (see Section 5).
5 Remove the coolant crossover manifold.
6 Remove the turbocharger(s) (see Chapter 4), if equipped.
7 Remove the exhaust manifold (non-turbocharged models) (see Section 6).
8 On turbocharged models, disconnect the high-pressure fuel line(s) from the fuel rail and high-pressure fuel pump (see Chapter 4).
Caution: *The manufacturer requires the fuel line from the fuel rail to the high-pressure fuel pump always be replaced once it has been removed.*
9 Remove the valve covers (see Section 4).
10 Remove the timing chain cover (see Section 9), timing chains and timing chain guides (see Section 10).
11 Remove the camshafts and tappets (see Section 11).
12 Remove the two Camshaft Position (CMP) sensor(s) at the rear of each cylinder head (see Chapter 6), then disconnect any other electrical connectors from the cylinder head and mark them with tape for correct reassembly.
13 Loosen the cylinder head bolts in 1/4-turn increments until they can be removed by hand. Work from bolt-to-bolt in a pattern that's the reverse of the tightening sequence (see illustration 8.22a or 8.22b).
Caution: *Remove the bolts and discard them - new bolts must be used when installing the cylinder head(s).*
14 Lift the cylinder head(s) off the engine. If resistance is felt, DO NOT pry between the cylinder head and engine block as damage to the mating surfaces will result. To dislodge the cylinder head, place a wood block against the end of it and strike the wood block with a hammer or pry against a casting protrusion. Store the cylinder heads on blocks of wood to prevent damage to the gasket sealing surfaces.
15 Cylinder head disassembly and inspection procedures should be performed by a qualified automotive machine shop.

Installation

16 The mating surfaces of the cylinder heads and engine block must be perfectly clean when the cylinder heads are installed.
Caution: *Do not use a scraper of any kind or abrasive discs to remove old gasket material, as the block, heads and intake manifold(s) are*

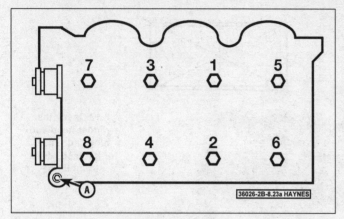

8.22a Cylinder head bolt tightening sequence - front cylinder bank

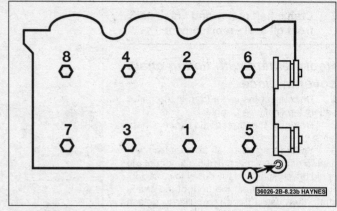

8.22b Cylinder head bolt tightening sequence - rear cylinder bank

aluminum. Cover all openings with shop rags to keep debris out of the engine. Use a vacuum cleaner to remove any debris that falls into the valley or intake ports.
17 Check the engine block and cylinder head mating surfaces for nicks, deep scratches and other damage. If damage is slight, it can be removed with a file - if it's excessive, machining may be the only alternative. Use a straightedge and feeler gauges to check for warpage (see illustration). If the warpage is beyond Specifications, have the head machined at an automotive machine shop.
18 Use a tap of the correct size to chase the threads in the block head bolt holes. Dirt, corrosion, sealant and damaged threads will affect torque readings.
19 Position the new gasket(s) over the dowel pins in the engine block. Make sure it's facing the right way. If the cylinder head is to be replaced, a new secondary timing chain tensioner will be required.
20 Carefully position the cylinder head(s) on the engine block without disturbing the gasket(s).
21 Before installing the new cylinder head bolts, lightly oil the threads.
22 Install the bolts and tighten them finger-tight. Tighten the bolts, in the recommended sequence, to the torque listed in this Chapter's Specifications (see illustrations).
23 Install and tighten the M6 bolt at the front of the cylinder head to the torque listed in this Chapter's Specifications.
24 The remaining installation steps are the reverse of removal.
25 Change the engine oil and filter and refill the cooling system (see Chapter 1), then start the engine and check carefully for oil and coolant leaks.

9 Timing chain cover - removal and installation

Warning: *The air conditioning system is under high pressure. DO NOT loosen any fittings or remove any components until after the system has been discharged. Air conditioning re-frigerant must be properly discharged into an EPA-approved container at a dealer service department or an automotive air conditioning repair facility. Always wear eye protection when disconnecting air conditioning system fittings.*
Warning: *The engine must be completely cool before beginning this procedure.*

Removal

1 Have the air conditioning system discharged by a licensed air conditioning technician.
2 Relieve the fuel system pressure (see Chapter 4), then disconnect the cable from the negative terminal of the battery (see Chapter 5).
3 Raise the vehicle and support it securely on jackstands. Remove the right inner fender splash shield (see Chapter 11), then remove the drivebelt (see Chapter 1).
4 Twist the lock retainers securing the underbody air duct (see illustration 6.3) and remove the air duct.
5 Drain the cooling system and engine oil, and remove the oil filter (see Chapter 1).
6 Disconnect the air conditioning tube connection at the rear of the engine. Cap the lines to prevent contamination.
7 Remove the expansion tank (see Chapter 3).
8 Remove the air filter housing and inlet and outlet ducts (see Chapter 4).
9 Disconnect the upper and lower radiator hoses from the engine, then secure the hoses out of the way.
10 On turbocharged models, remove the charge air cooler inlet and outlet pipes (see Chapter 4).
11 On turbocharged models, remove the turbocharger inlet and outlet tube (see Chapter 4).
12 Remove the valve covers (see Section 4).
13 Remove the crankshaft pulley and crankshaft front oil seal (see Section 7).
14 Remove all accessory brackets attached to the timing chain cover.
15 Remove drivebelt idler pulley bolt and pulley.
16 Remove the alternator (see Chapter 5).

17 Place a block of wood between the jack head and the oil pan, then carefully raise the engine or transaxle just enough to take the weight off the mounts. Remove the front engine mount and mounting bracket (see Section 16).
18 Remove the front cover bolts and separate the cover from the engine block.
Caution: *DO NOT use excessive force or you may crack the cover. If the cover is difficult to remove, double check to make sure all of the bolts have been removed.*
Note: *There are seven pry spots around the front cover. Work gently at each one until the cover comes off.*

Installation

19 Clean the gasket mating surfaces with brake system cleaner.
20 While the cover is off the engine, it's a good idea to install a new crankshaft front seal (see Section 7).
21 Apply a 1/8-inch bead of High Performance RTV sealant to the mating surface of the cover. Be sure to also apply the sealant to the bosses on the interior of the cover where the bolts go through.
22 Apply a 3/16-inch bead of High Performance RTV sealant to the areas where the oil pan and cylinder heads meet the engine block.
23 Slide the front cover onto the engine. The dowel pins will position it correctly.
Caution: *The front cover must be installed within 10 minutes of applying the RTV sealant or oil leaks may occur.*
24 Install all the bolts finger-tight within four minutes of applying the High performance RTV.
25 On 3.5L engine models, follow the correct sequence (see illustration), tighten the bolts to the torque listed in this Chapter's Specifications. All the bolts must be tightened within 35 minutes.
26 Install the remaining parts in the reverse order of removal.
27 Install a new oil filter, then add engine oil and coolant (see Chapter 1).
28 Run the engine and check for leaks.

Chapter 2B V6 engines

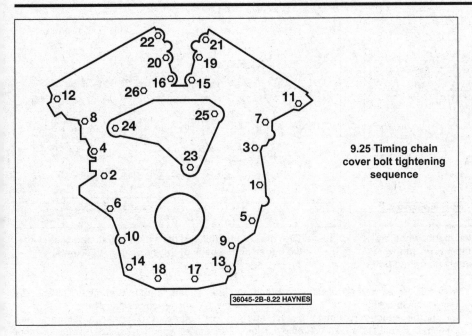

9.25 Timing chain cover bolt tightening sequence

10.5 Remove three of the camshaft cap bolts to remove the valvetrain oil tubes

10 Timing chains and sprockets - removal and installation

Warning: *The engine must be completely cool before beginning this procedure.*

Caution: *These engines are difficult to work on and require some special tools. On any procedure involving the timing chains, the Steps must be read carefully and disassembly must proceed using the special tools, otherwise damage to the engine will result.*

Caution: *Because this is an "interference" engine design, if the timing chain has broken, there will be damage to the valves (and possibly the pistons) and will require removal of the cylinder heads.*

Caution: *The timing system is complex. Severe engine damage will occur if you make any mistakes. Do not attempt this procedure unless you are highly experienced with this type of repair. If you are at all unsure of your abilities, consult an expert. Double-check all your work and be sure everything is correct before you attempt to start the engine.*

Removal

1 Disconnect the cable from the negative battery terminal (see Chapter 5, Section 3).
2 Remove the valve covers (see Section 4).
3 Remove the timing chain cover (see Section 9).

Non-turbocharged engines

4 Turn the engine clockwise until the VCT units (camshaft sprockets) on each intake camshaft are positioned at 90-degrees (12 o'clock) in relation to the top surface of the cylinder heads.
5 Remove the bolts that secure the valvetrain oil tubes to each cylinder head (see illustration).
6 Install the special camshaft locking tools (303-1248), with the tools holding the flats on each camshaft.
7 If you can't see the timing marks on the timing chain, add a paint dot to align with the marks on the VCT units and the crankshaft timing sprocket.
8 Remove the two bolts and the primary timing chain tensioner arm, then unbolt and remove the two tensioners.
9 Remove the mounting bolts and the lower-left chain guide, then the lower right chain guide.
10 Remove the bolts securing the VCT solenoids to the cylinder heads (see illustration). You may have to twist or wiggle the solenoids to disengage them.
Note: *The intake solenoids are white, the exhaust solenoids are black.*
11 Keep the VCT solenoids in clean plastic sandwich bags, marked with their destination.
12 Remove the primary timing chain.
13 The two camshafts on each cylinder head are jointly driven by a smaller secondary timing chain.
14 Depress and lock the secondary chain tensioner on each bank, using the factory tool (303-1530) inserted in the large hole at the camshaft first bearing cap, or a length of threaded rod and nuts to hold the tensioner down, by pushing from the camshaft bearing cap.
15 With the camshafts still locked at TDC, remove the mounting bolts from the VCT units (camshaft sprockets), and remove the sprockets and secondary timing chains from the camshafts.
16 If the primary timing chain sprocket (at the center of the front of the block) is to be replaced, remove the nine bolts securing the plate that mounts the gear to the front of the block.

Turbocharged engines

Note: *Two special camshaft holding tools (manufacturer tool no. 303-1248) are required for this procedure.*

Note: *Since the water pump on this engine is located behind the timing chain and components, it makes sense to replace the water pump while it is easily accessed during a timing chain replacement procedure.*

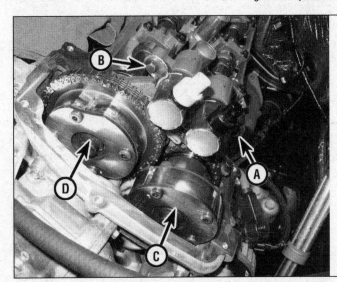

10.10 Remove the VCT oil control solenoid mounting bolts (A and B) - (C) is the exhaust VCT unit and (D) is the intake VCT unit

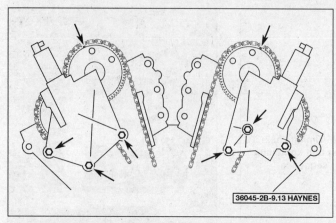

10.17 The position of the intake cam sprocket marks when the engine is set to TDC compression for cylinder no. 1 (upper arrows) - lower arrows indicate the VCT assembly mounting bolts

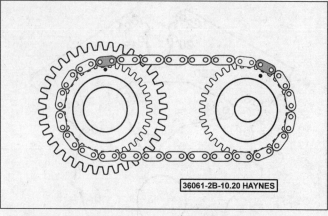

10.34 Align the colored links of the secondary timing chain with the marks on the back of the sprockets (VCT units)

17 Remove the timing chain cover (see Section 9). Reinstall the crankshaft sprocket bolt and set the engine to TDC for cylinder number 1 (see illustration). **Note:** *At this point the TDC position will be approximate; when the special tools are installed in Step 19, the TDC position will be exact.*

18 These engines have three timing chains: The long primary chain drives the intake cam VCT sprocket on each intake camshaft and the water pump, while a smaller chain on each head drives the exhaust camshafts.

19 Install the special camshaft holding tools (manufacturer tool no. 303-1248) on each cylinder head to lock the camshafts in the TDC position.

20 Remove the VCT assembly mounting bolts from the rear cylinder head, then the front cylinder head (see illustration 10.17).

21 Remove the bolts and the primary timing chain tensioner and tensioner arm. Remove the lower chain guide from the front cylinder bank. Mark the chain links opposite the timing marks on the camshafts (if not already marked), and remove the primary timing chain.

22 Remove the upper chain guide from the front cylinder bank. If necessary, remove the crankshaft sprocket.

23 If the secondary timing chains and camshaft sprockets are to be removed for inspection or another procedure, compress the tensioner(s) and insert a pin into the hole in the tensioner to hold the tensioner in the retracted position. Unscrew the bolts for the intake (VCT) and exhaust camshaft sprockets, then remove both sprockets and the chain as an assembly.

Caution: *Make sure the camshaft holding tool is in place and prevents the camshafts from turning while unscrewing the bolts. Also, new bolts must be used for reassembly.*

24 Remove the bolts and the secondary chain tensioners.

Note: *The camshaft holding tool must be loosened and tilted (toward the flywheel end of the engine) to access the rearmost bolt of the tensioner.*

Inspection

Note: *Do not mix parts from the front and rear timing chains and tensioners, and keep the primary and secondary tensioners separate.*

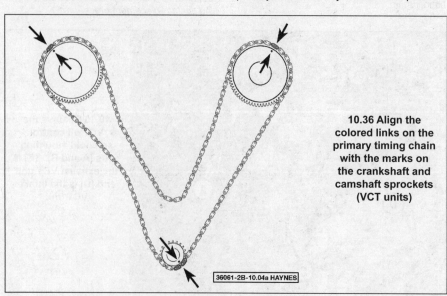

10.36 Align the colored links on the primary timing chain with the marks on the crankshaft and camshaft sprockets (VCT units)

25 Clean all parts with clean solvent. Dry with compressed air, if available.

26 Inspect the chain tensioners and tensioner arms for excessive wear or other damage.

27 Inspect the timing chain guides for deep grooves, excessive wear, or other damage.

28 Inspect the timing chain for excessive wear or damage.

29 Inspect the camshaft and crankshaft sprockets for chipped or broken teeth, excessive wear, or damage.

30 Replace any component that is in questionable condition.

Installation

Non-turbocharged engines

31 If the primary chain gear and its plate were removed, clean the mounting surface and the block and install the sprocket and plate with a new gasket. Prepare the primary chain tensioners by pushing in the release button and compressing the plunger until a nail or large paper clip can be inserted to hold the plunger.

32 If the secondary timing chain tensioners were removed, they cannot be reused. New tensioners must be installed.

Caution: *Do not remove the plastic clip that is keeping the new tensioner compressed.*

33 The tensioner is installed to the correct depth when you hear a "snap" sound. Install the tensioner shoe. The plastic clip can now be removed with pliers.

34 Align the colored links with the marks on the VCT units and install the chain on the backside of the VCT units (see illustration). Align the two VCT units with the dowel pins on the front of the camshafts.

35 New bolts must be used to install the VCT units. Tighten the bolts to the torque listed in this Chapter's Specifications.

36 When the secondary chains and VCT units are installed and aligned, install the primary timing chain, aligning the colored links with the marks on the sprockets (see illustration). When all components are installed and

Chapter 2B V6 engines

aligned, pull the pins holding back the two primary chain tensioners.

Turbocharged engines

37 Install the primary timing chain guide for the rear (right) cylinder bank and tighten the bolts to the torque listed in this Chapter's Specifications.

38 Tilt the rear (right) cylinder bank camshaft holding tool to the rear and install the secondary timing chain tensioner, tightening the bolts to the torque listed in this Chapter's Specifications. Don't remove the lock pin from the tensioner yet.

39 Install the secondary timing chains onto the camshaft sprockets. The colored links on the chain must align with the marks/camshaft keyway slots in the sprockets (see illustration).

40 Install the sprockets/VCT actuator (phaser)/secondary timing chain onto the rear camshafts. When installing the camshaft sprocket bolts, NEW bolts must be used. Tighten the bolts to the torque listed in this Chapter's Specifications.

41 Tilt the front (left) cylinder bank camshaft holding tool to the rear and install the secondary timing chain tensioner, tightening the bolts to the torque listed in this Chapter's Specifications. Don't remove the lock pin from the tensioner yet.

42 Install the sprockets/VCT actuator (phaser)/secondary timing chain (with the marks aligned as shown in illustration 10.39) onto the front camshafts. When installing the camshaft sprocket bolts, NEW bolts must be used. Tighten the bolts to the torque listed in this Chapter's Specifications. Remove the lock pin from the tensioner.

43 Install the crankshaft sprocket.

44 Install the primary timing chain onto the crankshaft sprocket and the VCT sprockets, making sure the colored links (or marks) on the chain align with the marks on the sprockets (see illustration 10.36).

Note: *Note: At TDC compression for cylinder no. 1, the crankshaft sprocket keyway will be in approximately the 11 o'clock position and the timing mark on the crankshaft sprocket will be in approximately the 4 o'clock position.*

45 Install the upper chain guide on the front (left) cylinder bank, tightening the bolts to the torque listed in this Chapter's Specifications.

46 Install the lower chain guide for the front cylinder bank, tightening the bolts to the torque listed in this Chapter's Specifications.

47 Install the primary timing chain tensioner arm to the rear cylinder bank.

48 Place the timing chain tensioner in a soft-jawed vise, with the jaws bearing on the tensioner body and the plunger. Swing the lever on the tensioner counterclockwise, then tighten the vise until the tensioner plunger has been retracted. Swing the lever clockwise, aligning the hole in the lever with the hole in the tensioner body, then install a lock pin made of heavy wire or a drill bit or small Allen wrench. Remove the tensioner from the vise.

49 Install the primary timing chain tensioner, tightening the bolts to the torque listed in this Chapter's Specifications. Remove the lock pin from the tensioner.

Note: *It might be necessary to rotate the crankshaft slightly to create a little slack in the chain to enable the tensioner to be installed.*

50 Recheck the timing marks to make sure they are all in proper alignment.

51 Check the seals on the VCT housing, replacing them if necessary.

52 Install the VCT housings, making sure the dowels engage completely with their corresponding holes in the cylinder head, then install the bolts and tighten them a little at a time to the torque listed in this Chapter's Specifications.

53 Install the timing chain cover (see Section 9).

All models

54 Reinstall the remaining parts in the reverse order of removal.

55 Carefully rotate the crankshaft by hand through at least two full revolutions (use a socket and breaker bar on the crankshaft pulley center bolt).

Caution: *If you feel any resistance, STOP! There is something wrong - most likely, valves are contacting the pistons. You must find the problem before proceeding.*

56 Install a new oil filter, then add engine oil and coolant (see Chapter 1).

57 Run the engine and check for leaks.

11 Camshafts and tappets/roller followers and lash adjusters - removal, inspection and installation

Warning: *The engine must be completely cool before beginning this procedure.*

Caution: *These engines are difficult to work on and require some special tools. On any procedure involving the timing chains, the Steps must be read carefully and disassembly must proceed using the special tools, otherwise damage to the engine will result.*

Caution: *Because this is an "interference" engine design, if the timing chain has broken, there will be damage to the valves (and possibly the pistons) and will require removal of the cylinder heads.*

Caution: *The timing system is complex. Severe engine damage will occur if you make any mistakes. Do not attempt this procedure unless you are highly experienced with this type of repair. If you are at all unsure of your abilities, consult an expert. Double-check all your work and be sure everything is correct before you attempt to start the engine.*

Warning: *Turbocharged models are equipped with Direct Injection (DI) and a high-pressure fuel pump. Fuel pressure in the high-pressure system on Direct Injection (DI) systems is under extremely high pressure. Be sure to correctly perform the fuel pressure relief procedure prior to servicing any of the high-pressure fuel system components to prevent injury (see Chapter 4, Section 3).*

Removal

1 Remove the engine and transaxle assembly from the vehicle (see Chapter 2C).

2 Before removing the camshafts, check the valve clearances (see Section 3).

3 On non-turbocharged models, remove the upper intake manifold and on turbocharged models the intake manifold (see Section 5).

4 When removing the right-side camshafts on turbocharged models, remove the brake vacuum pump (see Chapter 9).

5 Refer to Section 10 and remove the timings chain and camshaft VCT units also the tensioner holding tool and camshaft holding tools.

Note: *Make sure the camshafts are in the neutral positions (see illustrations 11.16a, 11.16b, 11.16c and 11.16d).*

6 Loosen the camshaft caps and "megacap" mounting bolts and remove the caps.

Caution: *Keep the caps in order and don't mix them up. Remove the camshafts from the cylinder head.*

7 With the camshafts removed, the tappets can be removed using a magnet.

Caution: *They must be stored in an egg carton or other divided and marked container, so they can be restored to their original locations during reassembly.*

10.39 Align the secondary timing chain colored links with the keyways of the sprockets

Chapter 2B V6 engines

11.10a Lobe lift can be obtained by measuring camshaft lobe height . . .

11.10b . . . and by measuring the camshaft base circle - the difference between the two measurements equals lobe lift

Inspection

8 After the camshafts have been removed, clean them with solvent, then inspect the bearing journals for uneven wear and pitting. If the journals are damaged, the bearing saddles in the cylinder heads and caps are probably damaged as well. The head and camshaft caps will have to be replaced.

9 Measure the bearing journals with a micrometer and compare to the Specifications in this Chapter to determine whether they are excessively worn or out-of-round.

10 Measure the camshaft lobe height and the base circle (see illustrations). The difference between the two measurements is the lobe lift (lobe height - base circle = lobe lift). Record this figure for future reference and repeat the check on the remaining camshaft lobes. Compare the results to the values listed in this Chapter's Specifications.

11 Inspect the camshaft lobes for heat discoloration, score marks, chipped areas, pitting and uneven wear. If the lobes are in good condition and if the lobe lift measurements are as specified, you can reuse the camshafts.

Installation

12 Rotate the crankshaft counterclockwise and position the keyway in the 9 o'clock position (this is the neutral position).
Caution: *The crankshaft must remain in the neutral position (crankshaft keyway at 9 o'clock position) until after the camshafts are installed and the valve clearance is checked and adjusted, if necessary. Do not turn the crankshaft until Step. Failure to follow this process will result in severe engine damage.*

13 Lubricate the tappets with clean engine oil and reinstall them in their original locations. If any valve clearances were out-of-specification, replace the tappet(s) with new ones of the proper thickness to achieve the desired clearance.

14 On non-turbocharged models, at the front of each camshaft, there are two grooves to hold seals. Obtain new seals and install them in the camshaft grooves.
Note: *The split where the ends of each seal come together must face UP (12 o'clock position) on the camshafts when they are installed.*

15 Lubricate the camshaft bearing journals and cam lobes with moly-based grease or camshaft installation lubricant.

16 Position the camshafts in the cylinder head in their neutral positions (see illustrations).

17 Install the camshaft bearing caps (in their original locations), and tighten the bolts, in sequence, to the torque listed in this Chapter's Specifications (see illustration).

18 Recheck the valve clearances (see Section 3). Since the crankshaft is in the neutral position, the camshafts can be turned without the valves contacting the pistons.

19 On non-turbocharged models, remove the camshaft cap bolts that secure the valve train oil tubes. Remove the oil tubes.

20 Rotate the camshafts to position the dowel pins in their proper positions for installing the VCT units and timing chains (see illustrations). Install the camshaft holding tools.

21 On turbocharged models, the dowel pins should be positioned pointing upwards.

22 Rotate the crankshaft clockwise to position the keyway in the 11 o'clock position.

23 Install the VCT units (see Section 10).

24 Install the timing chains (see Section 10).

25 Reinstall the valvetrain oil tubes and tighten the bearing cap bolts to the torque listed in this Chapter's Specifications.

26 The remaining installation steps are the reverse of removal.

27 Before starting and running the engine, refill the cooling system, change the oil and install new oil filter (see Chapter 1).

12 Oil pan - removal and installation

Removal

1 Disconnect the cable from the negative battery terminal (see Chapter 5).
2 Remove the oil dipstick.
3 Raise the vehicle and support it securely on jackstands.

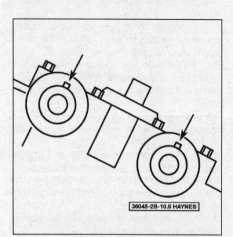

11.16a On turbocharged models, when installing the left (front) side cylinder head camshafts, position the dowel pins like this

11.16b On turbocharged models, when installing the right (rear) side cylinder head camshafts, position the dowel pins like this

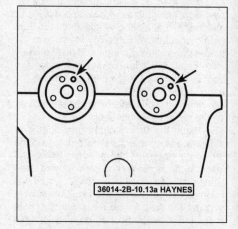

11.16c On non-turbocharged models, when installing the right (rear) side cylinder head camshafts, position the dowel pins like this

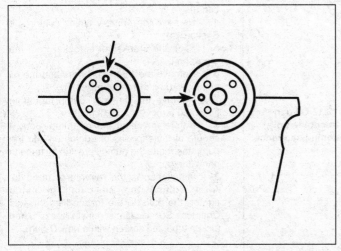

11.16d On non-turbocharged models, when installing the left (front) side cylinder head camshafts, position the dowel pins like this

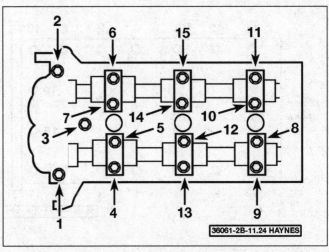

11.17a Non-turbocharged model, camshaft bearing cap tightening sequence - both cylinder banks

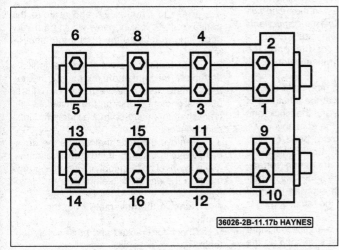

11.17b Turbocharged model, right (rear) side camshaft bearing cap tightening sequence

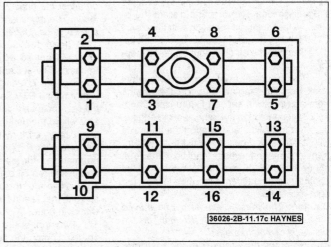

11.17c Turbocharged model, left (front) side camshaft bearing cap tightening sequence

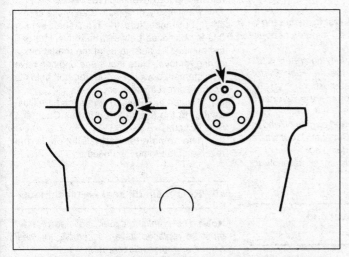

11.20a On non-turbocharged models, rotate the right (rear) side cylinder head camshafts so the dowel pins are positioned like this (TDC position), then install the camshaft holding tool

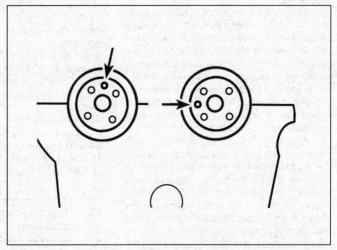

11.20b On non-turbocharged models, rotate the left (front) side cylinder head camshafts so the dowel pins are positioned like this (TDC position), then install the camshaft holding tool

Chapter 2B V6 engines

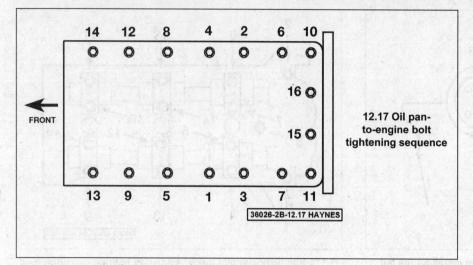

12.17 Oil pan-to-engine bolt tightening sequence

4   Drain the engine oil (see Chapter 1), then clean and install the engine oil drain plug, tightening it to the torque listed in the Chapter 1 Specifications. Remove and discard the oil filter, so that it can be replaced with the oil.
5   Remove the engine lower splash shield.
6   Remove the air filter housing (see Chapter 4).
7   The transaxle must be moved slightly back from the engine. Support the engine from above with a support fixture (see Chapter 2C) connected to the left end of the engine, near the transaxle, and support the transaxle with a floor jack. Loosen the upper engine-to-transaxle mounting bolts and back them off about 3/16-inch (5 mm), loosen the left-side engine-to-bellhousing bolts, then loosen the right-side bolts.
Caution: *The transaxle must not be moved back more than 3/16-inch (5 mm) or the transaxle can be damaged.*
8   Remove the transaxle roll restrictor mount through bolts and remove the mount (see Chapter 7A).
9   Remove the catalytic converters (see Chapter 4).
10   Remove the four bellhousing-to-pan bolts, and one oil pan-to-bellhousing bolt. Also remove the timing chain cover-to-oil pan fasteners.
11   Remove the a/c compressor and tie the compressor out of the way (see Chapter 3), without disconnecting the compressor lines.
12   Use a screwdriver to pry between the engine and transaxle until the bellhousing has moved away from the block to the limit of the loosened bolts (about 0.20-inch [5 mm]).
13   Progressively unscrew the oil pan retaining bolts evenly until all the bolts are removed. Use a rubber mallet to loosen the oil pan seal, then lower the oil pan, turning it as necessary to clear the exhaust system. Unfortunately, the use of sealant can make removal of the oil pan more difficult. Be careful when prying between the mating surfaces, otherwise they will be damaged, resulting in leaks when finished. With care, a putty knife can be used to cut through the sealant.

Installation

Caution: *Do not use abrasive discs or scrapers on either the engine or the pan mounting surfaces. The aluminum and plastic components can easily be gouged, leading to oil leakage.*
14   Clean the surfaces of the pan, the engine, front cover and transmission of all traces of RTV sealant.
15   Apply a 7/32-inch (5.5 mm) bead of High Performance RTV sealant at the junctures of the rear main seal housing and the block, and also where the pan meets the front cover.
16   Apply a 1/8-inch (3 mm) bead of High Performance RTV sealant around the inner perimeter of the oil pan's mounting surface and install the oil pan. Install the four corner bolts within four minutes of applying the RTV sealant. Install and tighten the remaining bolts within an hour.
17   When all of the pan-to-engine bolts are started, tighten the bolts, in sequence, to the torque listed in this Chapter's Specifications (see illustration). After the pan-to-engine bolts are tightened, install and tighten the pan-to-transmission bolts, and all fasteners loosened in Steps 10 and 11.
Caution: *You must work fast to secure the oil pan to the engine within 10 minutes of applying the RTV.*
Note: *All bolts must be fully tightened within 60 minutes of applying the the High Performance RTV or oil leaks can occur.*
18   The remaining steps are the reverse of removal.
Caution: *Don't forget to add engine oil and install a new oil filter (see Chapter 1), but wait at least 90 minutes before doing so.*
19   Start the engine and check carefully for oil leaks at the oil pan.

13 Oil pump - removal and installation

1   Drain the oil and remove the oil filter (see Chapter 1).
2   Remove the oil pan (see Section 12).
3   Remove the engine front cover (see Section 9).
4   Remove the primary timing chain (see Section 10).
5   Slide the crankshaft sprocket from the crankshaft.
6   Remove the three fasteners and the oil pump pickup and screen assembly.
7   Remove the three oil pump mounting bolts and remove the oil pump.
8   Before installing the oil pump, pour a couple of tablespoons of engine oil into the pump the rotate the pump sprocket by hand to prime the pump.
9   Installation is the reverse of removal. After bolting the oil pump in place tighten the bolts to the torque listed in this Chapter's Specifications, reinstall the oil pump pickup tube and screen with a new O-ring.

14 Driveplate - removal and installation

1   Refer to Chapter 7A and remove the transaxle. If it's leaking, now would be a very good time to replace the front pump seal/O-ring.
2   Look for factory paint marks that indicate driveplate-to-crankshaft alignment. If they aren't there, use a center-punch or paint to make alignment marks on the driveplate and crankshaft to ensure correct alignment during reinstallation.
3   Remove the bolts that secure the driveplate to the crankshaft. If the crankshaft turns, use a flywheel holding tool or wedge a screwdriver in the ring gear teeth to jam the driveplate.
4   Remove the driveplate from the crankshaft.
5   Check for cracked and broken ring gear teeth.
6   Clean and inspect the mating surfaces of the driveplate and the crankshaft. If the crankshaft rear seal is leaking, replace it before reinstalling the driveplate (see Section 15).
7   Position the driveplate against the crankshaft, but make sure the crankshaft sensor ring is positioned between the driveplate and the engine. Be sure to align the marks made during removal. Note that some engines have an alignment dowel or staggered bolt holes to ensure correct installation.
8   Tighten the bolts, in sequence (see illustration) to the torque listed in this Chapter's Specifications.
9   The remainder of installation is the reverse of the removal procedure.

15 Rear main oil seal - replacement

Note: *The crankshaft rear seal and retainer must be replaced as an assembly, the seal cannot be replaced separately.*
1   Remove the transaxle (see Chapter 7A).
2   Remove the driveplate (see Section 14) then remove the crankshaft sensor ring from the end of the crankshaft.

Chapter 2B V6 engines

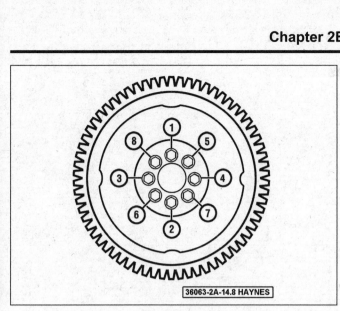

14.8 Driveplate bolt tightening sequence

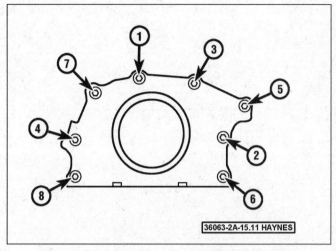

15.10 Rear main oil seal retainer tightening sequence

3   Disconnect the crankshaft sensor electrical connector, then remove the crankshaft sensor mounting bolts and the sensor from the seal housing (see Chapter 6).
4   Remove the oil pan-to-seal housing bolts.
5   Remove the seal retainer mounting bolts and the retainer from the rear of the engine.
6   Use only a plastic gasket scraper or putty knife to remove all traces of old gasket material and sealant from the pan and engine block.
Caution: *To prevent an oil leak after the new seal is installed, be very careful not to scratch or otherwise damage the crankshaft sealing surface or the engine block.*
7   Remove all traces of RTV sealant. Apply a 3/16-inch bead of high performance RTV along the sealing surfaces of the retainer where the block and oil pan flanges meet.
8   Apply a film of clean engine oil to the lips of the new seal and around the crankshaft where the seal makes contact.
9   Install the retainer assembly to the engine at an angle, with the bottom of the retainer touching the oil pan, tilt the retainer up and onto the rear of the cylinder block then install all the bolts within 10 minutes of applying the sealant.
10 Install the seal retainer-to-engine block bolts first, tighten the bolts, in sequence (see illustration) to the torque listed in this Chapter's Specifications, then install the oil pan-to-seal retainer bolts and tighten the bolts to the torque listed in this Chapter's Specifications.
11 Reinstall the crankshaft sensor (see Chapter 6), the crankshaft sensor ring and the driveplate, making sure the crankshaft sensor ring is between the crankshaft and the driveplate.

12 Reinstall the transmission (see Chapter 7A).

16 Engine mounts - check and replacement

Note: *The tightening torque required for installing the engine mount through-bolts is considerable. Make sure you have access to a high-torque torque wrench before beginning.*
Note: *The following is the factory-recommended procedure, using an engine support fixture from above. With some creativity and depending on tool selection, it may be possible to replace the engine mounts without performing some of the steps listed here.*
1   There are three powertrain mounts on the vehicles covered by this manual; left and right engine mounts attached to the engine block and the frame, and a rear mount attached to the transmission and the transmission crossmember. The rear transmission mount is covered in Chapter 7A. Engine mounts seldom require attention, but broken or deteriorated mounts should be replaced immediately or the added strain placed on the driveline components may cause damage or wear.

Check
2   During the check, the engine must be raised slightly to remove the weight from the mounts.
3   Raise the vehicle and support it securely on jackstands.
4   Position a jack under the engine oil pan. Place a large wood block between the jack head and the oil pan, then carefully raise the engine just enough to take the weight off the mounts.
Warning: *DO NOT place any part of your body under the engine when it's supported only by a jack!*
5   Check for relative movement between the inner and outer portions of the mount (use a large screwdriver or prybar to attempt to move the mounts). If movement is noted, lower the engine and tighten the mount fasteners.
6   Check the mounts to see if the rubber is cracked, hardened or separated from the metal casing which would indicate a need for replacement.

Replacement
7   Disconnect the negative battery cable (see Chapter 5). If you're replacing the right-side engine mount, remove the air filter housing (see Chapter 4), and the coolant expansion tank (see Chapter 3).
8   If you're replacing the front or rear transaxle mount, raise the vehicle and support it securely on jackstands.
9   Place a block of wood between the jack head and the oil pan, then carefully raise the engine or transaxle just enough to take the weight off the mounts.
Caution: *Do not disconnect more than one mount at a time unless the engine will be removed from the vehicle.*
10 Remove the fasteners holding the mount to the bracket.
11 Installation is the reverse of removal.

Notes

2C-1

Chapter 2 Part C
General engine overhaul procedures

Contents

	Section		Section
Balance shaft assembly (four-cylinder engine) - removal and installation	13	Engine rebuilding alternatives	5
Crankshaft - removal and installation	10	Engine removal - methods and precautions	6
Cylinder compression check	3	General information - engine overhaul	1
Engine - removal and installation	7	Initial start-up and break-in after overhaul	12
Engine overhaul - dissassembly sequence	8	Oil pressure check	2
Engine overhaul - reassembly sequence	11	Pistons and connecting rods - removal and installation	9
		Vacuum gauge diagnostic checks	4

Specifications

General

Cylinder compression	Lowest cylinder must be within 75% of the highest cylinder
Oil pressure	
Four-cylinder engines, hot, at 2000 rpm	
2.0L engine	29 to 39 psi (200 to 268 kPa)
2.3L engine	29 to 60 psi (200 to 414 kPa)
V6 engines, at 1500 rpm	30 psi (206 kPa)
Balance shaft assembly backlash (four-cylinder engines)	0.0008 to 0.0047 inch (0.020 to 0.120 mm)

Chapter 2 Part C General engine overhaul procedures

Torque specifications Ft-lbs (unless otherwise indicated) Nm

Note: *One foot-pound (ft-lb) of torque is equivalent to 12 inch-pounds (in-lbs) of torque. Torque values below approximately 15 foot-pounds are expressed in inch-pounds, because most foot-pound torque wrenches are not accurate at these smaller values.*

	Ft-lbs	Nm
Balance shaft assembly mounting bolts (four-cylinder engines)		
Step 1	18	25
Step 2	31	42
Connecting rod bolts*		
Four-cylinder engines		
Step 1	89 in-lbs	10
Step 2	21	29
Step 3	Tighten an additional 90 degrees	
V6 engines		
Step 1	17	23
Step 2	32	43
Step 3	Tighten an additional 90 degrees	
Main bearing bolts*		
Four-cylinder engines		
Step 1	44 in-lbs	5
Step 2	18	25
Step 3	Tighten an additional 90 degrees	
V6 engines		
Step 1, bolts 1 through 8		
2012 and earlier models	44	60
2013 and later models	24	33
Step 2, bolts 1 through 8		
2012 and earlier models	Tighten an additional 90 degrees	
2013 and later models	Tighten an additional 135 degrees	
Step 3, side bolts 10 through 16		
2012 and earlier models	33	45
2013 through 2015 models	18	24
2016 and later models	33	45
Step 4, side bolts 10 through 16		
2012 and earlier models	Tighten an additional 90 degrees	
2013 through 2015 models		
Non-turbocharged engines	Tighten an additional 180 degrees	
Turbocharged engines	Tighten an additional 90 degrees	
2016 and later models	Tighten an additional 90 degrees	
Main bearing cap support brace (V6 engines)		
Step 1	18	24
Step 2	Tighten an additional 180-degrees	

* Use new bolts

Chapter 2 Part C General engine overhaul procedures

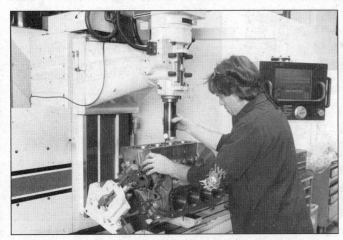

1.9a An engine block being bored. An engine rebuilder will use special machinery to recondition the cylinder bores

1.9b If the cylinders are bored, the machine shop will normally hone the engine on a machine like this

1 General information - engine overhaul

1 Included in this portion of Chapter 2A are general information and diagnostic testing procedures for determining the overall mechanical condition of your engine.

2 The information ranges from advice concerning preparation for an overhaul and the purchase of replacement parts and/or components to detailed, step-by-step procedures covering removal and installation.

3 The following Sections have been written to help you determine whether your engine needs to be overhauled and how to remove and install it once you've determined it needs to be rebuilt. For information concerning in-vehicle engine repair, see Chapter 2A or Chapter 2B.

4 It's not always easy to determine when, or if, an engine should be completely overhauled, because a number of factors must be considered.

5 High mileage is not necessarily an indication that an overhaul is needed, while low mileage doesn't preclude the need for an overhaul. Frequency of servicing is probably the most important consideration. An engine that's had regular and frequent oil and filter changes, as well as other required maintenance, will most likely give many thousands of miles of reliable service. Conversely, a neglected engine may require an overhaul very early in its service life.

6 Excessive oil consumption is an indication that piston rings, valve seals and/or valve guides are in need of attention. Make sure that oil leaks aren't responsible before deciding that the rings and/or guides are bad. Perform a cylinder compression check to determine the extent of the work required (see Section 3). Also check the vacuum readings under various conditions (see Section 4).

7 Check the oil pressure with a gauge installed in place of the oil pressure sending unit and compare it to this Chapter's Specifications (see Section 2). If it's extremely low, the bearings and/or oil pump are probably worn out.

8 Loss of power, rough running, knocking or metallic engine noises, excessive valve train noise and high fuel consumption rates may also point to the need for an overhaul, especially if they're all present at the same time. If a complete tune-up doesn't remedy the situation, major mechanical work is the only solution.

9 An engine overhaul involves restoring the internal parts to the specifications of a new engine. During an overhaul, the piston rings are replaced and the cylinder walls are reconditioned (rebored and/or honed) (see illustrations 1.9a and 1.9b). If a rebore is done by an automotive machine shop, new oversize pistons will also be installed. The main bearings, connecting rod bearings and camshaft bearings are generally replaced with new ones and, if necessary, the crankshaft may be reground to restore the journals (see illustration 1.9c). Generally, the valves are serviced as well, since they're usually in less-than-perfect condition at this point. While the engine is being overhauled, other components, such as the starter and alternator, can be rebuilt or replaced as well. The end result should be similar to a new engine that will give many trouble free miles.

Note: *Critical cooling system components such as the hoses, drivebelts, thermostat and water pump should be replaced with new parts when an engine is overhauled. The radiator should be checked carefully to ensure that it isn't clogged or leaking (see Chapter 3). If you purchase a rebuilt engine or short block, some rebuilders will not warranty their engines unless the radiator has been professionally flushed. Also, we don't recommend overhauling the oil pump - always install a new one when an engine is rebuilt.*

10 Overhauling the internal components on today's engines is a difficult and time-consuming task which requires a significant amount of specialty tools and is best left to a professional engine rebuilder (see illustrations 1.10a, 1.10b and 1.10c). A competent engine rebuilder will handle the inspection of your old parts and offer advice concerning the reconditioning or replacement of the original engine, never purchase parts or have machine work done on other components until the block has been thoroughly inspected by a profes-

1.9c A crankshaft having a main bearing journal ground

1.10a A machinist checks for a bent connecting rod, using specialized equipment

sional machine shop. As a general rule, time is the primary cost of an overhaul, especially since the vehicle may be tied up for a minimum of two weeks or more. Be aware that some engine builders only have the capability to rebuild the engine you bring them while other rebuilders have a large inventory of rebuilt exchange engines in stock. Also be aware that many machine shops could take as much as two weeks time to completely rebuild your engine depending on shop workload. Sometimes it makes more sense to simply exchange your engine for another engine that's already rebuilt to save time.

2 Oil pressure check

1 Low engine oil pressure can be a sign of an engine in need of rebuilding. A low oil pressure indicator (often called an "idiot light") is not a test of the oiling system. Such indicators only come on when the oil pressure is dangerously low. Even a factory oil pressure gauge in the instrument panel is only a relative indication, although much better for driver information than a warning light. A better test is with a mechanical (not electrical) oil pressure gauge.
2 Locate the oil pressure sending unit:
 a) On four-cylinder engines, the oil pressure sending unit is located on the front of the engine, threaded into the oil filter adapter (see illustration).
 b) On V6 engines, the sending unit is located on the left side of the block, just forward of the oil filter adapter.
3 Unscrew the oil pressure sending unit and screw in the hose for your oil pressure gauge (see illustration). If necessary, install an adapter fitting. Use Teflon tape or thread sealant on the threads of the adapter and/or the fitting on the end of your gauge's hose.
4 Connect an accurate tachometer to the engine, according to the tachometer manufacturer's instructions.

1.10b A bore gauge being used to check the main bearing bore

5 Check the oil pressure with the engine running (normal operating temperature) at the specified engine speed, and compare it to this Chapter's Specifications. If it's extremely low, the bearings and/or oil pump are probably worn out.

3 Cylinder compression check

1 A compression check will tell you what mechanical condition the upper end of your engine (pistons, rings, valves, head gaskets) is in. Specifically, it can tell you if the compression is down due to leakage caused by worn piston rings, defective valves and seats or a blown head gasket.
Note: *The engine must be at normal operating temperature and the battery must be fully charged for this check.*
2 Begin by cleaning the area around the ignition coils before you remove them (compressed air should be used, if available). The idea is to prevent dirt from getting into the cylinders as the compression check is being done.

1.10c Uneven piston wear like this indicates a bent connecting rod

3 Remove all of the spark plugs from the engine (see Chapter 1).
4 Disable the fuel pump (see Chapter 4, Section 3).
5 Install a compression gauge in the spark plug hole (see illustration).
6 Have an assistant depress the accelerator pedal to the floor and crank the engine over at least seven compression strokes, while you watch the gauge watch the gauge. The compression should build up quickly in a healthy engine. Low compression on the first stroke, followed by gradually increasing pressure on successive strokes, indicates worn piston rings. A low compression reading on the first stroke, which doesn't build up during successive strokes, indicates leaking valves or a blown head gasket (a cracked head could also be the cause). Deposits on the undersides of the valve heads can also cause low compression. Record the highest gauge reading obtained.
7 Repeat the procedure for the remaining cylinders and compare the results to this Chapter's Specifications.
8 Add some engine oil (about three squirts from a plunger-type oil can) to each cylinder,

2.2 The oil pressure sending unit is located on the oil filter adapter (2.0L four-cylinder engine shown)

2.3 The oil pressure can be checked by removing the sending unit and installing a pressure gauge in its place

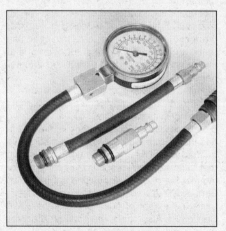

3.5 Use a compression gauge with a threaded fitting for the spark plug hole, not the type that requires hand pressure to maintain the seal

Chapter 2 Part C General engine overhaul procedures

4.4 A simple vacuum gauge can be handy in diagnosing engine condition and performance

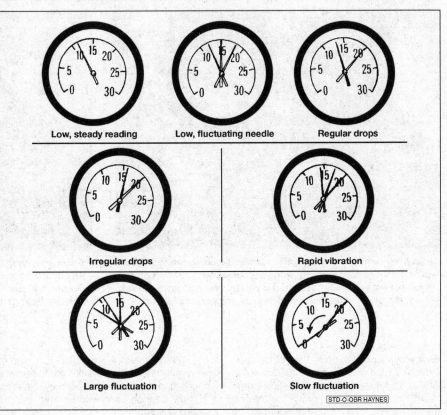

4.6 Typical vacuum gauge readings

through the spark plug hole, and repeat the test.

9 If the compression increases after the oil is added, the piston rings are definitely worn. If the compression doesn't increase significantly, the leakage is occurring at the valves or head gasket. Leakage past the valves may be caused by burned valve seats and/or faces or warped, cracked or bent valves.

10 If two adjacent cylinders have equally low compression, there's a strong possibility that the head gasket between them is blown. The appearance of coolant in the combustion chambers or the crankcase would verify this condition.

11 If one cylinder is slightly lower than the others, and the engine has a slightly rough idle, a worn lobe on the camshaft could be the cause.

12 If the compression is unusually high, the combustion chambers are probably coated with carbon deposits. If that's the case, the cylinder head(s) should be removed and decarbonized.

13 If compression is way down or varies greatly between cylinders, it would be a good idea to have a leak-down test performed by an automotive repair shop. This test will pinpoint exactly where the leakage is occurring and how severe it is.

4 Vacuum gauge diagnostic checks

1 A vacuum gauge provides inexpensive but valuable information about what is going on in the engine. You can check for worn rings or cylinder walls, leaking head or intake manifold gaskets, incorrect carburetor adjustments, restricted exhaust, stuck or burned valves, weak valve springs, improper ignition or valve timing and ignition problems.

2 Unfortunately, vacuum gauge readings are easy to misinterpret, so they should be used in conjunction with other tests to confirm the diagnosis.

3 Both the absolute readings and the rate of needle movement are important for accurate interpretation. Most gauges measure vacuum in inches of mercury (in-Hg). The following references to vacuum assume the diagnosis is being performed at sea level. As elevation increases (or atmospheric pressure decreases), the reading will decrease. For every 1,000 foot increase in elevation above approximately 2,000 feet, the gauge readings will decrease about one inch of mercury.

4 Connect the vacuum gauge directly to the intake manifold vacuum, not to ported (throttle body) vacuum (see illustration). Be sure no hoses are left disconnected during the test or false readings will result.

5 Before you begin the test, allow the engine to warm up completely. Block the wheels and set the parking brake. With the transaxle in Park, start the engine and allow it to run at normal idle speed.

Warning: *Keep your hands and the vacuum gauge clear of the fans.*

6 Read the vacuum gauge; an average, healthy engine should normally produce about 17 to 22 in-Hg with a fairly steady needle (see illustration). Refer to the following vacuum gauge readings and what they indicate about the engine's condition:

7 A low steady reading usually indicates a leaking gasket between the intake manifold and cylinder head(s) or throttle body, a leaky vacuum hose, late ignition timing or incorrect camshaft timing. Check ignition timing with a timing light and eliminate all other possible causes, utilizing the tests provided in this Chapter before you remove the timing chain cover to check the timing marks.

8 If the reading is three to eight inches below normal and it fluctuates at that low reading, suspect an intake manifold gasket leak at an intake port or a faulty fuel injector.

9 If the needle has regular drops of about two-to-four inches at a steady rate, the valves are probably leaking. Perform a compression check or leak-down test to confirm this.

10 An irregular drop or down-flick of the needle can be caused by a sticking valve or an ignition misfire. Perform a compression check or leak-down test and read the spark plugs.

11 A rapid vibration of about four in-Hg vibration at idle combined with exhaust smoke indicates worn valve guides. Perform a leak-down test to confirm this. If the rapid vibration occurs with an increase in engine speed, check for a leaking intake manifold gasket or head gasket, weak valve springs, burned valves or ignition misfire.

12 A slight fluctuation, say one inch up and down, may mean ignition problems. Check all the usual tune-up items and, if necessary, run the engine on an ignition analyzer.

13 If there is a large fluctuation, perform a compression or leak-down test to look for a weak or dead cylinder or a blown head gasket.

14 If the needle moves slowly through a wide range, check for a clogged PCV system, incorrect idle fuel mixture, throttle body or intake manifold gasket leaks.

15 Check for a slow return after revving the

6.3a After tightly wrapping water-vulnerable components, use a spray cleaner on everything, with particular concentration on the greasiest areas, usually around the valve cover and lower edges of the block. If one section dries out, apply more cleaner

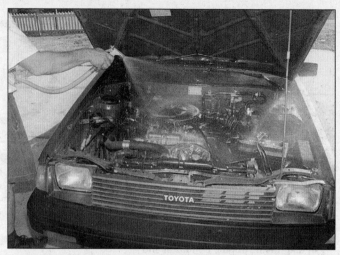

6.3b Depending on how dirty the engine is, let the cleaner soak in according to the directions and then hose off the grime and cleaner. Get the rinse water down into every area you can get at; then dry important components with a hair dryer or paper towels

engine by quickly snapping the throttle open until the engine reaches about 2,500 rpm and let it shut. Normally the reading should drop to near zero, rise above normal idle reading (about 5 in-Hg over) and then return to the previous idle reading. If the vacuum returns slowly and doesn't peak when the throttle is snapped shut, the rings may be worn. If there is a long delay, look for a restricted exhaust system (often the muffler or catalytic converter). An easy way to check this is to temporarily disconnect the exhaust ahead of the suspected part and redo the test.

5 Engine rebuilding alternatives

1   The do-it-yourselfer is faced with a number of options when purchasing a rebuilt engine. The major considerations are cost, warranty, parts availability and the time required for the rebuilder to complete the project. The decision to replace the engine block, piston/connecting rod assemblies and crankshaft depends on the final inspection results of your engine. Only then can you make a cost effective decision whether to have your engine overhauled or simply purchase an exchange engine for your vehicle.
2   Some of the rebuilding alternatives include:
3   **Individual parts** - If the inspection procedures reveal that the engine block and most engine components are in reusable condition, purchasing individual parts and having a rebuilder rebuild your engine may be the most economical alternative. The block, crankshaft and piston/connecting rod assemblies should all be inspected carefully by a machine shop first.

4   **Short block** - A short block consists of an engine block with a crankshaft and piston/connecting rod assemblies already installed. All new bearings are incorporated and all clearances will be correct. The existing camshafts, valve train components, cylinder head and external parts can be bolted to the short block with little or no machine shop work necessary.
5   **Long block** - A long block consists of a short block plus an oil pump, oil pan, cylinder head, valve cover, camshaft and valve train components, timing sprockets and chain or gears and timing cover. All components are installed with new bearings, seals and gaskets incorporated throughout. The installation of manifolds and external parts is all that's necessary.
6   **Low mileage used engines** - Some companies now offer low mileage used engines which are a very cost effective way to get your vehicle up and running again. These engines often come from vehicles which have been totaled in accidents or come from other countries which have a higher vehicle turn over rate. A low mileage used engine also usually has a similar warranty like the newly remanufactured engines.
7   Give careful thought to which alternative is best for you and discuss the situation with local automotive machine shops, auto parts dealers and experienced rebuilders before ordering or purchasing replacement parts.

6 Engine removal - methods and precautions

1   If you've decided that an engine must be removed for overhaul or major repair work, several preliminary steps should be taken. Read all removal and installation procedures carefully prior to committing to this job. These engines are removed by lowering the engine to the floor, along with the transaxle, and then raising the vehicle sufficiently to slide the assembly out; this will require a vehicle hoist as well as an engine hoist. Make sure the engine hoist is rated in excess of the combined weight of the engine and transaxle. A transmission jack is also very helpful. Safety is of primary importance, considering the potential hazards involved in removing the engine from the vehicle.
2   Locating a suitable place to work is extremely important. Adequate work space, along with storage space for the vehicle, will be needed. If a shop or garage isn't available, at the very least a flat, level, clean work surface made of concrete or asphalt is required.
3   Cleaning the engine compartment and engine before beginning the removal procedure will help keep tools clean and organized (see illustrations).
4   If you're a novice at engine removal, get at least one helper. One person cannot easily do all the things you need to do to remove a big heavy engine and transmission assembly from the engine compartment. Also helpful is to seek advice and assistance from someone who's experienced in engine removal.
5   Plan the operation ahead of time. Arrange for or obtain all of the tools and equipment you'll need prior to beginning the job (see illustration). Some of the equipment necessary to perform engine removal and installation safely and with relative ease are (in addition to a vehicle hoist and an engine hoist) a heavy duty floor jack (preferably fitted with a transmission jack head adapter), complete sets of wrenches and sockets as

described in the front of this manual, wooden blocks, plenty of rags and cleaning solvent for mopping up spilled oil, coolant and gasoline.
6   Plan for the vehicle to be out of use for quite a while. A machine shop can do the work that is beyond the scope of the home mechanic. Machine shops often have a busy schedule, so before removing the engine, consult the shop for an estimate of how long it will take to rebuild or repair the components that may need work.

7   Engine - removal and installation

Warning: *The models covered by this manual are equipped with a Supplemental Restraint system (SRS), more commonly known as airbags. Always disable the airbag system before working in the vicinity of airbag system components to avoid the possibility of accidental deployment of the airbag, which could cause personal injury (see Chapter 12).*
Warning: *Gasoline is extremely flammable, so take extra precautions when you work on any part of the fuel system. Don't smoke or allow open flames or bare light bulbs near the work area, and don't work in a garage where a gas-type appliance (such as a water heater or clothes dryer) is present. Since gasoline is carcinogenic, wear fuel-resistant gloves when there's a possibility of being exposed to fuel, and, if you spill any fuel on your skin, rinse it off immediately with soap and water. Mop up any spills immediately and do not store fuel-soaked rags where they could ignite. The fuel system is under constant pressure, so, if any fuel lines are to be disconnected, the fuel pressure in the system must be relieved first (see Chapter 4 for more information). When you perform any kind of work on the fuel system, wear safety glasses and have a Class B type fire extinguisher on hand.*
Warning: *The engine must be completely cool before beginning this procedure.*
Warning: *The air conditioning system is under high pressure. DO NOT loosen any fittings or remove any components until after the system has been discharged. Air conditioning refrigerant should be properly discharged into an EPA-approved container at a dealership service department or an automotive air conditioning repair facility. Always wear eye protection when disconnecting air conditioning system fittings.*
Note: *Engine removal on these models is a difficult job, especially for the do-it-yourself mechanic working at home. Because of the vehicle's design, the manufacturer states that the engine and transaxle have to be removed as a unit from the bottom of the vehicle, not the top. With a floor jack and jackstands the vehicle can't be raised high enough and supported safely enough for the engine/transaxle assembly to slide out from underneath. The manufacturer recommends that removal of the engine/transaxle assembly only be performed on a vehicle hoist.*

6.5 Get an engine stand sturdy enough to firmly support the engine while you're working on it. Stay away from three-wheeled models; they have a tendency to tip over more easily, so get a four-wheeled unit

Removal

1   Have the air conditioning system discharged by an automotive air conditioning technician.
2   Park the vehicle on a frame-contact type vehicle hoist. The pads of the hoist arms must contact the body welt along each side of the vehicle.
3   Relieve the fuel system pressure (see Chapter 4), then disconnect the negative cable from the battery (see Chapter 5)..
4   Place protective covers on the fenders and cowl and remove the hood (see Chapter 11).
5   Remove the air filter housing (see Chapter 4).
6   Remove the cowl panels (see Chapter 11).
7   Remove the battery and battery tray (see Chapter 5). On 2.3L models, remove the battery tray support bracket once the tray has been removed.
8   Loosen the front wheel lug nuts and the driveaxle/hub nuts, then raise the vehicle on the hoist. Drain the cooling system and engine oil and remove the drivebelt (see Chapter 1).
9   Clearly label, then disconnect all vacuum lines, coolant and emissions hoses, wiring harness connectors, ground straps and fuel lines. Masking tape and/or a touch up paint applicator work well for marking items. Take instant photos or sketch the locations of components and brackets.
10 On non-turbocharged V6 models, remove the upper intake manifold (see Chapter 2B).
11 On four-cylinder models, disconnect the high-pressure fuel pump quick-connect line (see Chapter 4).
12 Remove the alternator and the starter (see Chapter 5).
13 Remove the power steering pump and all hose retainers (see Chapter 10).
14 Remove the heater hoses and radiator hoses (see Chapter 3).
15 On four-cylinder models, disconnect and remove the charge air cooler inlet and outlet pipes (see Chapter 4).
16 Remove the coolant expansion tank, engine cooling fan and radiator (see Chapter 3).
Note: *This step is not absolutely necessary, but it will help avoid damage to the cooling fans and radiator as the engine is lowered out of the vehicle. If the radiator is not taken out it will still be necessary to detach the transaxle oil cooler lines from the bottom of the radiator.*
17 Disconnect the shift cable(s) from the transaxle (see Chapter 7A).
18 Disconnect any wiring harness connectors from the transaxle.
19 Remove the air conditioning compressor (see Chapter 3), and cap all the lines to prevent dirt and moisture from entering the system.
20 Place the wheels in a straight position, then lock the steering wheel in place and detach the steering column shaft from the steering gear (see Chapter 10).
Note: *It may be necessary to use a steering wheel holding tool to prevent the wheel from moving.*
21 Disconnect the PCM harness connectors and place the connectors out of the way (see Chapter 6).
22 On four-cylinder models, disconnect the Transmission Control Module (TCM) electrical connector and harness retainers, then place the harness out of the way.
23 Detach the exhaust pipe(s) from the exhaust manifold(s) (see Chapter 4).
24 Detach all wiring harnesses and hoses from between the engine/transaxle and the chassis. Be sure to mark all connectors to facilitate reassembly.
25 Remove the wheels, then remove the brake calipers. Tie the calipers out of the way with a piece of wire (see Chapter 9).
26 Remove the driveaxles (see Chapter 8). On AWD vehicles, remove the rear driveshaft.
27 Disconnect the EVAP line quick disconnect connectors at the rear of the subframe.
28 Detach the stabilizer bar links from the bar and separate the lower control arms from the steering knuckles. Also detach the tie-rod ends from the steering knuckles (see Chapter 10).
29 Mark the relationship of the torque converter to the driveplate, then remove the torque converter-to-driveplate nuts (see Chapter 7A).
30 Remove the roll restrictor between the lower rear part of the engine and the subframe.
31 Remove the subframe-to-bumper nuts and separate the bumper from the subframe.
32 Remove the subframe brace bolts and remove the brace.
33 Attach a lifting sling or chain to the engine. Position an engine hoist and connect the sling to it. If no lifting hooks or brackets are present, you'll have to fasten the chains or

9.1 Before you try to remove the pistons, use a ridge reamer to remove the raised material (ridge) from the top of the cylinders

9.3 Checking the connecting rod endplay (side clearance)

slings to some substantial part of the engine - ones that are strong enough to take the weight, but in locations that will provide good balance. Take up the slack until there is slight tension on the sling or chain. Position the chain on the hoist so it balances the engine and the transaxle level with the vehicle.

34 Mark the position of each corner of the subframe to the chassis. Support the subframe with two floor jacks, then remove the fasteners and lower the subframe. Remove it from underneath the vehicle.

35 Remove the transaxle mount (see Chapter 7A) and the right-side engine mount (see Chapter 2A).

36 Recheck to be sure nothing is still connecting the engine or transaxle to the vehicle. Disconnect and label anything still remaining.

37 Slowly lower the engine/transaxle from the vehicle.

Note: *Placing a sheet of hardboard or paneling between the engine and the floor makes moving the powertrain easier.*

38 Once the powertrain is on the floor, disconnect the engine lifting hoist and raise the vehicle hoist until the powertrain can be slid out from underneath.

Note: *A helper will be needed to move the powertrain.*

39 Reconnect the chain or sling and raise the engine/transaxle with the hoist. Support the transmission with a jack (preferably one with a transmission jack head adapter). Separate the engine from the transaxle (see Chapter 7A).

40 Remove the driveplate and the crankshaft sensor ring, then mount the engine on a stand.

Installation

Note: *The manufacturer recommends replacing all subframe and suspension fasteners with new ones whenever they are loosened or removed.*

41 Installation is the reverse of removal, noting the following points:

a) Check the engine/transaxle mounts. If they're worn or damaged, replace them.
b) Attach the transaxle to the engine following the procedure described in Chapter 7A.
c) Add coolant, oil, power steering and transmission fluids as needed (see Chapter 1).
d) Align the subframe reference marks before tightening the bolts.
e) Tighten the subframe mounting fasteners to the torque listed in the Chapter 10 Specifications.
f) Reconnect the negative battery cable (see Chapter 5).
g) Run the engine and check for proper operation and leaks. Shut off the engine and recheck fluid levels.
h) Have the air conditioning system recharged and leak tested by the shop that discharged it.

8 Engine overhaul - disassembly sequence

1 It's much easier to remove the external components if it's mounted on a portable engine stand. A stand can often be rented quite cheaply from an equipment rental yard. Before the engine is mounted on a stand, the driveplate should be removed from the engine.

2 If a stand isn't available, it's possible to remove the external engine components with it blocked up on the floor. Be extra careful not to tip or drop the engine when working without a stand.

3 If you're going to obtain a rebuilt engine, all external components must come off first, to be transferred to the replacement engine. These components include:

Turbocharger(s)
Driveplate
Emissions-related components
Engine mounts and mount brackets
Intake/exhaust manifolds
Fuel injection components
Oil filter
Ignition coils and spark plugs
Thermostat and housing assembly
Water pump

Note: *When removing the external components from the engine, pay close attention to details that may be helpful or important during installation. Note the installed position of gaskets, seals, spacers, pins, brackets, washers, bolts and other small items.*

4 If you're going to obtain a short block (assembled engine block, crankshaft, pistons and connecting rods), remove the timing chain, cylinder head, oil pan, oil pump pick-up tube, oil pump and water pump from your engine so that you can turn in your old short block to the rebuilder as a core. See *Engine rebuilding alternatives* for additional information regarding the different possibilities to be considered.

9 Pistons and connecting rods - removal and installation

Removal

Note: *Prior to removing the piston/connecting rod assemblies, remove the cylinder head and oil pan (see Chapter 2A or).*

1 Use your fingernail to feel if a ridge has formed at the upper limit of ring travel (about 1/4-inch down from the top of each cylinder). If carbon deposits or cylinder wear have produced ridges, they must be completely removed with a special tool (see illustration). Follow the manufacturer's instructions provided with the tool. Failure to remove the ridges before attempting to remove the piston/connecting rod assemblies may result in piston breakage.

2 After the cylinder ridges have been removed, turn the engine so the crankshaft is facing up. On four-cylinder engines, remove the balance shaft assembly (see Section 13). On V6 engines, remove the main bearing cap support brace.

3 Before the main bearing caps or main-bearing support bridge and connecting rods are removed, check the connecting rod endplay with feeler gauges. Slide them between the first connecting rod and the crankshaft throw until the play is removed (see illustration). Repeat this procedure for each connecting rod. The endplay is equal to the thickness of the feeler gauge(s). Check with an automotive machine shop for the endplay service limit (a typical end play limit should measure between 0.005 to 0.015 inch [0.127 to 0.381 mm]). If the play exceeds the service limit, new connecting rods will be required. If new rods (or a new crankshaft) are installed, the endplay may fall under the minimum allowable. If it does, the rods will have to be machined to restore it. If necessary, consult an automotive machine shop for advice.

Chapter 2 Part C General engine overhaul procedures

9.4 If the connecting rods and caps are not marked, use permanent ink or paint to mark the caps to the rods by cylinder number (for example, this would be the No. 4 connecting rod)

9.13 Install the piston ring into the cylinder then push it down into position using a piston so the ring will be square in the cylinder

4   Check the connecting rods and caps for identification marks. If they aren't plainly marked, use paint or marker to clearly identify each rod and cap (1, 2, 3, etc., depending on the cylinder they're associated with) (see illustration).

5   Loosen each of the connecting rod cap bolts 1/2-turn at a time until they can be removed by hand.

Note: *New connecting rod cap bolts must be used when reassembling the engine, but save the old bolts for use when checking the connecting rod bearing oil clearance.*

6   Remove the number one connecting rod cap and bearing insert. Don't drop the bearing insert out of the cap.

7   Remove the bearing insert and push the connecting rod/piston assembly out through the top of the engine. Use a wooden or plastic hammer handle to push on the upper bearing surface in the connecting rod. If resistance is felt, double-check to make sure that all of the ridge was removed from the cylinder.

8   Repeat the procedure for the remaining cylinders.

9   After removal, reassemble the connecting rod caps and bearing inserts in their respective connecting rods and install the cap bolts finger tight. Leaving the old bearing inserts in place until reassembly will help prevent the connecting rod bearing surfaces from being accidentally nicked or gouged.

10   The pistons and connecting rods are now ready for inspection and overhaul at an automotive machine shop.

Piston ring installation

11   Before installing the new piston rings, the ring end gaps must be checked. It's assumed that the piston ring side clearance has been checked and verified correct.

12   Lay out the piston/connecting rod assemblies and the new ring sets so the ring sets will be matched with the same piston and cylinder during the end gap measurement and engine assembly.

13   Insert the top (number one) ring into the first cylinder and square it up with the cylinder walls by pushing it in with the top of the piston (see illustration). The ring should be near the bottom of the cylinder, at the lower limit of ring travel.

14   To measure the end gap, slip feeler gauges between the ends of the ring until a gauge equal to the gap width is found (see illustration). The feeler gauge should slide between the ring ends with a slight amount of drag. A typical ring gap should fall between 0.010 and 0.020 inch (0.25 to 0.50 mm) for compression rings and up to 0.030 inch (0.76 mm) for the oil ring steel rails. If the gap is larger or smaller than specified, double-check to make sure you have the correct rings before proceeding.

15   If the gap is too small, it must be enlarged or the ring ends may come in contact with each other during engine operation, which can cause serious damage to the engine. If necessary, increase the end gaps by filing the ring ends very carefully with a fine file. Mount the file in a vise equipped with soft jaws, slip the ring over the file with the ends contacting the file face and slowly move the ring to remove material from the ends. When performing this operation, file only by pushing the ring from the outside end of the file towards the vise (see illustration).

9.14 With the ring square in the cylinder, measure the ring end gap with a feeler gauge

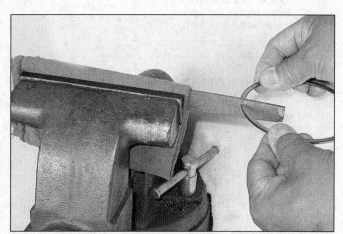

9.15 If the ring end gap is too small, clamp a file in a vise as shown and file the piston ring ends - be sure to remove all raised material

ENGINE BEARING ANALYSIS

Debris

Babbitt bearing embedded with debris from machinings

Microscopic detail of debris

Microscopic detail of gouges

Overplated copper alloy bearing gouged by cast iron debris

Aluminum bearing embedded with glass beads

Microscopic detail of glass beads

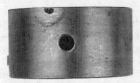

Damaged lining caused by dirt left on the bearing back

Misassembly

Result of a lower half assembled as an upper - blocking the oil flow

Excessive oil clearance is indicated by a short contact arc

Polished and oil-stained backs are a result of a poor fit in the housing bore

Result of a wrong, reversed, or shifted cap

Overloading

Damage from excessive idling which resulted in an oil film unable to support the load imposed

Damaged upper connecting rod bearings caused by engine lugging; the lower main bearings (not shown) were similarly affected

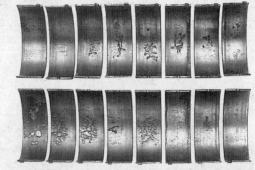

The damage shown in these upper and lower connecting rod bearings was caused by engine operation at a higher-than-rated speed under load

Misalignment

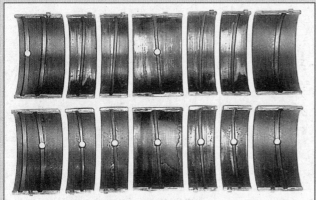

A warped crankshaft caused this pattern of severe wear in the center, diminishing toward the ends

A poorly finished crankshaft caused the equally spaced scoring shown

A bent connecting rod led to the damage in the "V" pattern

A tapered housing bore caused the damage along one edge of this pair

Lubrication

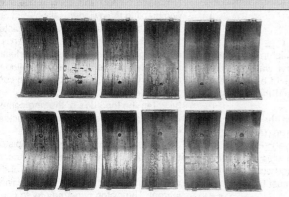

Result of dry start: The bearings on the left, farthest from the oil pump, show more damage

Result of a low oil supply or oil starvation

Severe wear as a result of inadequate oil clearance

Corrosion

Microscopic detail of corrosion

Corrosion is an acid attack on the bearing lining generally caused by inadequate maintenance, extremely hot or cold operation, or inferior oils or fuels

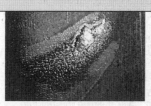

Microscopic detail of cavitation

Example of cavitation - a surface erosion caused by pressure changes in the oil film

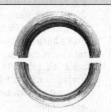

Damage from excessive thrust or insufficient axial clearance

Bearing affected by oil dilution caused by excessive blow-by or a rich mixture

© 1986 Federal-Mogul Corporation
Copy and photographs courtesy of Federal Mogul Corporation

9.19a Installing the spacer/expander in the oil ring groove

9.19b DO NOT use a piston ring installation tool when installing the oil control side rails

9.22 Use a piston ring installation tool to install the number 2 and the number 1 (top) rings - be sure the directional mark on the piston ring(s) is facing toward the top of the piston

16 Excess end gap isn't critical unless it's greater than 0.040 inch (1.01 mm). Again, double-check to make sure you have the correct ring type.

17 Repeat the procedure for each ring that will be installed in the first cylinder and for each ring in the remaining cylinders. Remember to keep rings, pistons and cylinders matched up.

18 Once the ring end gaps have been checked/corrected, the rings can be installed on the pistons.

19 The oil control ring (lowest one on the piston) is usually installed first. It's composed of three separate components. Slip the spacer/expander into the groove (see illustration). If an anti-rotation tang is used, make sure it's inserted into the drilled hole in the ring groove. Next, install the upper side rail in the same manner (see illustration). Don't use a piston ring installation tool on the oil ring side rails, as they may be damaged. Instead, place one end of the side rail into the groove between the spacer/expander and the ring land, hold it firmly in place and slide a finger around the piston while pushing the rail into the groove. Finally, install the lower side rail.

20 After the three oil ring components have been installed, check to make sure that both the upper and lower side rails can be rotated smoothly inside the ring grooves.

21 The number two (middle) ring is installed next. It's usually stamped with a mark which must face up, toward the top of the piston. Do not mix up the top and middle rings, as they have different cross-sections.

Note: *Always follow the instructions printed on the ring package or box - different manufacturers may require different approaches.*

22 Use a piston ring installation tool and make sure the identification mark is facing the top of the piston, then slip the ring into the middle groove on the piston (see illustration). Don't expand the ring any more than necessary to slide it over the piston.

23 Install the number one (top) ring in the same manner. Make sure the mark is facing up. Be careful not to confuse the number one and number two rings.

24 Repeat the procedure for the remaining pistons and rings.

Installation

25 Before installing the piston/connecting rod assemblies, the cylinder walls must be perfectly clean, the top edge of each cylinder bore must be chamfered, and the crankshaft must be in place.

26 Remove the cap from the end of the number one connecting rod (refer to the marks made during removal). Remove the original bearing inserts and wipe the bearing surfaces of the connecting rod and cap with a clean, lint-free cloth. They must be kept spotlessly clean.

Connecting rod bearing oil clearance check

27 Clean the back side of the new upper bearing insert, then lay it in place in the connecting rod.

28 Make sure the tab on the bearing fits into the recess in the rod. Don't hammer the bearing insert into place and be very careful not to nick or gouge the bearing face. Don't lubricate the bearing at this time.

29 Clean the back side of the other bearing insert and install it in the rod cap. Again, make sure the tab on the bearing fits into the recess in the cap, and don't apply any lubricant. It's critically important that the mating surfaces of the bearing and connecting rod are perfectly clean and oil free when they're assembled.

30 Position the piston ring gaps at 90-degree intervals around the piston as shown (see illustrations).

31 Lubricate the piston and rings with clean engine oil and attach a piston ring compressor to the piston. Leave the skirt protruding about 1/4-inch to guide the piston into the cylinder. The rings must be compressed until they're flush with the piston.

32 Rotate the crankshaft until the number one connecting rod journal is at BDC (bottom dead center) and apply a liberal coat of engine oil to the cylinder walls.

33 With the mark on top of the piston facing the front (timing belt or chain end) of the engine, gently insert the piston/connecting rod assembly into the number one cylinder bore and rest the bottom edge of the ring compressor on the engine block.

34 Tap the top edge of the ring compressor to make sure it's contacting the block around its entire circumference.

35 Gently tap on the top of the piston with the end of a wooden or plastic hammer handle (see illustration) while guiding the end of the connecting rod into place on the crankshaft journal. The piston rings may try to pop out of the ring compressor just before entering the cylinder bore, so keep some downward pressure on the ring compressor. Work slowly, and if any resistance is felt as the piston enters the cylinder, stop immediately. Find out what's hanging up and fix it before proceeding. Do not, for any reason, force the piston into the cylinder - you might break a ring and/or the piston.

36 Once the piston/connecting rod assembly is installed, the connecting rod bearing oil clearance must be checked before the rod cap is permanently installed.

37 Cut a piece of the appropriate size Plastigage slightly shorter than the width of the connecting rod bearing and lay it in place on the number one connecting rod journal, parallel with the journal axis (see illustration).

38 Clean the connecting rod cap bearing face and install the rod cap. Make sure the mating mark on the cap is on the same side as the mark on the connecting rod (see illustration 9.4).

39 Install the old rod bolts, at this time, and tighten them to the torque listed in this Chapter's Specifications.

Note: *Use a thin-wall socket to avoid erro-*

Chapter 2 Part C General engine overhaul procedures

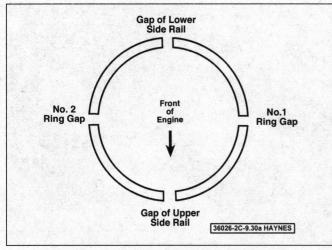

9.30a Position the piston ring end gaps as shown - four-cylinder engines

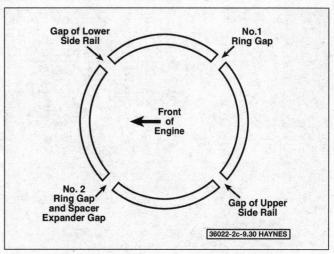

9.30b V6 engine piston ring end gaps positions

9.35 Use a plastic or wooden hammer handle to push the piston into the cylinder

9.37 Place Plastigage on each connecting rod bearing journal parallel to the crankshaft centerline

9.41 Use the scale on the Plastigage package to determine the bearing oil clearance - be sure to measure the widest part of the Plastigage and use the correct scale; it comes with both standard and metric scales

neous torque readings that can result if the socket is wedged between the rod cap and the bolt head. If the socket tends to wedge itself between the fastener and the cap, lift up on it slightly until it no longer contacts the cap. DO NOT rotate the crankshaft at any time during this operation.

40 Remove the fasteners and detach the rod cap, being very careful not to disturb the Plastigage.

41 Compare the width of the crushed Plastigage to the scale printed on the Plastigage envelope to obtain the oil clearance (see illustration). The connecting rod oil clearance is usually about 0.001 to 0.002 inch (0.025 to 0.05 mm). Consult an automotive machine shop for the clearance specified for the rod bearings on your engine. One the covered vehicles, code numbers on the block and crankshaft are used to select the proper bearings, using a chart at a dealership service/parts department.

42 If the clearance is not as specified, the bearing inserts may be the wrong size (which means different ones will be required). Before deciding that different inserts are needed, make sure that no dirt or oil was between the bearing inserts and the connecting rod or cap when the clearance was measured. Also, recheck the journal diameter. If the Plastigage was wider at one end than the other, the journal may be tapered. If the clearance still exceeds the limit specified, the bearing will have to be replaced with an undersize bearing.

Caution: *When installing a new crankshaft always use a standard size bearing.*

Final installation

43 Carefully scrape all traces of the Plastigage material off the rod journal and/or bearing face. Be very careful not to scratch the bearing - use your fingernail or the edge of a plastic card.

44 Make sure the bearing faces are perfectly clean, then apply a uniform layer of clean moly-base grease or engine assembly lube to both of them. You'll have to push the piston into the cylinder to expose the face of the bearing insert in the connecting rod.

45 Slide the connecting rod back into place on the journal, install the rod cap, install the new bolts and tighten them to the torque listed in this Chapter's Specifications.

Caution: *Install new connecting rod cap bolts. Do NOT reuse old bolts - they have stretched and cannot be reused. Again, work up to the torque in three steps.*

46 Repeat the entire procedure for the remaining pistons/connecting rods.

10.1 Checking crankshaft endplay with a dial indicator

10.3 Checking the crankshaft endplay with feeler gauges at the thrust bearing journal

47 The important points to remember are:
a) *Keep the back sides of the bearing inserts and the insides of the connecting rods and caps perfectly clean when assembling them.*
b) *Make sure you have the correct piston/rod assembly for each cylinder.*
c) *The mark on the piston must face the front (timing chain) of the engine.*
d) *Lubricate the cylinder walls liberally with clean oil.*
e) *Lubricate the bearing faces when installing the rod caps after the oil clearance has been checked.*

48 After all the piston/connecting rod assemblies have been correctly installed, rotate the crankshaft a number of times by hand to check for any obvious binding.

49 As a final step, check the connecting rod endplay, as described in Step 3. If it was correct before disassembly and the original crankshaft and rods were reinstalled, it should still be correct. If new rods or a new crankshaft were installed, the endplay may be inadequate. If so, the rods will have to be removed and taken to an automotive machine shop for resizing.

10 Crankshaft - removal and installation

Removal

Note: *The crankshaft can be removed only after the engine has been removed from the vehicle. It's assumed that the driveplate, crankshaft pulley, timing chain, oil pan, oil pump body, oil filter and piston/connecting rod assemblies have already been removed. The rear main oil seal retainer must be unbolted and separated from the block before proceeding with crankshaft removal.*

1 Before the crankshaft is removed, measure the endplay. Mount a dial indicator with the indicator in line with the crankshaft and just touching the end of the crankshaft as shown (see illustration).

2 Pry the crankshaft all the way to the rear and zero the dial indicator. Next, pry the crankshaft to the front as far as possible and check the reading on the dial indicator. The distance traveled is the endplay. A typical crankshaft endplay will fall between 0.003 to 0.010 inch (0.076 to 0.254 mm). If it is greater than that, check the crankshaft thrust surfaces for wear after it's removed. If no wear is evident, new main bearings should correct the endplay.

3 If a dial indicator isn't available, feeler gauges can be used. Gently pry the crankshaft all the way to the front of the engine. Slip feeler gauges between the crankshaft and the front face of the thrust bearing or washer to determine the clearance (see illustration).

4 Loosen the main bearing cap beam bolts (four-cylinder engines) or main bearing cap support brace and main bearing cap bolts (V6 engines) 1/4-turn at a time each, until they can be removed by hand. Loosen the bolts in the reverse of the tightening sequence (see illustration(s) 10.20a or 10.20b and 10.20c).

Note: *New main bearing cap bolts must be used when reassembling the engine, but save the old bolts for use when checking the main bearing oil clearance.*

5 If you're working on a four-cylinder engine, remove the main bearing cap support beam. If you're working on a V6 engine, remove the main bearing caps. Try not to drop the bearing inserts if they come out with the assembly.

6 Carefully lift the crankshaft out of the engine. It may be a good idea to have an assistant available, since the crankshaft is quite heavy and awkward to handle. With the bearing inserts in place inside the engine block and main bearing caps or lower cylinder block, reinstall the main bearing caps or lower cylinder block onto the engine block and tighten the bolts finger tight.

Installation

7 Crankshaft installation is the first step in engine reassembly. It's assumed at this point that the engine block and crankshaft have been cleaned, inspected and repaired or reconditioned.

8 Position the engine block with the bottom facing up.

9 Remove the bolts and lift off the support beam, main bearing caps or lower cylinder block, as applicable.

10 If they're still in place, remove the original bearing inserts. Wipe the bearing surfaces of the block and main bearing cap assembly with a clean, lint-free cloth. They must be kept spotlessly clean. This is critical for determining the correct bearing oil clearance.

Main bearing oil clearance check

11 Without mixing them up, clean the back sides of the new upper main bearing inserts (with grooves and oil holes) and lay one in each main bearing saddle in the engine block. Each upper bearing (engine block) has an oil groove and oil hole in it.

Caution: *The oil holes in the block must line up with the oil holes in the upper bearing inserts.*

Note: *The thrust bearing on the four-cylinder engine is located on the engine block number 3 (center) journal. The thrust washers on the V6 engines are located on the 4th journal (two upper on each side of the main saddle, one lower on the rear of the main bearing cap). The grooves on shim-type thrust washers must face the crankshaft (not the main bearing saddle). Clean the back sides of the lower main bearing inserts and lay them in the corresponding location in the main bearing caps or the lower cylinder block. Make sure the tab on the bearing insert fits into its corresponding recess.*

Caution: *Do not hammer the bearing insert into place and don't nick or gouge the bearing faces. DO NOT apply any lubrication at this time.*

Chapter 2 Part C General engine overhaul procedures

10.17 Place the Plastigage onto the crankshaft bearing journal as shown

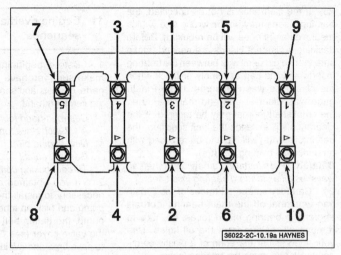

10.20a Main bearing cap beam bolt tightening sequence (four-cylinder engines)

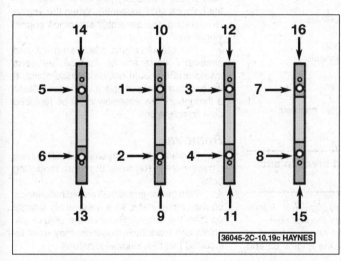

10.20b Main bearing cap bolt tightening sequence (V6 engines)

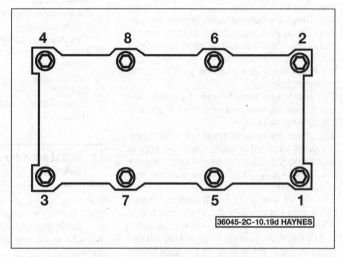

10.20c Main bearing cap support brace, bolt tightening sequence (V6 engines)

12 Clean the faces of the bearing inserts in the block and the crankshaft main bearing journals with a clean, lint-free cloth.
13 Check or clean the oil holes in the crankshaft, as any dirt here can go only one way - straight through the new bearings.
14 Once you're certain the crankshaft is clean, carefully lay it in position in the cylinder block.
15 Before the crankshaft can be permanently installed, the main bearing oil clearance must be checked.
16 Cut several strips of the appropriate size of Plastigage. They must be slightly shorter than the width of the main bearing journal.
17 Place one piece on each crankshaft main bearing journal, parallel with the journal axis as shown (see illustration).
18 Clean the faces of the bearing inserts in the main bearing caps or the lower cylinder block. Hold the bearing inserts in place and install the caps or the lower cylinder block onto the crankshaft and cylinder block. DO NOT disturb the Plastigage.
19 Apply clean engine oil to all bolt threads prior to installation, then install all bolts finger-tight.
Note: *Use the old bolts at this time.*
20 Tighten the bolts in the sequence shown (see illustrations) progressing in steps, to the torque listed in this Chapter's Specifications. DO NOT rotate the crankshaft at any time during this operation.
21 Remove the bolts in the reverse order of the tightening sequence and carefully lift the caps or the lower cylinder block straight up and off the block. Do not disturb the Plastigage or rotate the crankshaft.
22 Compare the width of the crushed Plastigage on each journal to the scale printed on the Plastigage envelope to determine the main bearing oil clearance (see illustration). Check with an automotive machine shop for the oil clearance for your engine.

10.22 Use the scale on the Plastigage package to determine the bearing oil clearance - be sure to measure the widest part of the Plastigage and use the correct scale; it comes with both standard and metric scales

23 If the clearance is not as specified, the bearing inserts may be the wrong size (which means different ones will be required). Before deciding if different inserts are needed, make sure that no dirt or oil was between the bearing inserts and the cap assembly or block when the clearance was measured. If the Plastigage was wider at one end than the other, the crankshaft journal may be tapered. If the clearance still exceeds the limit specified, the bearing insert(s) will have to be replaced with an undersize bearing insert(s).
Caution: *When installing a new crankshaft always install a standard bearing insert set.*
24 Carefully scrape all traces of the Plastigage material off the main bearing journals and/or the bearing insert faces. Be sure to remove all residue from the oil holes. Use your fingernail or the edge of a plastic card - don't nick or scratch the bearing faces.

Final installation

25 Carefully lift the crankshaft out of the cylinder block.
26 Clean the bearing insert faces in the cylinder block, then apply a thin, uniform layer of moly-base grease or engine assembly lube to each of the bearing surfaces. Be sure to coat the thrust faces as well as the journal face of the thrust bearing.
27 Make sure the crankshaft journals are clean, then lay the crankshaft back in place in the cylinder block.
28 Clean the bearing insert faces and apply the same lubricant to them. Clean the engine block and the bearing caps/lower cylinder block thoroughly. The surfaces must be free of oil residue.
29 On V6 engines, install the lower cylinder block onto the crankshaft and cylinder block. If you're working on a four-cylinder engine, install the main bearing caps in their proper locations, with the arrows on the caps facing the front of the engine.
30 Prior to installation, apply clean engine oil to all of the NEW bolt threads, wiping off any excess, then install all bolts finger-tight.
31 Tighten the support beam (four-cylinder engine) or caps (V6 engines), following the correct torque sequence (see illustration 10.20a or 10.20b). Torque the bolts to the Specifications listed in this Chapter. On V6 engines, install the main bearing cap support brace, tightening the bolts to the torque listed in this Chapter's Specifications in the proper sequence (see illustration 10.20c).
32 Recheck the crankshaft endplay with a feeler gauge or a dial indicator. The endplay should be correct if the crankshaft thrust faces aren't worn or damaged and if new bearings have been installed.
33 Rotate the crankshaft a number of times by hand to check for any obvious binding. It should rotate with a running torque of 50 in-lbs or less. If the running torque is too high, correct the problem at this time.
34 Install a new rear main oil seal (see Chapter 2A or 2B).

11 Engine overhaul - reassembly sequence

1 Before beginning engine reassembly, make sure you have all the necessary new parts, gaskets and seals as well as the following items on hand:

Common hand tools
A 1/2-inch drive torque wrench
New engine oil
Gasket sealant
Thread locking compound

2 If you obtained a short block it will be necessary to install the cylinder head, the oil pump and pick-up tube, the oil pan, the water pump, the timing belt and timing cover, and the valve cover (see Chapter 2A or). In order to save time and avoid problems, the external components must be installed in the following general order:

Thermostat and housing cover
Water pump
Intake and exhaust manifolds
Turbocharger(s)
Fuel injection components
Emission control components
Spark plugs
Ignition coils
Oil filter
Engine mounts and mount brackets
Driveplate

12 Initial start-up and break-in after overhaul

Warning: *Have a fire extinguisher handy when starting the engine for the first time.*
1 Once the engine has been installed in the vehicle, double-check the engine oil and coolant levels.
2 With the spark plugs out of the engine and the fuel pump disabled (see Chapter 4, Section 3), crank the engine until oil pressure registers on the gauge or the light goes out.
3 Install the spark plugs, and restore the ignition system and fuel pump functions.
4 Start the engine. It may take a few moments for the fuel system to build up pressure, but the engine should start without a great deal of effort.
5 After the engine starts, it should be allowed to warm up to normal operating temperature. While the engine is warming up, make a thorough check for fuel, oil and coolant leaks.
6 Shut the engine off and recheck the engine oil and coolant levels.
7 Drive the vehicle to an area with minimum traffic, accelerate from 30 to 50 mph, then allow the vehicle to slow to 30 mph with the throttle closed. Repeat the procedure 10 or 12 times. This will load the piston rings and cause them to seat properly against the cylinder walls. Check again for oil and coolant leaks.
8 Drive the vehicle gently for the first 500 miles (no sustained high speeds) and keep a constant check on the oil level. It is not unusual for an engine to use oil during the break-in period.
9 At approximately 500 to 600 miles, change the oil and filter.
10 For the next few hundred miles, drive the vehicle normally. Do not pamper it or abuse it.
11 After 2,000 miles, change the oil and filter again and consider the engine broken in.

13 Balance shaft assembly (four-cylinder engine) - removal and installation

1 The balance shaft assembly is bolted to the crankshaft main bearing support beam. A gear on the crankshaft meshes with a gear on the balance shaft assembly. When the engine is operating, the assembly smoothes engine vibrations.
2 The balancer is a precision-machined assembly, there are no serviceable parts inside and it should not be disassembled. If the backlash is out of Specification (see Steps 5 through 8), the assembly must be replaced as a complete unit.

Removal

3 Position the crankshaft at TDC (see Chapter 2A). Remove the four mounting bolts.
4 With the engine turned crankshaft side up on the engine stand, lift the assembly straight up from the engine. Remove the adjustment shims and mark their locations, they must be installed into their original locations.

Inspection

5 When the balance shaft assembly is in place, the backlash between the drive gear (on the crankshaft) and the driven gear on the assembly can be checked. Remove the timing peg.
6 Attach a 5 mm Allen wrench to the top of the driveshaft with the long end of the wrench pointing straight up using a hose clamp. Secure a dial-indicator fixture to the engine so that the tip of the indicator is against the top of the wrench (see illustration).
7 Use a pry tool against the crankshaft front counterweight to apply thrust pressure, while turning the crankshaft back-and-forth. Record the measurements of the dial-indicator. Measurements should be taken at the following degrees of engine rotation: 10, 30, 100, 190 and 210 degrees. Compare your results with the allowable range of backlash given in this Chapter's Specifications.
8 If the backlash is out of range, the assembly must be replaced.

Chapter 2 Part C General engine overhaul procedures

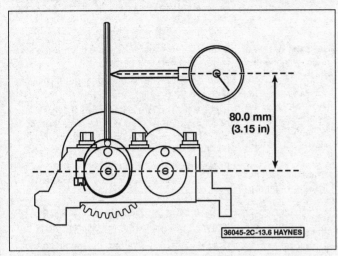

13.6 Method of checking the backlash in the balancer assembly

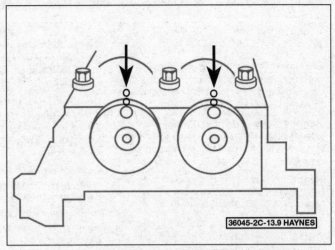

13.9 Remove/install the balance shaft assembly bolts only when the crank is at TDC and these balancer dots align

Installation

9   Install the adjustment shims in their original positions on the seat faces of the balance shaft assembly. With the balance shaft assembly, the engine must set to TDC for cylinder number 1 using special tool (303-507) (see Chapter 2A, Section 3), if necessary. Before installing the assembly, rotate it to align the timing marks on both shafts of the assembly (see illustration). Bolt the assembly to the engine and recheck that the timing marks are still aligned and that the crankshaft has not moved. Tighten the bolts in a criss-cross pattern to the torque listed in this Chapter's Specifications.

10   The crankshaft timing peg should remain installed to keep the engine at TDC until installation of the timing chain and sprockets is completed.

COMMON ENGINE OVERHAUL TERMS

B

Backlash - The amount of play between two parts. Usually refers to how much one gear can be moved back and forth without moving the gear with which it's meshed.

Bearing Caps - The caps held in place by nuts or bolts which, in turn, hold the bearing surface. This space is for lubricating oil to enter.

Bearing clearance - The amount of space left between shaft and bearing surface. This space is for lubricating oil to enter.

Bearing crush - The additional height which is purposely manufactured into each bearing half to ensure complete contact of the bearing back with the housing bore when the engine is assembled.

Bearing knock - The noise created by movement of a part in a loose or worn bearing.

Blueprinting - Dismantling an engine and reassembling it to EXACT specifications.

Bore - An engine cylinder, or any cylindrical hole; also used to describe the process of enlarging or accurately refinishing a hole with a cutting tool, as to bore an engine cylinder. The bore size is the diameter of the hole.

Boring - Renewing the cylinders by cutting them out to a specified size. A boring bar is used to make the cut.

Bottom end - A term which refers collectively to the engine block, crankshaft, main bearings and the big ends of the connecting rods.

Break-in - The period of operation between installation of new or rebuilt parts and time in which parts are worn to the correct fit. Driving at reduced and varying speed for a specified mileage to permit parts to wear to the correct fit.

Bushing - A one-piece sleeve placed in a bore to serve as a bearing surface for shaft, piston pin, etc. Usually replaceable.

C

Camshaft - The shaft in the engine, on which a series of lobes are located for operating the valve mechanisms. The camshaft is driven by gears or sprockets and a timing chain. Usually referred to simply as the cam.

Carbon - Hard, or soft, black deposits found in combustion chamber, on plugs, under rings, on and under valve heads.

Cast iron - An alloy of iron and more than two percent carbon, used for engine blocks and heads because it's relatively inexpensive and easy to mold into complex shapes.

Chamfer - To bevel across (or a bevel on) the sharp edge of an object.

Chase - To repair damaged threads with a tap or die.

Combustion chamber - The space between the piston and the cylinder head, with the piston at top dead center, in which air-fuel mixture is burned.

Compression ratio - The relationship between cylinder volume (clearance volume) when the piston is at top dead center and cylinder volume when the piston is at bottom dead center.

Connecting rod - The rod that connects the crank on the crankshaft with the piston. Sometimes called a con rod.

Connecting rod cap - The part of the connecting rod assembly that attaches the rod to the crankpin.

Core plug - Soft metal plug used to plug the casting holes for the coolant passages in the block.

Crankcase - The lower part of the engine in which the crankshaft rotates; includes the lower section of the cylinder block and the oil pan.

Crank kit - A reground or reconditioned crankshaft and new main and connecting rod bearings.

Crankpin - The part of a crankshaft to which a connecting rod is attached.

Crankshaft - The main rotating member, or shaft, running the length of the crankcase, with offset throws to which the connecting rods are attached; changes the reciprocating motion of the pistons into rotating motion.

Cylinder sleeve - A replaceable sleeve, or liner, pressed into the cylinder block to form the cylinder bore.

D

Deburring - Removing the burrs (rough edges or areas) from a bearing.

Deglazer - A tool, rotated by an electric motor, used to remove glaze from cylinder walls so a new set of rings will seat.

E

Endplay - The amount of lengthwise movement between two parts. As applied to a crankshaft, the distance that the crankshaft can move forward and back in the cylinder block.

F

Face - A machinist's term that refers to removing metal from the end of a shaft or the face of a larger part, such as a flywheel.

Fatigue - A breakdown of material through a large number of loading and unloading cycles. The first signs are cracks followed shortly by breaks.

Feeler gauge - A thin strip of hardened steel, ground to an exact thickness, used to check clearances between parts.

Free height - The unloaded length or height of a spring.

Freeplay - The looseness in a linkage, or an assembly of parts, between the initial application of force and actual movement. Usually perceived as slop or slight delay.

Freeze plug - See Core plug.

G

Gallery - A large passage in the block that forms a reservoir for engine oil pressure.

Glaze - The very smooth, glassy finish that develops on cylinder walls while an engine is in service.

H

Heli-Coil - A rethreading device used when threads are worn or damaged. The device is installed in a retapped hole to reduce the thread size to the original size.

I

Installed height - The spring's measured length or height, as installed on the cylinder head. Installed height is measured from the spring seat to the underside of the spring retainer.

J

Journal - The surface of a rotating shaft which turns in a bearing.

K

Keeper - The split lock that holds the valve spring retainer in position on the valve stem.

Key - A small piece of metal inserted into matching grooves machined into two parts fitted together - such as a gear pressed onto a shaft - which prevents slippage between the two parts.

Knock - The heavy metallic engine sound, produced in the combustion chamber as a result of abnormal combustion - usually detonation. Knock is usually caused by a loose or worn bearing. Also referred to as detonation, pinging and spark knock. Connecting rod or main bearing knocks are created by too much oil clearance or insufficient lubrication.

L

Lands - The portions of metal between the piston ring grooves.

Lapping the valves - Grinding a valve face and its seat together with lapping compound.

Lash - The amount of free motion in a gear train, between gears, or in a mechanical assembly, that occurs before movement can begin. Usually refers to the lash in a valve train.

Lifter - The part that rides against the cam to transfer motion to the rest of the valve train.

M

Machining - The process of using a machine to remove metal from a metal part.

Main bearings - The plain, or babbit, bearings that support the crankshaft.

Main bearing caps - The cast iron caps, bolted to the bottom of the block, that support the main bearings.

O

O.D. - Outside diameter.

Oil gallery - A pipe or drilled passageway in the engine used to carry engine oil from one area to another.

Oil ring - The lower ring, or rings, of a piston; designed to prevent excessive amounts of oil from working up the cylinder walls and into the combustion chamber. Also called an oil-control ring.

Oil seal - A seal which keeps oil from leaking out of a compartment. Usually refers to a dynamic seal around a rotating shaft or other moving part.

O-ring - A type of sealing ring made of a special rubberlike material; in use, the O-ring is compressed into a groove to provide the sealing action.

Overhaul - To completely disassemble a unit, clean and inspect all parts, reassemble it with the original or new parts and make all adjustments necessary for proper operation.

P

Pilot bearing - A small bearing installed in the center of the flywheel (or the rear end of the crankshaft) to support the front end of the input shaft of the transmission.

Pip mark - A little dot or indentation which indicates the top side of a compression ring.

Piston - The cylindrical part, attached to the connecting rod, that moves up and down in the cylinder as the crankshaft rotates. When the fuel charge is fired, the piston transfers the force of the explosion to the connecting rod, then to the crankshaft.

Piston pin (or wrist pin) - The cylindrical and usually hollow steel pin that passes through the piston. The piston pin fastens the piston to the upper end of the connecting rod.

Piston ring - The split ring fitted to the groove in a piston. The ring contacts the sides of the ring groove and also rubs against the cylinder wall, thus sealing space between piston and wall. There are two types of rings: Compression rings seal the compression pressure in the combustion chamber; oil rings scrape excessive oil off the cylinder wall.

Piston ring groove - The slots or grooves cut in piston heads to hold piston rings in position.

Piston skirt - The portion of the piston below the rings and the piston pin hole.

Plastigage - A thin strip of plastic thread, available in different sizes, used for measuring clearances. For example, a strip of plastigage is laid across a bearing journal and mashed as parts are assembled. Then parts are disassembled and the width of the strip is measured to determine clearance between journal and bearing. Commonly used to measure crankshaft main-bearing and connecting rod bearing clearances.

Press-fit - A tight fit between two parts that requires pressure to force the parts together. Also referred to as drive, or force, fit.

Prussian blue - A blue pigment; in solution, useful in determining the area of contact between two surfaces. Prussian blue is commonly used to determine the width and location of the contact area between the valve face and the valve seat.

R

Race (bearing) - The inner or outer ring that provides a contact surface for balls or rollers in bearing.

Ream - To size, enlarge or smooth a hole by using a round cutting tool with fluted edges.

Ring job - The process of reconditioning the cylinders and installing new rings.

Runout - Wobble. The amount a shaft rotates out-of-true.

S

Saddle - The upper main bearing seat.

Scored - Scratched or grooved, as a cylinder wall may be scored by abrasive particles moved up and down by the piston rings.

Scuffing - A type of wear in which there's a transfer of material between parts moving against each other; shows up as pits or grooves in the mating surfaces.

Seat - The surface upon which another part rests or seats. For example, the valve seat is the matched surface upon which the valve face rests. Also used to refer to wearing into a good fit; for example, piston rings seat after a few miles of driving.

Short block - An engine block complete with crankshaft and piston and, usually, camshaft assemblies.

Static balance - The balance of an object while it's stationary.

Step - The wear on the lower portion of a ring land caused by excessive side and back-clearance. The height of the step indicates the ring's extra side clearance and the length of the step projecting from the back wall of the groove represents the ring's back clearance.

Stroke - The distance the piston moves when traveling from top dead center to bottom dead center, or from bottom dead center to top dead center.

Stud - A metal rod with threads on both ends.

T

Tang - A lip on the end of a plain bearing used to align the bearing during assembly.

Tap - To cut threads in a hole. Also refers to the fluted tool used to cut threads.

Taper - A gradual reduction in the width of a shaft or hole; in an engine cylinder, taper usually takes the form of uneven wear, more pronounced at the top than at the bottom.

Throws - The offset portions of the crankshaft to which the connecting rods are affixed.

Thrust bearing - The main bearing that has thrust faces to prevent excessive endplay, or forward and backward movement of the crankshaft.

Thrust washer - A bronze or hardened steel washer placed between two moving parts. The washer prevents longitudinal movement and provides a bearing surface for thrust surfaces of parts.

Tolerance - The amount of variation permitted from an exact size of measurement. Actual amount from smallest acceptable dimension to largest acceptable dimension.

U

Umbrella - An oil deflector placed near the valve tip to throw oil from the valve stem area.

Undercut - A machined groove below the normal surface.

Undersize bearings - Smaller diameter bearings used with re-ground crankshaft journals.

V

Valve grinding - Refacing a valve in a valve-refacing machine.

Valve train - The valve-operating mechanism of an engine; includes all components from the camshaft to the valve.

Vibration damper - A cylindrical weight attached to the front of the crankshaft to minimize torsional vibration (the twist-untwist actions of the crankshaft caused by the cylinder firing impulses). Also called a harmonic balancer.

W

Water jacket - The spaces around the cylinders, between the inner and outer shells of the cylinder block or head, through which coolant circulates.

Web - A supporting structure across a cavity.

Woodruff key - A key with a radiused backside (viewed from the side).

Notes

Chapter 3
Cooling, heating and air conditioning systems

Contents

Section		Section
Air conditioning and heating system - check and maintenance.....	3	Engine cooling fans - removal and installation 5
Air conditioning compressor - removal and installation	13	General information 1
Air conditioning condenser - removal and installation	14	Heater core - replacement... 12
Air conditioning pressure cycling switch - replacement.................	16	Heater/air conditioner control assembly and HVAC module -
Air conditioning receiver-drier - removal and installation..............	15	removal and installation 11
Air conditioning thermostatic expansion valve (TXV) -		Radiator - removal and installation... 7
general information	17	Thermostat - replacement 4
Blower motor resistor and blower motor - replacement.................	10	Troubleshooting... 2
Coolant expansion tank - removal and installation........................	6	Water pump - replacement.. 8
Coolant temperature sending unit - check and replacement.........	9	

Specifications

General

Expansion tank cap pressure rating	14 to 18 psi (95 to 124 kPa)
Cooling system capacity...	See Chapter 1
Refrigerant type..	R-134a
Refrigerant capacity...	Refer to HVAC specification tag

Torque specifications

Ft-lbs (unless otherwise indicated) **Nm**

Note: *One foot-pound (ft-lb) of torque is equivalent to 12 inch-pounds (in-lbs) of torque. Torque values below approximately 15 ft-lbs are expressed in inch-pounds, because most foot-pound torque wrenches are not accurate at these smaller values.*

Condenser inlet and outlet nuts...	80 in-lbs	9 Nm
Receiver-drier plug (3.5L non-turbocharged models)..........................	27 in-lbs	3 Nm
Receiver drier mounting bolts (four-cylinder and		
3.5L turbocharged models)	80 in-lbs	9 Nm
Thermostat		
Four-cylinder engines (thermostat/housing bolts).............................	89 in-lbs	10 Nm
V6 engines		
Coolant inlet cover bolts	89 in-lbs	10 Nm
Thermostat housing bolts (non-turbocharged models)		
Step 1	71 in-lbs	8 Nm
Step 2	Tighten an additional 90 degrees	

Chapter 3 Cooling, heating and air conditioning systems

Torque specifications

Note: *One foot-pound (ft-lb) of torque is equivalent to 12 inch-pounds (in-lbs) of torque. Torque values below approximately 15 ft-lbs are expressed in inch-pounds, because most foot-pound torque wrenches are not accurate at these smaller values.*

	Ft-lbs (unless otherwise indicated)	Nm
Water pump fasteners		
Four-cylinder engines	89 in-lbs	10 Nm
V6 engines		
Step 1	89 in-lbs	10 Nm
Step 2	Tighten an additional 45 degrees	
Water pump pulley bolts (four-cylinder engines)	177 in-lbs	20 Nm

1 General Information

Warning: *Do not allow antifreeze to come in contact with your skin or painted surfaces of the vehicle. Rinse off spills immediately with plenty of water. Antifreeze is highly toxic if ingested. Never leave antifreeze lying around in an open container or in puddles on the floor; children and pets are attracted by it's sweet smell and may drink it. Check with local authorities about disposing of used antifreeze. Many communities have collection centers which will see that antifreeze is disposed of safely. Never dump used antifreeze on the ground or pour it into drains.*

Engine cooling system

1 All modern vehicles employ a pressurized engine cooling system with thermostatically controlled coolant circulation. The cooling system consists of a radiator, an expansion tank or coolant reservoir, a pressure cap (located on the expansion tank or radiator), a thermostat, a cooling fan, and a water pump.

2 The water pump circulates coolant through the engine. The coolant flows around each cylinder and around the intake and exhaust ports, near the spark plug areas and in close proximity to the exhaust valve guides.

3 A thermostat controls engine coolant temperature. During warm up, the closed thermostat prevents coolant from circulating through the radiator. As the engine nears normal operating temperature, the thermostat opens and allows hot coolant to travel through the radiator, where it's cooled before returning to the engine.

Heating system

4 The heating system consists of a blower fan and heater core located in a housing under the dash, the hoses connecting the heater core to the engine cooling system and the heater/air conditioning control head on the dashboard. Hot engine coolant is circulated through the heater core. When the heater mode is activated, a flap door in the housing opens to expose the heater core to the passenger compartment through air ducts. A fan switch on the control head activates the blower motor, which forces air through the core, heating the air.

Air conditioning system

5 The air conditioning system consists of a condenser mounted in front of the radiator, an evaporator mounted adjacent to the heater core, a compressor mounted on the engine, a receiver-drier or accumulator and the plumbing connecting all of the above components.

6 A blower fan forces the warmer air of the passenger compartment through the evaporator core (sort of a radiator-in-reverse), transferring the heat from the air to the refrigerant. The liquid refrigerant boils off into low pressure vapor, taking the heat with it when it leaves the evaporator.

2 Troubleshooting

Coolant leaks

1 A coolant leak can develop anywhere in the cooling system, but the most common causes are:

A loose or weak hose clamp
A defective hose
A faulty pressure cap
A damaged radiator
A bad heater core
A faulty water pump
A leaking gasket at any joint that carries coolant

2 Coolant leaks aren't always easy to find. Sometimes they can only be detected when the cooling system is under pressure. Here's where a cooling system pressure tester comes in handy. After the engine has cooled completely, the tester is attached in place of the pressure cap, then pumped up to the pressure value equal to that of the pressure cap rating (see illustration). Now, leaks that only exist when the engine is fully warmed up will become apparent. The tester can be left connected to locate a nagging slow leak.

Coolant level drops, but no external leaks

3 If you find it necessary to keep adding coolant, but there are no external leaks, the probable causes include:

a) *A blown head gasket*
b) *A leaking intake manifold gasket (only on engines that have coolant passages in the manifold)*
c) *A cracked cylinder head or cylinder block*

2.2 The cooling system pressure tester is connected in place of the pressure cap, then pumped up to pressurize the system

Chapter 3 Cooling, heating and air conditioning systems

2.5a The combustion leak detector consists of a bulb, syringe and test fluid

2.5b Place the tester over the cooling system filler neck and use the bulb to draw a sample into the tester

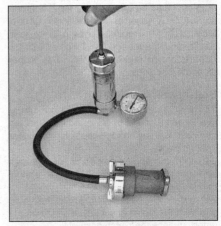

2.8 Checking the cooling system pressure cap with a cooling system pressure tester

4 Any of the above problems will also usually result in contamination of the engine oil, which will cause it to take on a milkshake-like appearance. A bad head gasket or cracked head or block can also result in engine oil contaminating the cooling system.

5 Combustion leak detectors (also known as block testers) are available at most auto parts stores. These work by detecting exhaust gases in the cooling system, which indicates a compression leak from a cylinder into the coolant. The tester consists of a large bulb-type syringe and bottle of test fluid (see illustration). A measured amount of the fluid is added to the syringe. The syringe is placed over the cooling system filler neck and, with the engine running, the bulb is squeezed and a sample of the gases present in the cooling system are drawn up through the test fluid (see illustration). If any combustion gases are present in the sample taken, the test fluid will change color.

6 If the test indicates combustion gas is present in the cooling system, you can be sure that the engine has a blown head gasket or a crack in the cylinder head or block, and will require disassembly to repair.

Pressure cap

Warning: *Wait until the engine is completely cool before beginning this check.*

7 The cooling system is sealed by a spring-loaded cap, which raises the boiling point of the coolant. If the cap's seal or spring are worn out, the coolant can boil and escape past the cap. With the engine completely cool, remove the cap and check the seal; if it's cracked, hardened or deteriorated in any way, replace it with a new one.

8 Even if the seal is good, the spring might not be; this can be checked with a cooling system pressure tester (see illustration). If the cap can't hold a pressure within approximately 1-1/2 lbs of its rated pressure (which is marked on the cap), replace it with a new one.

9 The cap is also equipped with a vacuum relief spring. When the engine cools off, a vacuum is created in the cooling system. The vacuum relief spring allows air back into the system, which will equalize the pressure and prevent damage to the radiator (the radiator tanks could collapse if the vacuum is great enough). If, after turning the engine off and allowing it to cool down you notice any of the cooling system hoses collapsing, replace the pressure cap with a new one.

Thermostat

10 Before assuming the thermostat (see illustration) is responsible for a cooling system problem, check the coolant level (see Chapter 1), drivebelt tension (see Chapter 1) and temperature gauge (or light) operation.

11 If the engine takes a long time to warm up (as indicated by the temperature gauge or heater operation), the thermostat is probably stuck open. Replace the thermostat with a new one.

12 If the engine runs hot or overheats, a thorough test of the thermostat should be performed.

13 Definitive testing of the thermostat can only be made when it is removed from the vehicle. If the thermostat is stuck in the open position at room temperature, it is faulty and must be replaced.

Caution: *Do not drive the vehicle without a thermostat. The computer may stay in open loop and emissions and fuel economy will suffer.*

14 To test a thermostat, suspend the (closed) thermostat on a length of string or wire in a pot of cold water.

15 Heat the water on a stove while observing the thermostat. The thermostat should fully open before the water boils.

16 If the thermostat doesn't open and close as specified, or sticks in any position, replace it.

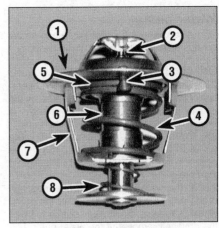

2.10 Typical thermostat:

1	Flange	5	Valve seat
2	Piston	6	Valve
3	Jiggle valve	7	Frame
4	Main coil spring	8	Secondary coil spring

Cooling fan

Electric cooling fan

17 If the engine is overheating and the cooling fan is not coming on when the engine temperature rises to an excessive level, unplug the fan motor electrical connector(s) and connect the motor directly to the battery with fused jumper wires. If the fan motor doesn't come on, replace the motor.

18 If the fan motor is okay, but it isn't coming on when the engine gets hot, the cooling fan control module or the Powertrain Control Module (PCM) might be defective.

19 These control circuits are fairly complex, and checking them should be left to a qualified automotive technician.

20 Check all wiring and connections to the fan motor. Refer to the wiring diagrams at the end of Chapter 12.

21 If no obvious problems are found, the

problem could be the Cylinder Head Temperature (CHT) sensor or the Powertrain Control Module (PCM). Have the cooling fan system and circuit diagnosed by a dealer service department or repair shop with the proper diagnostic equipment.

Belt-driven cooling fan

22 Disconnect the cable from the negative terminal of the battery and rock the fan back and forth by hand to check for excessive bearing play.
23 With the engine cold (and not running), turn the fan blades by hand. The fan should turn freely.
24 Visually inspect for substantial fluid leakage from the clutch assembly. If problems are noted, replace the clutch assembly.
25 With the engine completely warmed up, turn off the ignition switch and disconnect the negative battery cable from the battery. Turn the fan by hand. Some drag should be evident. If the fan turns easily, replace the fan clutch.

Water pump

26 A failure in the water pump can cause serious engine damage due to overheating.

Drivebelt-driven water pump

27 There are two ways to check the operation of the water pump while it's installed on the engine. If the pump is found to be defective, it should be replaced with a new or rebuilt unit.
28 Water pumps are equipped with weep (or vent) holes (see illustration). If a failure occurs in the pump seal, coolant will leak from the hole.
29 If the water pump shaft bearings fail, there may be a howling sound at the pump while it's running. Shaft wear can be felt with the drivebelt removed if the water pump pulley is rocked up and down (with the engine off). Don't mistake drivebelt slippage, which causes a squealing sound, for water pump bearing failure.

Timing chain or timing belt-driven water pump

30 Water pumps driven by the timing chain or timing belt are located underneath the timing chain or timing belt cover.
31 Checking the water pump is limited because of where it is located. However, some basic checks can be made before deciding to remove the water pump. If the pump is found to be defective, it should be replaced with a new or rebuilt unit.
32 One sign that the water pump may be failing is that the heater (climate control) may not work well. Warm the engine to normal operating temperature, confirm that the coolant level is correct, then run the heater and check for hot air coming from the ducts.
33 Check for noises coming from the water pump area. If the water pump impeller shaft or bearings are failing, there may be a howling sound at the pump while the engine is running.
Note: *Be careful not to mistake drivebelt noise (squealing) for water pump bearing or shaft failure.*
34 It you suspect water pump failure due to noise, wear can be confirmed by feeling for play at the pump shaft. This can be done by rocking the drive sprocket on the pump shaft up and down. To do this you will need to remove the tension on the timing chain or belt as well as access the water pump.

All water pumps

35 In rare cases or on high-mileage vehicles, another sign of water pump failure may be the presence of coolant in the engine oil. This condition will adversely affect the engine in varying degrees.
Note: *Finding coolant in the engine oil could indicate other serious issues besides a failed water pump, such as a blown head gasket or a cracked cylinder head or block.*
36 Even a pump that exhibits no outward signs of a problem, such as noise or leakage, can still be due for replacement. Removal for close examination is the only sure way to tell. Sometimes the fins on the back of the impeller can corrode to the point that cooling efficiency is diminished significantly.

Heater system

37 Little can go wrong with a heater. If the fan motor will run at all speeds, the electrical part of the system is okay. The three basic heater problems fall into the following general categories:
a) Not enough heat
b) Heat all the time
c) No heat

38 If there's not enough heat, the control valve or door is stuck in a partially open position, the coolant coming from the engine isn't hot enough, or the heater core is restricted. If the coolant isn't hot enough, the thermostat in the engine cooling system is stuck open, allowing coolant to pass through the engine so rapidly that it doesn't heat up quickly enough. If the vehicle is equipped with a temperature gauge instead of a warning light, watch to see if the engine temperature rises to the normal operating range after driving for a reasonable distance.
39 If there's heat all the time, the control valve or the door is stuck wide open.
40 If there's no heat, coolant is probably not reaching the heater core, or the heater core is plugged. The likely cause is a collapsed or plugged hose, core, or a frozen heater control valve. If the heater is the type that flows coolant all the time, the cause is a stuck door or a broken or kinked control cable.

Air conditioning system

41 If the cool air output is inadequate:
a) *Inspect the condenser coils and fins to make sure they're clear*
b) *Check the compressor clutch for slippage.*
c) *Check the blower motor for proper operation.*
d) *Inspect the blower discharge passage for obstructions.*
e) *Check the system air intake filter for clogging.*

42 If the system provides intermittent cooling air:
a) *Check the circuit breaker, blower switch and blower motor for a malfunction.*
b) *Make sure the compressor clutch isn't slipping.*
c) *Inspect the plenum door to make sure it's operating properly.*
d) *Inspect the evaporator to make sure it isn't clogged.*
e) *If the unit is icing up, it may be caused by excessive moisture in the system, incorrect super heat switch adjustment or low thermostat adjustment.*

43 If the system provides no cooling air:
a) *Inspect the compressor drivebelt. Make sure it's not loose or broken.*
b) *Make sure the compressor clutch engages. If it doesn't, check for a blown fuse.*
c) *Inspect the wire harness for broken or disconnected wires.*
d) *If the compressor clutch doesn't engage, bridge the terminals of the A/C pressure switch(es) with a jumper wire; if the clutch now engages, and the system is properly charged, the pressure switch is bad.*
e) *Make sure the blower motor is not disconnected or burned out.*
f) *Make sure the compressor isn't partially or completely seized.*
g) *Inspect the refrigerant lines for leaks.*
h) *Check the components for leaks.*
i) *Inspect the receiver-drier/accumulator or expansion valve/tube for clogged screens.*

44 If the system is noisy:
a) *Look for loose panels in the passenger compartment.*
b) *Inspect the compressor drivebelt. It may be loose or worn.*
c) *Check the compressor mounting bolts. They should be tight.*
d) *Listen carefully to the compressor. It may be worn out.*
e) *Listen to the idler pulley and bearing and the clutch. Either may be defective.*

2.28 The water pump weep hole is generally located on the underside of the pump

Chapter 3 Cooling, heating and air conditioning systems

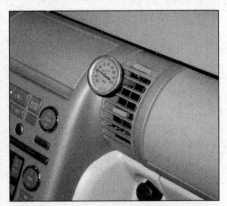

3.9 Insert a thermometer in the center vent, turn on the air conditioning system and wait for it to cool down; depending on the humidity, the output air should be 35 to 40 degrees cooler than the ambient air temperature

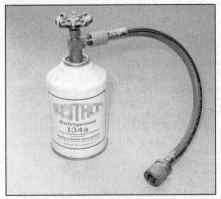

3.11 R-134a automotive air conditioning charging kit

3.13 Location of the low-side charging port

f) The winding in the compressor clutch coil or solenoid may be defective.
g) The compressor oil level may be low.
h) The blower motor fan bushing or the motor itself may be worn out.
i) If there is an excessive charge in the system, you'll hear a rumbling noise in the high pressure line, a thumping noise in the compressor, or see bubbles or cloudiness in the sight glass.
j) If there's a low charge in the system, you might hear hissing in the evaporator case at the expansion valve, or see bubbles or cloudiness in the sight glass.

3 Air conditioning and heating system - check and maintenance

Air conditioning system

Warning: *The air conditioning system is under high pressure. Do not loosen any hose fittings or remove any components until after the system has been discharged. Air conditioning refrigerant should be properly discharged into an EPA-approved recovery/recycling unit at a dealer service department or an automotive air conditioning repair facility. Always wear eye protection when disconnecting air conditioning system fittings.*
Caution: *All models covered by this manual use environmentally friendly R-134a. This refrigerant (and its appropriate refrigerant oils) are not compatible with R-12 refrigerant system components and must never be mixed or the components will be damaged.*
Caution: *When replacing entire components, additional refrigerant oil should be added equal to the amount that is removed with the component being replaced. Read the can before adding any oil to the system, to make sure it is compatible with the R-134a system.*

1 The following maintenance checks should be performed on a regular basis to ensure that the air conditioning continues to operate at peak efficiency.

a) Inspect the condition of the compressor drivebelt. If it is worn or deteriorated, replace it (see Chapter 1).
b) Check the drivebelt tension (see Chapter 1).
c) Inspect the system hoses. Look for cracks, bubbles, hardening and deterioration. Inspect the hoses and all fittings for oil bubbles or seepage. If there is any evidence of wear, damage or leakage, replace the hose(s).
d) Inspect the condenser fins for leaves, bugs and any other foreign material that may have embedded itself in the fins. Use a fin comb or compressed air to remove debris from the condenser.
e) Make sure the system has the correct refrigerant charge.

2 It's a good idea to operate the system for about ten minutes at least once a month. This is particularly important during the winter months because long term non-use can cause hardening, and subsequent failure, of the seals. Note that using the Defrost function operates the compressor.
3 If the air conditioning system is not working properly, proceed to Step 6 and perform the general checks outlined below.
4 Because of the complexity of the air conditioning system and the special equipment necessary to service it, in-depth troubleshooting and repairs beyond checking the refrigerant charge and the compressor clutch operation are not included in this manual. However, simple checks and component replacement procedures are provided in this Chapter. For more complete information on the air conditioning system, refer to the *Haynes Automotive Heating and Air Conditioning Manual*.
5 The most common cause of poor cooling is simply a low system refrigerant charge. If a noticeable drop in system cooling ability occurs, one of the following quick checks will help you determine if the refrigerant level is low.

Checking the refrigerant charge

6 Warm the engine up to normal operating temperature.
7 Place the air conditioning temperature selector at the coldest setting and put the blower at the highest setting.
8 After the system reaches operating temperature, feel the larger pipe exiting the evaporator at the firewall. The outlet pipe should be cold (the tubing that leads back to the compressor). If the evaporator outlet pipe is warm, the system probably needs a charge.
9 Insert a thermometer in the center air distribution duct (see illustration) while operating the air conditioning system at its maximum setting - the temperature of the output air should be 35 to 40 degrees F below the ambient air temperature (down to approximately 40 degrees F). If the ambient (outside) air temperature is very high, say 110 degrees F, the duct air temperature may be as high as 60 degrees F, but generally the air conditioning is 35 to 40 degrees F cooler than the ambient air.
10 Further inspection or testing of the system requires special tools and techniques and is beyond the scope of the home mechanic.

Adding refrigerant

Caution: *Make sure any refrigerant, refrigerant oil or replacement component you purchase is designated as compatible with R-134a systems.*
11 Purchase an R-134a automotive charging kit at an auto parts store (see illustration). A charging kit includes a can of refrigerant, a tap valve and a short section of hose that can be attached between the tap valve and the system low side service valve.
Caution: *Never add more than one can of refrigerant to the system. If more refrigerant than that is required, the system should be evacuated and leak tested.*
12 Back off the valve handle on the charging kit and screw the kit onto the refrigerant can, making sure first that the O-ring or rubber seal inside the threaded portion of the kit is in place.
Warning: *Wear protective eyewear when dealing with pressurized refrigerant cans.*
13 Remove the dust cap from the low-side charging port and attach the hose's quick-connect fitting to the port (see illustration).
Warning: *DO NOT hook the charging kit hose to the system high side! The fittings on the charging kit are designed to fit only on the low side of the system.*

3.24 Insert the nozzle of the disinfectant can into the return-air intake behind the glove box

14 Warm up the engine and turn On the air conditioning. Keep the charging kit hose away from the fan and other moving parts.
Note: *The charging process requires the compressor to be running. If the clutch cycles off, you can put the air conditioning switch on High and leave the car doors open to keep the clutch on and compressor working. The compressor can be kept on during the charging by removing the connector from the pressure switch and bridging it with a paper clip or jumper wire during the procedure.*
15 Turn the valve handle on the kit until the stem pierces the can, then back the handle out to release the refrigerant. You should be able to hear the rush of gas. Keep the can upright at all times, but shake it occasionally. Allow stabilization time between each addition.
Note: *The charging process will go faster if you wrap the can with a hot-water-soaked rag to keep the can from freezing up.*
16 If you have an accurate thermometer, you can place it in the center air conditioning duct inside the vehicle and keep track of the output air temperature. A charged system that is working properly should cool down to approximately 40 degrees F. If the ambient (outside) air temperature is very high, say 110 degrees F, the duct air temperature may be as high as 60 degrees F, but generally the air conditioning is 35 to 40 degrees F cooler than the ambient air.
17 When the can is empty, turn the valve handle to the closed position and release the connection from the low-side port. Reinstall the dust cap.
18 Remove the charging kit from the can and store the kit for future use with the piercing valve in the UP position, to prevent inadvertently piercing the can on the next use.

Heating systems

19 If the carpet under the heater core is damp, or if antifreeze vapor or steam is coming through the vents, the heater core is leaking. Remove it (see Section 12) and install a new unit (most radiator shops will not repair a leaking heater core).
20 If the air coming out of the heater vents isn't hot, the problem could stem from any of the following causes:
   a) The thermostat is stuck open, preventing the engine coolant from warming up enough to carry heat to the heater core. Replace the thermostat (see Section 4).
   b) There is a blockage in the system, preventing the flow of coolant through the heater core. Feel both heater hoses at the firewall. They should be hot. If one of them is cold, there is an obstruction in one of the hoses or in the heater core, or the heater control valve is shut. Detach the hoses and back flush the heater core with a water hose. If the heater core is clear but circulation is impeded, remove the two hoses and flush them out with a water hose.
   c) If flushing fails to remove the blockage from the heater core, the core must be replaced (see Section 12).

Eliminating air conditioning odors

21 Unpleasant odors that often develop in air conditioning systems are caused by the growth of a fungus, usually on the surface of the evaporator core. The warm, humid environment there is a perfect breeding ground for mildew to develop.
22 The evaporator core on most vehicles is difficult to access, and factory dealerships have a lengthy, expensive process for eliminating the fungus by opening up the evaporator case and using a powerful disinfectant and rinse on the core until the fungus is gone. You can service your own system at home, but it takes something much stronger than basic household germ-killers or deodorizers.
23 Aerosol disinfectants for automotive air conditioning systems are available in most auto parts stores, but remember when shopping for them that the most effective treatments are also the most expensive. The basic procedure for using these sprays is to start by running the system in the RECIRC mode for ten minutes with the blower on its highest speed. Use the highest heat mode to dry out the system and keep the compressor from engaging by disconnecting the wiring connector at the compressor.
24 The disinfectant can usually comes with a long spray hose. Insert the nozzle into an intake port inside the cabin, and spray according to the manufacturer's recommendations (see illustration). Try to cover the whole surface of the evaporator core, by aiming the spray up, down and sideways. Follow the manufacturer's recommendations for the length of spray and waiting time between applications.

Automatic heating and air conditioning systems

25 Some vehicles are equipped with an optional automatic climate control system. This system has its own computer that receives inputs from various sensors in the heating and air conditioning system. This computer, like the PCM, has self-diagnostic capabilities to help pinpoint problems or faults within the system. Vehicles equipped with automatic heating and air conditioning systems are very complex and considered beyond the scope of the home mechanic. Vehicles equipped with automatic heating and air conditioning systems should be taken to dealer service department or other qualified facility for repair.

4 Thermostat - replacement

Warning: *Wait until the engine is completely cool before performing this procedure.*

Removal

Note: *The thermostat and thermostat housing on four-cylinder models are replaced as an assembly.*
1 Disconnect the cable from the negative battery terminal (see Chapter 5).
2 Remove the engine cover by pulling straight up on the cover to disengage the grommets from the cover studs, if equipped.
Caution: *On 2.3L models, the engine cover must be pulled straight up, do not pull the cover forward or sideways to remove it. The cover mounting points and studs are easily damaged if the cover is not pulled straight upward from the underside of the cover.*
3 Drain the cooling system (see Chapter 1). If the coolant is relatively new and still in good condition, save it and reuse it.
4 Follow the lower radiator hose to the engine to locate the thermostat housing.
5 Loosen the hose clamp, then detach the hose from the fitting. If it's stuck, grasp it near the end with a pair of adjustable pliers and twist it to break the seal, then pull it off. If the hose is old or if it has deteriorated, cut it off and install a new one.
6 If the outer surface of the thermostat housing cover, which mates with the hose, is already corroded, pitted, or otherwise deteriorated, it might be damaged even more by hose removal. If it is, replace the thermostat housing cover.

Four-cylinder models

7 Disconnect the hoses from the thermostat housing.
8 Remove the thermostat housing bolts (see illustration), then detach the housing from the engine.

V6 models

Thermostat

9 Remove the air filter housing and duct (see Chapter 4).
10 Remove the two bolts and detach the coolant inlet cover (see illustration). If the cover is stuck, tap it with a soft-face hammer to jar it loose. Be prepared for some coolant

Chapter 3 Cooling, heating and air conditioning systems

4.8 Remove the thermostat housing bolts and housing - 2.0L four-cylinder engine shown

4.10 Coolant inlet cover bolts

to spill as the gasket seal is broken.
11 Note how it's installed, which end is facing up, or out, then remove the thermostat and O-ring (see illustration).

Thermostat housing (3.5L non-turbocharged engines)
12 Remove the air filter housing and duct (see Chapter 4).
13 Disconnect and remove the upper and lower radiator hoses, heater hoses and expansion tank hose from the housing.
14 Remove the mounting bolts and separate the housing from the coolant tube.
Caution: *The other end of the coolant tube is a slip fit into the engine block; if the tube is pulled when separating the thermostat housing it may come out of the block.*

Installation
15 Clean the sealing surfaces. Also inspect the hoses, replacing them as necessary.

Four-cylinder models
16 Install the thermostat housing, using a new gasket. Tighten the bolts to the torque listed in this Chapter's Specifications. Reconnect the radiator hose and heater hose.

V6 models

Thermostat
17 Install the thermostat in the housing, spring-end first, then install the O-ring. If the replacement thermostat is equipped with a jiggle valve, make sure the valve is positioned at 12 o'clock.
18 Install the coolant inlet cover and bolts, then tighten the bolts to the torque listed in this Chapter's Specifications.

Thermostat housing - 3.5L non-turbocharged engines
19 Install a new thermostat-to-coolant pipe O-ring, then lubricate the O-ring with clean coolant.
20 Install a new O-ring gasket, then install

the housing.
21 Install the thermostat housing bolts, then tighten the bolts to the torque listed in this Chapter's Specifications.
22 Reattach the radiator hoses, heater hoses and expansion tank hose to the thermostat housing. Make sure that the hose clamp is still tight. If it isn't, replace it.

All models
23 Refill the cooling system and bleed the air from the system (see Chapter 1).
24 Start the engine and allow it to reach normal operating temperature, then check for leaks and proper thermostat operation.

5 Engine cooling fans - removal and installation

Warning: *To avoid possible injury or damage, DO NOT operate the engine with a damaged fan. Do not attempt to repair fan blades - replace a damaged fan with a new one.*
Warning: *Wait until the engine is completely cool before performing this procedure.*
Warning: *These models have an airbag sensor mounted near the radiator support bracket. It will be necessary to disarm the system prior to performing any work around the radiator, fans or other components in this area (see Chapter 12).*
Note: *Always be sure to check for blown fuses before attempting to diagnose an electrical circuit problem.*
Note: *All models are equipped with two fan motors and are controlled by a fan module which is controlled by the PCM.*
1 Remove the air filter housing and duct (see Chapter 4).

Four-cylinder models
2 Remove the engine cover by pulling straight up to disengage the grommets from the cover studs.

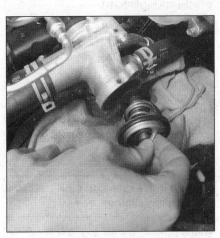

4.11 Remove the thermostat, noting how it's oriented

Caution: *The engine cover must be pulled straight up - do not pull the cover forward or sideways to remove it. The cover mounting points and studs are easily damaged if the cover is not pulled straight upward from the underside of the cover.*
3 Remove the intake air ducts and charge air cooler tubes (see Chapter 4).
4 Disconnect the electrical connector to the fan controller and and the cooling fan electrical connectors then move the harness out of the way.
5 Remove the fan shroud fastener from the upper sides of the fan shroud and lift the fan shroud assembly out of the vehicle

V6 models
6 On turbocharged models, remove the charge air cooler tubes.
7 Disconnect the block heater harness from the top of the fan shroud, if equipped.
8 Disconnect the fan motor electrical connectors and harness connectors.

Chapter 3 Cooling, heating and air conditioning systems

5.9a Fan shroud bolt - right side

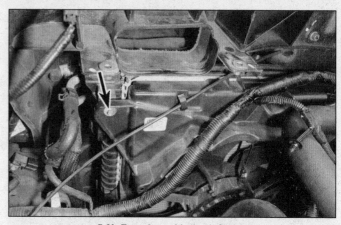

5.9b Fan shroud bolt - left side

9 Remove the fan shroud mounting bolts and lift the fan shroud assembly out of the vehicle (see illustrations).

All models

10 Installation is the reverse of removal.
Note: *When reinstalling the fan assembly, make sure the rubber air shields around the shroud are still in place - without them, the cooling system may not work efficiently.*
11 Reconnect the battery (see Chapter 5).
12 Start the engine and allow it to reach normal operating temperature and check for proper operation.

6 Coolant expansion tank - removal and installation

Warning: *Wait until the engine is completely cool before beginning this procedure.*
1 Drain the cooling system (see Chapter 1).
2 Disengage the receiver-drier line clip from the expansion tank.
3 Remove the fasteners securing the expansion tank to the fender (see illustration).
4 Lift the tank up enough to disconnect the hoses.
5 Clean out the tank with soapy water and a brush to remove any deposits inside. Inspect the reservoir carefully for cracks. If you find a crack, replace the reservoir.
6 Installation is the reverse of removal.

7 Radiator - removal and installation

Warning: *Wait until the engine is completely cool before beginning this procedure.*
Warning: *These models have an airbag sensor mounted near the radiator support. It will be necessary to disarm the airbag system prior to performing any work around the radiator, fans or other components in this area (see Chapter 12).*

Removal

1 Disconnect the cable from the negative battery terminal (see Chapter 5).
2 Raise the vehicle and support it securely on jackstands. Remove the lower splash shield.
3 Drain the cooling system (see Chapter 1). If the coolant is relatively new and in good condition, save it and reuse it. Detach the radiator hose from the bottom of the radiator.
4 Disconnect the expansion tank hose from the radiator.
5 Remove the cooling fan/shroud assembly from the radiator (see Section 5).
6 Remove the upper and lower radiator hoses.
7 Remove the front bumper cover (see Chapter 11).

Four-cylinder models

8 Remove the charge air cooler-to-radiator bolts (see Chapter 4).
9 Remove the radiator mount bolts from the front side (see illustration 7.16a), then remove the mounts from the back side (see illustration 7.16b).
10 Move the radiator towards the engine.
11 Disengage the clip for the upper air deflector and remove the deflector (see illustration 7.17a).
12 Remove the charge air cooler mounting bolts and charge air cooler (see Chapter 4).
Note: *The left side bolt is longer than the right side.*
13 Remove the a/c condenser bolts (see illustrations 7.18a and 7.18b) and move the condenser out of the way.
14 Unclip the air deflector from the top of the radiator (see illustration 7.19a).
15 Carefully lift out the radiator. Don't spill coolant on the vehicle or scratch the paint. Make sure the rubber radiator insulators that fit on the bottom and top of the radiator and into the sockets in the body remain in place in the body for proper reinstallation of the radiator.

V6 models

16 On turbocharged models, remove the charge air cooler-to-radiator bolts (see Chapter 4). On all models, remove the radiator mounting bolts from the front side (see illustration) then remove the mount from the back side (see illustration).
17 Remove the air deflector push-pin type retainers, then remove the deflectors (see illustrations).
18 Remove the condenser mounting bolts (see illustrations) and lift the condenser tab up and out of the mounting clip (see illustration).
19 Remove the deflector clip from the top of the radiator then remove the deflector (see illustrations).

6.3 Remove the expansion tank fasteners (A), disengage the receiver-drier line clip from the expansion tank (B), then disconnect the hoses from the tank

Chapter 3 Cooling, heating and air conditioning systems 3-9

7.16a Remove the radiator mount bolts from the front side…

7.16b … then remove the mounts from the back side

7.17a Remove the air deflector push-pins…

7.17b … then remove the left and right deflectors

7.18a Remove the condenser right side mounting bolt…

7.18b … left side mounting bolt…

7.18c … then lift the condenser up to disengage the tab from the radiator mount - right side shown

7.19a Use a small screwdriver to pop the retaining clip up...

7.19b ... then remove the upper deflector

8.7 Remove the bolts and separate the water pump pulley from the pump

20 Carefully lift out the radiator. Don't spill coolant on the vehicle or scratch the paint. Make sure the rubber radiator insulators that fit on the bottom of the radiator and into the sockets in the body remain in place in the body for proper reinstallation of the radiator.

Installation

21 Remove bugs and dirt from the radiator with compressed air and a soft brush. Don't bend the cooling fins. Inspect the radiator for leaks and damage. If it needs repair, have a radiator shop or a dealer service department do the work.

22 Inspect the rubber insulators in the lower crossmember for cracks and deterioration. Make sure that they're free of dirt and gravel. When installing the radiator, make sure that it's correctly seated on the insulators before fastening the top brackets.

23 Installation is otherwise the reverse of the removal procedure. After installation, fill the cooling system with the correct mixture of antifreeze and water (see Chapter 1).

24 Start the engine and check for leaks. Allow the engine to reach normal operating temperature, indicated by the upper radiator hose becoming hot. Recheck the coolant level and add more if required.

8 Water pump - replacement

Warning: *Wait until the engine is completely cool before beginning this procedure.*

Removal

1 Disconnect the cable from the negative battery terminal (see Chapter 5).
2 Drain the cooling system (see Chapter 1).
3 On V6 models remove the drivebelt (see Chapter 1).

Four-cylinder models

4 Remove the engine cover by pulling straight up on the cover to disengage the grommets from the cover studs.
Caution: *The engine cover must be pulled straight up - do not pull the cover forward or sideways to remove it. The cover mounting points and studs are easily damaged if the cover is not pulled straight upward from the underside of the cover.*

5 Loosen the right front wheel lug nuts. Raise the vehicle and support it securely on jackstands.
6 Remove the right front wheel and the inner fender splash shield (see Chapter 11).
7 Remove the water pump pulley (see illustration).
8 Remove the bolts attaching the water pump to the engine block and remove the pump from the engine (see illustration). If the water pump is stuck, gently tap it with a soft-faced hammer to break the seal.
9 Clean the bolt threads and the threaded holes in the engine and remove all corrosion and sealant. Remove all traces of old gasket material from the sealing surfaces.

V6 models

10 The water pump is mounted to the front of the block but is driven by the timing chain and thus not accessible without removing the timing chain (see Chapter 2B for timing chain removal). Remove the pump mounting bolts (see illustration).

Installation

11 Compare the new pump to the old one to make sure that they're identical.
12 Apply a thin film of RTV sealant to hold the new gasket in place during installation. O-rings should be coated with clean coolant.
Caution: *Make sure that the gasket is correctly positioned on the water pump and the mating surfaces are clean and free of old gasket material.*

13 Carefully mate the pump to the water pump housing.
14 Install the water pump bolts (four-cylinder models, see illustration 8.8; V6 models see illustration 8.10) and tighten them to the torque listed in this Chapter's Specifications.
15 The remainder of installation is the reverse of removal. Refill and bleed the cooling system (see Chapter 1).
16 Operate the engine to check for leaks.

8.8 Remove the water pump bolts - four-cylinder engines

Chapter 3 Cooling, heating and air conditioning systems

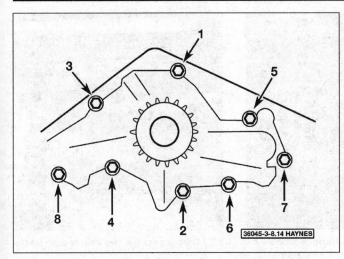

8.10 Water pump bolt locations and tightening sequence - V6 engines

10.6 Location of the motor resistor connector (A), the blower motor connector (B) and the blower motor vent tube (C)

9 Coolant temperature sending unit - check and replacement

Warning: *Wait until the engine is completely cool before beginning this procedure.*

Check

1 The coolant temperature indicator system consists of a warning light or a temperature gauge on the dash and a Cylinder Head Temperature (CHT) sensor mounted on the engine. On the models covered by this manual, the Cylinder Head Temperature (CHT), which is an information sensor for the Powertrain Control Module (PCM), also functions as the coolant temperature sending unit.
2 If an overheating indication occurs, check the coolant level in the system and then make sure all connectors in the wiring harness between the sending unit and the indicator light or gauge are tight.
3 When the ignition switch is turned to START and the starter motor is turning, the indicator light (if equipped) should come on. This doesn't mean the engine is overheated; it just means that the bulb is good.
4 If the light doesn't come on when the ignition key is turned to START, the bulb might be burned out, the ignition switch might be faulty or the circuit might be open.
5 As soon as the engine starts, the indicator light should go out and remain off, unless the engine overheats. If the light doesn't go out, the wire between the cylinder head temperature (CHT) sensor and the light could be grounded; the cylinder head temperature (CHT) might be defective (have it checked by a dealer service department); or the ignition switch might be faulty (see Chapter 12). Check the coolant to make sure it's correctly mixed; plain water, with no antifreeze, or coolant that's mainly water, might have too low a boiling point to activate the cylinder head temperature (CHT) sensor (see Chapter 6).

Replacement

6 See Chapter 6 for the Cylinder Head Temperature (CHT) sensor replacement procedure.

10 Blower motor resistor and blower motor - replacement

Warning: *The models covered by this manual are equipped with Supplemental Restraint Systems (SRS), more commonly known as airbags. Always disable the airbag system before working in the vicinity of any airbag system component to avoid the possibility of accidental deployment of the airbag, which could cause personal injury (see Chapter 12).*

Blower motor resistor

Note: *The blower motor resistor also called the blower motor speed control works in conjunction with the HVAC module to vary the speed of the blower motor.*

1 Remove the glovebox (see Chapter 11).
2 Disconnect the electrical connector from the blower motor resistor (see illustration 10.6).
3 Remove the blower motor resistor mounting screws and remove the resistor from the evaporator housing.
4 Installation is the reverse of removal.

Blower motor

5 Remove the lower trim panel fasteners and panel from below the glovebox (see Chapter 11).
6 Disconnect the blower motor electrical connector (see illustration).
7 Depress the retaining tabs and remove the blower motor vent tube from the heater/evaporator core housing.
8 Rotate the blower motor counterclockwise approximately 45 degrees, and remove it from the heater/evaporator housing.
9 Remove the blower motor mounting screws and remove the blower motor.
10 Installation is the reverse of removal.

11 Heater/air conditioner control assembly and HVAC module - removal and installation

Warning: *The models covered by this manual are equipped with Supplemental Restraint Systems (SRS), more commonly known as airbags. Always disable the airbag system before working in the vicinity of any airbag system component to avoid the possibility of accidental deployment of the airbag, which could cause personal injury (see Chapter 12).*

Note: *The heater/air conditioner control assembly is incorporated in the Front Controls Interface Module (FCIM); if there is a problem, the entire module must be replaced. The HVAC module is mounted behind the radio ACM.*

Note: *The radio consists of the Front Controls Interface Module (FCIM) and the Audio Control Module (ACM). The FCIM (the controls) is what you see in the center trim panel; the ACM (the radio) is a separate component in the dash behind the FCIM. You can replace the FCIM at home, but not the ACM. You can remove the ACM to access something else, or to remove the instrument panel, but if the ACM must be replaced, it will have to be done by a dealer service department because module configuration must be programmed into the new ACM unit. Without the correct module configuration, the new ACM will not work.*

1 Disconnect the cable from the negative battery terminal (see Chapter 5).
2 Using a plastic trim tool, start from the

11.2 Using a plastic trim tool, start from the sides of the panel and work your way to the top

11.3 Front Controls Interface Module (FCIM) retaining screw locations

12.3 Disconnect the two heater core hoses

sides of the panel and work your way to the top, disengaging the retaining clips (see illustration).
3   Remove the FCIM retaining screws (see illustration) and remove the module.
4   Remove the FCIM from the instrument panel.
5   Remove the ACM (see Chapter 12, Section 12).
6   Remove the HVAC module mounting fasteners and remove the module from the instrument panel.
7   Disconnect the electrical connectors at the HVAC module.
8   Installation is the reverse of removal.

12 Heater core - replacement

Warning: *The models covered by this manual are equipped with Supplemental Restraint Systems (SRS), more commonly known as airbags. Always disarm the airbag system before working in the vicinity of any airbag system component to avoid the possibility of accidental deployment of the airbag, which could cause personal injury (see Chapter 12).*
Warning: *The air conditioning system is under high pressure. DO NOT loosen any fittings or remove any components until after the system has been discharged. Air conditioning refrigerant must be properly discharged into an EPA-approved container at a dealer service department or an automotive air conditioning repair facility. Always wear eye protection when disconnecting air conditioning system fittings.*
Warning: *Wait until the engine is completely cool before beginning this procedure.*
Note: *This is a difficult procedure for the home mechanic, involving numerous hard-to-find fasteners, clips and electrical connectors.*
1   Have the air conditioning system recovered by a dealer service department or by an automotive air conditioning shop before proceeding (see Warning above).
2   Disconnect the cable from the negative battery terminal (see Chapter 5).
3   Drain the cooling system (see Chapter 1). Disconnect the heater hoses from the heater core inlet and outlet pipes at the firewall (see illustration), then disconnect the thermostatic expansion valve and the two refrigerant pipes.
4   Remove the instrument panel (see Chapter 11). On V6 models, remove the lower cowl grille panel to access some HVAC unit fasteners.
5   Tag and disconnect the electrical connectors from the HVAC unit. Remove the housing-to-firewall nuts.
6   Remove the heater core cover screws and pull the heater core out of the housing.
7   Installation is the reverse of removal. Reconnect the heater core inlet and outlet hoses at the firewall. Use a new gasket where the two air conditioning pipes meet the expansion valve at the firewall.
8   Refill and bleed the cooling system (see Chapter 1). Have the air conditioning system evacuated, recharged and leak-tested by the shop that discharged it.

13 Air conditioning compressor - removal and installation

Warning: *The air conditioning system is under high pressure. DO NOT loosen any fittings or remove any components until after the system has been discharged. Air conditioning refrigerant must be properly discharged into an EPA-approved container at a dealer service department or an automotive air conditioning repair facility. Always wear eye protection when disconnecting air conditioning system fittings.*
Caution: *If the compressor is being replaced due to failure, the rest of the system should be flushed by a technician to remove particles or contaminants.*

Removal

1   Have the air conditioning system recovered by a dealer service department or by an automotive air conditioning shop before proceeding (see Warning above).
2   Loosen the right front wheel lug nuts. Raise the vehicle and secure it on jackstands. Remove the right front wheel.
3   Remove the right front fender splash shield and the engine splash shield.
4   Remove the drivebelt (see Chapter 1).

Four-cylinder models

5   Remove the mounting nut from the rear of the compressor, then remove the stud.
6   Disconnect the electrical connector from the compressor clutch field coil (see illustration).
7   Disconnect the inlet and outlet lines from the compressor. Remove and discard the old O-rings.

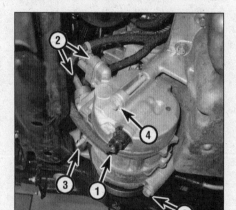

13.6 Compressor details (2.0L four-cylinder engine shown, 2.3L engine similar)

1   Compressor clutch field coil electrical connector
2   Compressor inlet and outlet lines
3   Mounting stud/nut
4   Mounting bolts

Chapter 3 Cooling, heating and air conditioning systems

8 Remove the compressor mounting bolts and remove the compressor.

V6 models
9 Remove the mounting nut from the rear of the compressor, then remove the stud.
10 Disconnect the electrical connector from the compressor clutch field coil (see illustration).
11 Release the twist lock retainers and remove the underbody air duct from under the vehicle.
12 Remove the compressor mounting bolts and lower the compressor until the inlet and outlet lines to the compressor can be reached.
13 Disconnect the inlet and outlet lines from the compressor. Remove and discard the old O-rings.
14 Remove the compressor from the vehicle.

Installation
15 If a new compressor is being installed, follow the directions that came with the compressor regarding the draining of excess oil prior to installation.
16 The clutch may have to be transferred from the original to the new compressor.
17 Before reconnecting the inlet and outlet lines to the compressor, replace all O-rings and lubricate them with refrigerant oil.
18 Installation is otherwise the reverse of removal.
19 Have the system evacuated, recharged and leak-tested by the shop that discharged it.

14 Air conditioning condenser - removal and installation

Warning: *The air conditioning system is under high pressure. DO NOT loosen any fittings or remove any components until after the system has been discharged. Air conditioning refrigerant must be properly discharged into an EPA-approved container at a dealer service department or an automotive air conditioning repair facility. Always wear eye protection when disconnecting air conditioning system fittings.*
Note: *The air conditioning condenser has an integrated receiver/drier, equipped with a receiver/drier desiccant bag. The receiver/drier desiccant bag may be removed and replaced with a new one (available from the dealer).*

1 Have the air conditioning system discharged by a dealer service department or an automotive air conditioning shop before proceeding (see Warning above).
2 Remove the radiator (see Section 7). Carefully lift out the radiator. Don't spill coolant on the vehicle or scratch the paint.
3 Disconnect the refrigerant inlet and outlet pipes from the condenser. Remove the bolt or nut securing the pipes to the block that is part of the condenser. Cap the lines to prevent contamination.

13.10 Compressor details (3.5L non-turbocharged engine shown)

1 *Compressor clutch field coil electrical connector*
2 *Compressor inlet and outlet lines*
3 *Mounting stud/nut*
4 *Mounting bolts*

4 Disconnect the transaxle cooler hoses from the left side of the condenser.
5 Remove the condenser. If reinstalling the same condenser, store it with the line fittings facing up to prevent oil from draining out.
6 If installing a new condenser, pour one ounce of refrigerant oil of the correct type into it prior to installation.
7 Before reconnecting the refrigerant lines to the condenser, coat a pair of new O-rings with refrigerant oil, install them in the refrigerant line fittings and tighten the condenser inlet and outlet nuts to the torque listed in this Chapter's Specifications.
8 Installation is otherwise the reverse of removal.
9 Have the system evacuated, recharged and leak tested by the shop that discharged it.

15 Air conditioning receiver-drier - removal and installation

Warning: *The air conditioning system is under high pressure. DO NOT loosen any fittings or remove any components until after the system has been discharged. Air conditioning refrigerant must be properly discharged into an EPA-approved container at a dealer service department or an automotive air conditioning repair facility. Always wear eye protection when disconnecting air conditioning system fittings.*
Note: *On 3.5L non-turbocharged models, the air conditioning condenser has an integrated receiver-drier, equipped with a receiver-drier desiccant bag. The receiver/drier desiccant bag may be removed and replaced with a new one (available from the dealer).*

1 Have the air conditioning system discharged by a dealer service department or an automotive air conditioning shop before proceeding (see Warning above).

Four-cylinder and 3.5L turbocharged V6 models
2 Raise the vehicle and support it securely on jackstands.

3 Remove the front bumper cover (see Chapter 11).
4 Disconnect the refrigerant inlet and outlet pipes from the condenser. Cap the lines to prevent contamination.
5 Disengage the receiver drier line clip from the expansion tank, then remove the receiver drier outlet fitting nut and disconnect the fitting from the drier. Secure the tube assembly out of the way and cap the openings to prevent contamination.
6 Remove the expansion tank (see Section 6).
7 Remove the receiver drier mounting bolts and remove the drier.
8 Tighten the receiver drier mounting bolts to the torque listed in this Chapter's Specifications, installation is otherwise the reverse of removal.

3.5L non-turbocharged V6 models

2015 and earlier models
9 Remove the radiator (see Section 7).
10 Disconnect the refrigerant inlet and outlet pipes from the condenser. Remove the bolt or nut securing the pipes to the block that is part of the condenser. Cap the lines to prevent contamination. The refrigerant pipe block is retained by a bolt and a rivet that must be drilled out.
11 Disconnect the transaxle cooler line clamps and remove the hoses from the fittings. Cap the lines to prevent contamination.
12 Remove the condenser (Section 14).

2016 and later models
13 Raise the vehicle and support it securely on jackstands.
14 Remove the air deflector panel fasteners and remove the panel from under the radiator.

All models
15 Remove the threaded plug at the bottom left of the condenser.
16 Using needle-nose pliers, pull the desiccant bag out of the receiver-drier.

16.3 Disconnect the pressure cycling switch (pressure transducer) electrical connector

17 Insert a new bag into the receiver-drier.
18 Install the plug and tighten it to the torque listed in this Chapter's Specifications.
19 Have the system evacuated, recharged and leak tested by the shop that discharged it.

16 Air conditioning pressure cycling switch - replacement

Warning: *The air conditioning system is under high pressure. DO NOT loosen any fittings or remove any components until after the system has been discharged. Air conditioning refrigerant must be properly discharged into an EPA-approved container at a dealer service department or an automotive air conditioning repair facility. Always wear eye protection when disconnecting air conditioning system fittings.*

1 The pressure cycling switch is located on top of the hard line from the condenser to the air conditioning thermostatic expansion valve (TXV). The pressure cycling switch detects low refrigerant line pressure, and switches the A/C system off, then back on again to provide higher pressure. If the pressure increases too high, the pressure cut-off switch (located in the high-pressure side of the system) shuts the system off.
2 Have the air conditioning system recovered by a dealer service department or by an automotive air conditioning shop before proceeding (see Warning above).
3 Unplug the electrical connector from the pressure cycling switch (see illustration).
4 Unscrew the pressure cycling switch.
5 Lubricate the switch O-ring with clean refrigerant oil of the correct type.
6 Screw the new switch into place until hand tight, then tighten it securely.
7 Reconnect the electrical connector.
8 Have the system evacuated, recharged and leak tested by the shop that discharged it.

17 Air conditioning thermostatic expansion valve (TXV) - general information

Warning: *The air conditioning system is under high pressure. DO NOT loosen any hose fittings or remove any components until the system has been discharged. Air conditioning refrigerant must be properly discharged into an EPA-approved recovery/recycling unit by a dealer service department or an automotive air conditioning repair facility. Always wear eye protection when disconnecting air conditioning system fittings.*

1 There are several ways that air conditioning systems convert the high-pressure liquid refrigerant from the compressor to lower-pressure vapor. The conversion takes place at the air conditioning evaporator; the evaporator is chilled as the refrigerant passes through, cooling the airflow through the evaporator for delivery to the vents. The conversion is usually accomplished by a sudden change in the tubing size. Many vehicles have a removable controlled orifice in one of the AC pipes at the firewall.
2 The models covered by this manual use a thermostatic expansion valve (TXV) that accomplishes the same purpose as a controlled orifice. To remove the TXV, have the air conditioning system discharged by a licensed air conditioning technician, then disconnect the refrigerant lines from the TXV at the firewall, remove the two bolts securing the valve, then remove the valve.

Chapter 4
Fuel and exhaust systems

Contents

	Section		Section
Air filter housing - removal and installation	9	Fuel rail and injectors - removal and installation	11
Charge air cooler - removal and installation	13	Fuel tank - removal and installation	8
Exhaust system servicing - general information	6	General information and precautions	1
Fuel lines and fittings - general information and disconnection	5	High-pressure fuel pump (turbocharged models) - removal and installation	12
Fuel pressure - check	4	Throttle body - removal and installation	10
Fuel pressure relief procedure	3	Troubleshooting	2
Fuel Pump Control Module (FPCM) - replacement	15	Turbocharger - removal and installation	14
Fuel pump module - removal and installation	7		

Specifications

General

Fuel system pressure
  2.0L, 2.3L, and 3.5L turbocharged engines (direct injection, low-pressure side)
    Key On, Engine Off (KOEO) .... 58 psi (400 kPa)
    Engine Running at Idle .... 55 to 75 psi (380 to 520 kPa)
  3.5L non-turbocharged engine
    Key On, Engine Off (KOEO) .... 58 psi (400 kPa)
    Engine Running at Idle .... 58 psi (400 kPa)

Chapter 4 Fuel and exhaust systems

Torque specifications — Ft-lbs (unless otherwise indicated) — Nm

Note: One foot-pound (ft-lb) of torque is equivalent to 12 inch-pounds (in-lbs) of torque. Torque values below approximately 15 ft-lbs are expressed in inch-pounds, since most foot-pound torque wrenches are not accurate at these smaller values.

Fastener	Ft-lbs (unless otherwise indicated)	Nm
Throttle body mounting fasteners		
2.0L, 2.3L, and 3.5L turbocharged engines and 2015 and earlier 3.5L non-turbocharged engines	89 in-lbs	10
2016 and later 3.5L non-turbocharged engines		
Step 1	89 in-lbs	10
Step 2	Tighten an additional 90 degrees	
Fuel rail mounting bolts		
Four-cylinder engines		
Step 1	71 in-lbs	8
Step 2	Tighten an additional 26-degrees	
V6 engines		
Step 1	89 in-lbs	10
Step 2	Tighten an additional 45 degrees	
High-pressure fuel pump mounting bolts		
Four-cylinder engines		
Step 1	44 in-lbs	5
Step 2	Tighten an additional 55-degrees	
V6 engine		
Step 1	89 in-lbs	10
Step 2	Tighten an additional 45-degrees	
High-pressure fuel tube flare nuts		
2012 and earlier models		
Step 1	133 in-lbs	15
Step 2	Tighten an additional 30-degrees	
2013 and later models		
Step 1	24	32
Step 2	Wait 10 minutes	
Step 3	24	32
High-pressure fuel tube bracket bolts	89 in-lbs	10
Turbocharger		
Coolant line banjo bolts		
Four-cylinder engines	21	28
V6 engine		
Step 1	106 in-lbs	12
Step 2	Tighten an additional 60-degrees	
Mounting flange		
Bolts (V6 engine)	33	45
Nuts (four-cylinder engines)		
2.0L	37	50
2.3L		
Step 1	18	25
Step 2	37	50
Oil supply line		
Banjo bolts		
Four-cylinder engine		
2.0L	18	25
2.3L	22	30
V6 engine		
Step 1	159 in-lbs.	18
Step 2	Tighten an additional 90-degrees	
Bolts	89 in-lbs	10
Oil drain tube bolts	89 in-lbs	10

Chapter 4 Fuel and exhaust systems

2.3 The fuel pump fuse (A) is located in the engine compartment fuse box; (B) is the fuel pump relay

3.9 The FPCM is mounted behind the right-side quarter trim ("C-pillar") panel

1 General information and precautions

Fuel system warnings

1 Gasoline is extremely flammable and repairing fuel system components can be dangerous. Consider your automotive repair knowledge and experience before attempting repairs which may be better suited for a professional mechanic.

a) Don't smoke or allow open flames or bare light bulbs near the work area
b) Don't work in a garage with a gas-type appliance (water heater, clothes dryer)
c) Use fuel-resistant gloves. If any fuel spills on your skin, wash it off immediately with soap and water
d) Clean up spills immediately
e) Do not store fuel-soaked rags where they could ignite
f) Prior to disconnecting any fuel line, you must relieve the fuel pressure (see Section 3)
g) Wear safety glasses
h) Have a proper fire extinguisher on hand

Fuel system

2 The fuel system consists of the fuel tank, electric fuel pump/fuel level sending unit (located in the fuel tank), fuel rail, fuel injectors and, on turbocharged models, a high-pressure fuel pump mounted to the end of the cylinder head. The fuel injection system is a multi-port system; multi-port fuel injection uses timed impulses to inject the fuel directly into the intake port of each cylinder. The Powertrain Control Module (PCM) controls the injectors. The PCM monitors various engine parameters and delivers the exact amount of fuel required into the intake ports.

3 Fuel is circulated from the fuel pump to the fuel rail through fuel lines running along the underside of the vehicle. Various sections of the fuel line are either rigid metal or nylon, or flexible fuel hose. The various sections of the fuel hose are connected either by quick-connect fittings or threaded metal fittings.

Exhaust system

4 The exhaust system consists of the exhaust manifold(s), catalytic converter(s), muffler(s), tailpipe and all connecting pipes, flanges and clamps. The catalytic converters are an emission control device added to the exhaust system to reduce pollutants.

2 Troubleshooting

Electric fuel pump

1 The electric fuel pump is located inside the fuel tank. Sit inside the vehicle with the windows closed, turn the ignition key to ON (not START) and listen for the sound of the fuel pump as it's briefly activated. You will only hear the sound for a second or two, but that sound tells you that the pump is working. Alternatively, have an assistant listen at the fuel filler cap.

2 A fuel-cut off system is used in case of an accident. If there has been an accident, the restraint system sends a signal to the Fuel Pump Control Module (FPCM) and shuts the module off. To reset the FPCM, turn the ignition key to the "OFF" position, then to the "ON" position and start the engine.
Note: *This may take a few attempts.*

3 Check the fuel pump fuse (F63, 30A) and relay (see illustration). If the fuse and relay are okay, check the wiring back to the fuel pump. If the fuse, relay, wiring and inertia switch are okay, the fuel pump is probably defective. If the pump runs continuously with the ignition key in the ON position, the Powertrain Control Module (PCM) is probably defective. Have the PCM checked by a professional mechanic.

Fuel injection system

Note: *The following procedure is based on the assumption that the fuel pump is working and the fuel pressure is adequate (see Section 4).*

4 Check all electrical connectors that are related to the system. Check the ground wire connections for tightness. Verify that the battery is fully charged (see Chapter 5).

5 Inspect the air filter element (see Chapter 1).

6 Check all fuses related to the fuel system (see Chapter 12).

7 Check the air induction system between the throttle body and the intake manifold for air leaks. Also inspect the condition of all vacuum hoses connected to the intake manifold and to the throttle body.

8 Remove the air intake duct from the throttle body and look for dirt, carbon, varnish, or other residue in the throttle body, particularly around the throttle plate. If it's dirty, clean it with carb cleaner, a toothbrush and a clean shop towel.

9 Check to see if any trouble codes are stored in the PCM (see Chapter 6).

3 Fuel pressure relief procedure

Warning: *Gasoline is extremely flammable. See Fuel system warnings in Section 1.*

Low-pressure side fuel pressure relief

2015 and earlier models

1 Locate and remove the fuel pump relay from the underhood fuse box (see illustration 2.3).

2 Start the engine an allow it to run until it dies.

3 After the engine stalls, crank it over for a few more seconds to make sure the pressure has been relieved.

4 On vehicles with hi-pressure direct injection fuel systems, perform the high-pressure fuel pump pressure release procedure.

5 Disconnect the cable from the negative terminal of the battery before working on the fuel system (see Chapter 5).

6 When the repair is complete, install the fuel pump relay.

7 Turn the ignition to RUN and check for leaks before starting the engine.

2016 and later models

8 Remove the passenger rear quarter trim panel to access the Fuel Pump Control Module (FPCM) (see Chapter 11, Section 26).

9 Disconnect the Fuel Pump Control Module (FPCM) electrical connector (see illustration).

10 Start the engine and allow to idle until the engine stalls, then crank the engine a few times to make sure all the pressure is released.

11 Disconnect the cable from the negative terminal of the battery before working on the fuel system (see Chapter 5).

12 When the repair is complete, reconnect the FPCM.

13 Turn the ignition to RUN and check for leaks before starting the engine.

3.15 Loosen the high pressure fuel tube with a flare-nut wrench while covering it with a shop towel (2.0L four-cylinder engine shown)

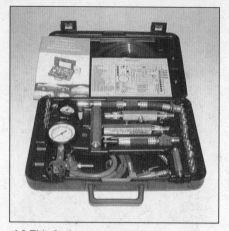

4.3 This fuel pressure testing kit contains all the necessary fittings and adapters, along with the fuel pressure gauge, to test most automotive systems

4.4 On turbocharged models, detach the fuel feed line from the high-pressure fuel pump and connect a fuel pressure gauge between the line and the pump (2.0L turbocharged engine shown)

High-pressure side fuel pressure relief (turbocharged models)

14 Relieve the fuel system pressure, then wait two hours before proceeding.
15 Place a flare nut wrench on the high-pressure fuel tube (see illustration), then wrap the wrench and nut with a shop towel to absorb any remaining fuel pressure while loosening the nut.
16 Loosen the high pressure fuel tube-to-fuel injection pump flare nut and release the remaining pressure.

4 Fuel pressure - check

Warning: *Gasoline is extremely flammable. See Fuel system warnings in Section 1.*
Note: *The following procedure assumes that the fuel pump is receiving voltage and runs.*

1 Release the fuel system pressure (see Section 3).
2 Disconnect the negative battery cable (see Chapter 5, Section 3).
3 A fuel pressure gauge, with the proper adapter(s) and capable of reading fuel pressures within the range listed in this Chapter's Specifications, will be required (see illustration).
4 Disconnect the fuel feed line from the fuel rail (non-turbocharged V6 models) or from the high-pressure fuel pump (turbocharged models) (see illustration) and tee-in the fuel pressure gauge using the proper adapters.
5 Reconnect the negative battery cable.
Note: *On 2015 and earlier models, ensure the fuel pump relay is installed. On 2016 and later models, ensure the FPCM is connected.*
6 Turn the ignition to RUN to pressurize the fuel system and check for leaks at the fuel pressure adapter.
7 Note the gauge reading as soon as the pressure stabilizes, and compare it with the pressure listed in this Chapter's Specifications.
8 Start the engine and allow it to idle. Note the gauge reading as soon as the pressure stabilizes, and compare it with the pressure listed in this Chapter's Specifications.
9 If the fuel pressure is not within specifications:
Check for a restriction in the fuel system (kinked fuel line, plugged fuel pump inlet strainer or clogged fuel filter). If no restrictions are found, replace the fuel pump module (see Section 7). If the fuel pressure is higher than specified, replace the fuel pump module (see Section 7).
10 Turn off the engine and monitor the fuel pressure. Fuel pressure should not fall more than 8 psi over five minutes. If it does, the problem could be a leaky fuel injector, fuel line leak, or faulty fuel pump module.
11 After completing the fuel pressure check, perform the fuel pressure relief procedure.
12 Disconnect the fuel pressure gauge and reconnect the fuel line. Wipe up any spilled gasoline.

5 Fuel lines and fittings - general information and disconnection

Warning: *Gasoline is extremely flammable. See Fuel system warnings in Section 1.*

1 Relieve the fuel pressure before servicing fuel lines or fittings (see Section 3), then disconnect the cable from the negative battery terminal (see Chapter 5) before proceeding.
2 The fuel supply line connects the fuel pump in the fuel tank to the fuel rail on the engine. The Evaporative emission (EVAP) system lines connect the fuel tank to the EVAP canister and connect the canister to the intake manifold.
3 Whenever you're working under the vehicle, be sure to inspect all fuel and evaporative emission lines for leaks, kinks, dents and other damage. Always replace a damaged fuel or EVAP line immediately.
4 If you find signs of dirt in the lines during disassembly, disconnect all lines and blow them out with compressed air. Inspect the fuel strainer on the fuel pump pick-up unit for damage and deterioration.

Steel tubing

5 It is critical that the fuel lines be replaced with lines of equivalent type and specification.
6 Some steel fuel lines have threaded fittings. When loosening these fittings, hold the stationary fitting with a wrench while turning the tube nut.

Plastic tubing

7 When replacing fuel system plastic tubing, use only original equipment replacement plastic tubing.
Caution: *When removing or installing plastic fuel line tubing, be careful not to bend or twist it too much, which can damage it. Also, plastic fuel tubing is NOT heat resistant, so keep it away from excessive heat.*

Flexible hoses

8 When replacing fuel system flexible hoses, use only original equipment replacements.
9 Don't route fuel hoses (or metal lines) within four inches of the exhaust system or within ten inches of the catalytic converter. Make sure that no rubber hoses are installed directly against the vehicle, particularly in places where there is any vibration. If allowed to touch some vibrating part of the vehicle, a hose can easily become chafed and it might start leaking. A good rule of thumb is to maintain a minimum of 1/4-inch clearance around a hose (or metal line) to prevent contact with the vehicle underbody.

Disconnecting Fuel Line Fittings

Two-tab type fitting; depress both tabs with your fingers, then pull the fuel line and the fitting apart

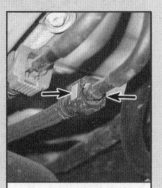

On this type of fitting, depress the two buttons on opposite sides of the fitting, then pull it off the fuel line

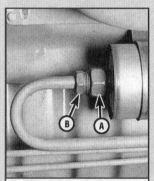

Threaded fuel line fitting; hold the stationary portion of the line or component (A) while loosening the tube nut (B) with a flare-nut wrench

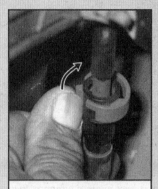

Plastic collar-type fitting; rotate the outer part of the fitting

Metal collar quick-connect fitting; pull the end of the retainer off the fuel line and disengage the other end from the female side of the fitting . . .

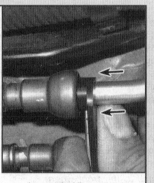

. . . insert a fuel line separator tool into the female side of the fitting, push it into the fitting and pull the fuel line off the pipe

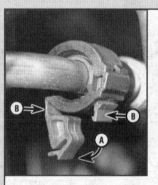

Some fittings are secured by lock tabs. Release the lock tab (A) and rotate it to the fully-opened position, squeeze the two smaller lock tabs (B) . . .

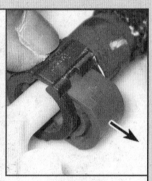

. . . then push the retainer out and pull the fuel line off the pipe

Spring-lock coupling; remove the safety cover, install a coupling release tool and close the tool around the coupling . . .

. . . push the tool into the fitting, then pull the two lines apart

Hairpin clip type fitting: push the legs of the retainer clip together, then push the clip down all the way until it stops and pull the fuel line off the pipe

4-6　Chapter 4　Fuel and exhaust systems

6.1 Exhaust system hangers. Inspect regularly and replace at the first sign of damage or deterioration

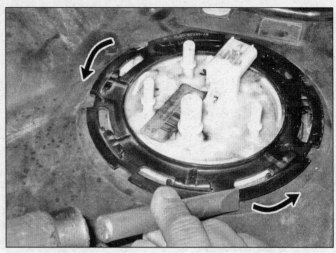

7.4 Using a brass punch and a hammer to unscrew the fuel pump module lock ring

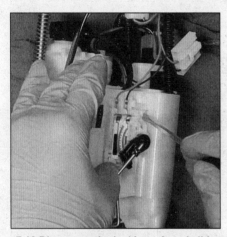

7.10 Disengage the locking tab and slide the sending unit off of the fuel pump module body

6 Exhaust system servicing - general information

Warning: *Allow exhaust system components to cool before inspection or repair. Also, when working under the vehicle, make sure it is securely supported on jackstands.*

1　The exhaust system consists of the exhaust manifolds, catalytic converter, muffler, tailpipe and all connecting pipes, flanges and clamps. The exhaust system is isolated from the vehicle body and from chassis components by a series of rubber hangers (see illustration). Periodically inspect these hangers for cracks or other signs of deterioration, replacing them as necessary.

Note: *On non-turbocharged V6 models, the front manifold and front catalytic converter have been combined into a single unit.*

2　Conduct regular inspections of the exhaust system to keep it safe and quiet. Look for any damaged or bent parts, open seams, holes, loose connections, excessive corrosion or other defects which could allow exhaust fumes to enter the vehicle. Do not repair deteriorated exhaust system components; replace them with new parts.

3　If the exhaust system components are extremely corroded, or rusted together, a cutting torch is the most convenient tool for removal. Consult a properly-equipped repair shop. If a cutting torch is not available, you can use a hacksaw, or if you have compressed air, there are special pneumatic cutting chisels that can also be used. Wear safety goggles to protect your eyes from metal chips and wear work gloves to protect your hands.

4　Here are some simple guidelines to follow when repairing the exhaust system:

a)　*Work from the back to the front when removing exhaust system components.*
b)　*Apply penetrating oil to the exhaust system component fasteners to make them easier to remove.*
c)　*Use new gaskets, hangers and clamps.*
d)　*Apply anti-seize compound to the threads of all exhaust system fasteners during reassembly.*
e)　*Be sure to allow sufficient clearance between newly installed parts and all points on the underbody to avoid overheating the floor pan and possibly damaging the interior carpet and insulation. Pay particularly close attention to the catalytic converter and heat shield.*

7 Fuel pump module - removal and installation

Warning: *Gasoline is extremely flammable. See Fuel system warnings in Section 1.*

Fuel pump module

The *fuel pump module includes the fuel pump, the fuel level sending unit and a fuel filter. The fuel level sending unit is the only item that can be serviced separately.*

1　Relieve the fuel system pressure (see Section 3). Disconnect the cable from the negative battery terminal (see Chapter 5).
2　Remove the fuel tank (see Section 8).
3　Place reference marks on the fuel pump module and the fuel tank.
4　Remove the pump module lock ring using a lock ring removal tool (manufacturer tool no. 310-123, or equivalent) or tap the fuel pump module lock ring counterclockwise with a hammer and brass punch (to avoid sparks) (see illustration).
5　Carefully pull the fuel pump module out of the tank enough to disconnect the fuel vapor hose (top of module) and fuel transfer tube (bottom of module) quick-connect fittings, and separate the hoses. Lift the module out of tank completely, angling it as necessary to protect the fuel level sensor float arm.
6　Inspect the O-ring and replace it if it shows any sign of deterioration.
7　Installation is the reverse of removal noting the following points:
　Ensure the vapor line and transfer tube are connected during module installation.
　Align the fuel pump module and the fuel tank in the correct position.
　Ensure the O-ring does not become distorted or shift during installation.

Fuel level sending unit

Note: *The fuel level sending unit is part of the fuel pump module, but can be replaced.*
8　Remove the fuel pump module as described above.
Note: *On 2016 and later models, the fuel level sensor contains the same fuel level sending unit as the fuel pump module and can be replaced.*
9　Disconnect the fuel level sending unit connector.
10　Press the locking tab inwards and slide the fuel level sender up and off of the fuel pump module (or fuel level sensor) housing (see illustration).
11　Installation is the reverse of removal.

Chapter 4 Fuel and exhaust systems

8.6a Loosen the clamp on the slip joint at the front of the rear portion of the exhaust system...

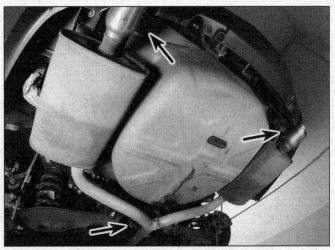

8.6b ... then detach the exhaust system hangers from the pipe and mufflers and slide the exhaust system from the front exhaust pipe

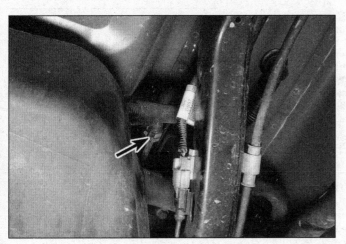

8.8 Disconnect the vapor hose

8.9 Fuel tank filler neck hose clamp

Fuel level sensor

Note: *The fuel level sensor is on the opposite side of the fuel tank from the fuel pump module.*

12 Remove the fuel tank (see Section 8).

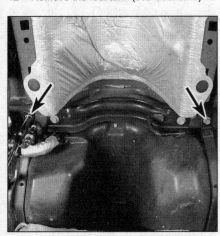

8.11 Fuel tank strap bolts

13 Remove the lock ring using a lock ring removal tool or tap the lock ring counterclockwise with a hammer and brass punch (to avoid sparks) (see illustration 7.4).
14 Carefully pull the fuel level sensor out of the tank enough to disconnect the fuel transfer tube quick-connect fitting, and separate the hose. Lift the module out of tank completely, angling it as necessary to protect the fuel level sensor float arm.
15 Discard and replace the O-ring.
16 On 2016 and later models, replace the fuel level sending unit if necessary following the fuel level sending unit replacement procedure.
17 Installation is the reverse of removal.

8 Fuel tank - removal and installation

Warning: *Gasoline is extremely flammable. See Fuel system warnings in Section 1.*
Note: *The following procedure is much easier to perform if the fuel tank is empty.*

1 Remove the fuel tank filler cap to relieve fuel tank pressure.
2 Relieve the fuel system pressure (see Section 3).
3 Disconnect the cable from the negative battery terminal (see Chapter 5).
4 Raise the rear of the vehicle and support it securely on jackstands.
5 On AWD models, remove the driveshaft (see Chapter 8).
6 Remove the rear portion of the exhaust system (see illustrations).
7 Remove the EVAP canister (see Chapter 6, Section 22).
8 Disconnect the EVAP vapor hose quick-connect fitting.
9 Loosen the hose clamp and disconnect the fuel filler neck hose from the tank (see illustration).
10 Disconnect the fuel lines at the fuel pump module.
11 Support the fuel tank securely, then remove the fuel tank retaining strap bolts and remove the straps (see illustration).
12 Carefully lower the fuel tank enough to

Chapter 4 Fuel and exhaust systems

8.12 Fuel pump module electrical connector

10.4 Loosen the charge air cooler pipe clamp (2.0L four-cylinder model shown)

10.6a Throttle body mounting fasteners (four-cylinder model shown) - manifold removed for clarity

10.6b Throttle body mounting fasteners (non-turbocharged V6 model shown)

disconnect the electrical connectors and harnesses on top of the tank (see illustration).
13 Disconnect the fresh air hose vent cap from the body, Lower and remove the tank.
14 Installation is the reverse of removal. Tighten the fuel tank strap bolts securely.
15 Reconnect the cable to the negative battery terminal (see Chapter 5), then start the engine and check for fuel leaks.

9 Air filter housing - removal and installation

Air intake duct

1 Disconnect the PCV fresh air hose(s) from the air intake duct.
2 On turbocharged models, loosen the clamp at the air intake duct and the charge air cooler inlet tube and remove the intake duct.
3 On non-turbocharged V6 models, loosen the clamps at the air filter housing and the throttle body and remove the air intake duct.
4 Installation is the reverse of removal.

Air filter housing

5 Loosen the air intake duct-to-air filter housing clamp.
6 Disconnect the electrical connector from the MAF sensor (see Chapter 6).
7 Remove the air filter housing cover and filter element (see Chapter 1).
8 Remove the two bolts attaching the air filter housing bracket to the radiator support and remove the housing.
9 Installation is the reverse of removal.

10 Throttle body - removal and installation

Warning: *Wait until the engine is completely cool before beginning this procedure.*

1 Disconnect the cable from the negative battery terminal (see Chapter 5).
2 On non-turbocharged V6 models, remove the air intake duct (see Section 9).
3 On 2016 and later 2.3L turbocharged engines, raise and support the vehicle on jackstands and remove the engine undercover.
Note: *On 2016 and later 2.3L turbocharged models, the throttle body is removed from under the vehicle.*
4 On all turbocharged four-cylinder and V6 models, loosen the clamp and remove the charge air cooler pipe from the throttle body (see illustration).
5 On all models, disconnect the electrical connector from the throttle body.
6 Remove the throttle body mounting fasteners and detach the throttle body from the intake manifold (see illustrations). Discard the gasket; it should be replaced with a new one. Cover the intake manifold opening with a clean shop towel.
7 Installation is the reverse of removal. Use a new gasket and tighten the throttle body fasteners to the torque listed in this Chapter's Specifications.

11 Fuel rail and injectors - removal and installation

Warning: *Gasoline is extremely flammable. See Fuel system warnings in Section 1.*

1 Relieve the fuel system pressure (see Section 3).
2 Disconnect the cable from the negative battery terminal (see Chapter 5).

Turbocharged models

Note: *This procedure applies to 2.0L, 2.3L and 3.5L turbocharged models with high-pressure direct injection systems.*

Removal

3 Remove the intake manifold (see Chapter 2A).
4 Disconnect the high-pressure fuel pump electrical connector and remove the sound insulator from the pump.
5 Disconnect the high-pressure fuel lines from the fuel pump and fuel rail (see Section 12).
Note: *Wrap the fittings in a rag to prevent fuel spray or spill. The manufacturer recommends replacing the fuel lines anytime they are removed.*
6 Disconnect the electrical connectors for all wiring that shares the harness with the fuel injectors.
7 On four-cylinder models, disconnect the injector electrical connectors. Using a trim removal tool, disengage the two pin-type retainers that secure the wiring harness and harness insulator from the fuel rail, then push the harness aside.
Note: *On V6 models, cut the harness-to-fuel rail tie wraps. The fuel injectors are not disconnected from the harness until the fuel rails are removed.*
8 On four-cylinder models, remove the fuel rail noise insulator from the fuel rail.
9 On all models, clean the area around each injector using compressed air.

Chapter 4 Fuel and exhaust systems   4-9

11.10 Fuel rail mounting bolts (four-cylinder model shown)

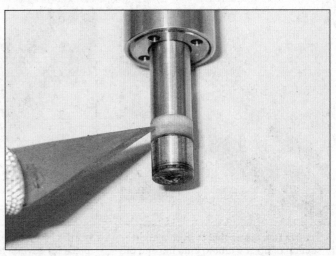

11.14 To remove the Teflon sealing ring, cut it off with a hobby knife (be careful not to scratch the injector groove)

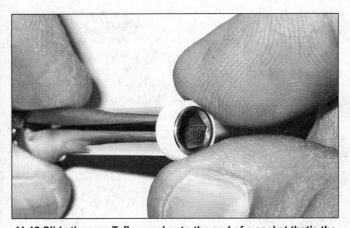

11.19 Slide the new Teflon seal onto the end of a socket that's the same diameter as the end of the fuel injector . . .

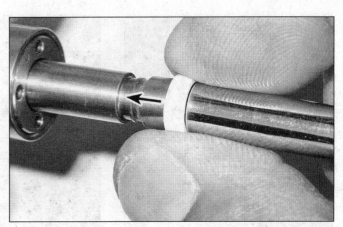

11.20 . . . align the socket with the end of the injector and slide the seal onto the injector and into its mounting groove

10   Remove the fuel rail mounting fasteners (see illustration) and discard them, then remove the fuel rail and injectors. Make sure to use new fuel rail bolts on installation.
Note: *The fuel injectors may remain in the fuel rails when the rail is removed, but normally they remain in the cylinder heads and require the use of a removal tool.*
11   On V6 models, disconnect the fuel injector electrical connectors and remove the harness from the fuel rail.
12   On all models, remove each fuel injector retaining clip with a pair of needle-nose pliers, then remove the injector from its bore in the fuel rail. Remove and discard the upper injector O-rings. Repeat this procedure for each injector.
Note: *Even if you only removed the fuel rail assembly to replace a single injector or a leaking O-ring, replace all of the fuel injector retaining clips and O-rings.*
13   If any injectors stick in the cylinder head, use special tool #310-206 attached to a slide hammer to remove the injector(s).
14   Remove the old combustion chamber Teflon sealing ring, upper O-ring and support ring from each injector (see illustration).
Caution: *Be extremely careful not to damage the groove for the seal or the rib in the floor of the groove. If you damage the groove or the rib, you must replace the injector.*
15   Before installing the new Teflon seal on each injector, thoroughly clean the groove for the seal and the injector shaft. Remove all combustion residue and varnish with a clean shop rag.

Teflon seal installation using the special tools
16   The manufacturer recommends that you use the tools included in the special injector tool set to install the Teflon lower seals on the injectors: Install the special seal assembly cone on the injector, install the special sleeve on the injector and use the sleeve to push on the assembly cone, which pushes the Teflon seal into place on its groove. Do NOT use any lubricants to do so.
17   Pushing the Teflon seal into place in its groove expands it slightly. There are three sizing sleeves in the special tool set with progressively smaller inside diameters. Using a clockwise rotating motion of about 180 degrees, install the slightly larger sleeve onto the injector and over the Teflon seal until the sleeve hits its stop, then carefully turn the sleeve counterclockwise as you pull it off the injector. Use the slightly smaller sizing sleeve the same way, followed by the smallest sizing ring. The seal is now sized. Repeat this step for each injector.

Teflon seal installation without special tools
18   If you don't have the special injector tool set, the Teflon seal can be installed using this method: First, find a socket that is equal or very close in diameter to the diameter of the end of the fuel injector.
19   Work the new Teflon seal onto the end of the socket (see illustration).
20   Place the socket against the end of the injector (see illustration) and slide the seal from the socket onto the injector. Do NOT use any lubricants to do so. Continue pushing the seal onto the injector until it seats into its mounting groove.
21   Because the inside diameter of the seal

4-10   Chapter 4 Fuel and exhaust systems

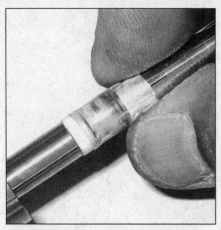

11.21a Use the socket to push a short section of plastic tubing onto the end of the injector and over the new seal . . .

11.21b . . . then leave the plastic tubing in place for several hours to compress the new seal

11.22 Note that the upper O-ring (1) is installed above the support ring (2)

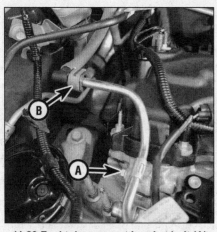

11.29 Fuel tube support bracket bolt (A) and fuel supply line fitting (B)

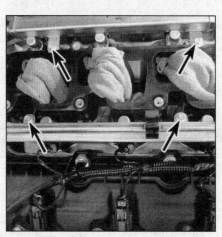

11.32 Fuel rail mounting bolts (non-turbocharged V6 engine)

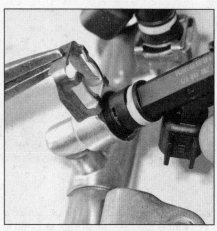

11.33 Before removing the injector retaining clips, note how they're installed to ensure that they're correctly reinstalled

has to be stretched open to fit over the bore of the socket and the injector, its outside diameter is now slightly too large - it is no longer flush with the surface of the injector. It must be shrunk back to its original size. To do so, push a piece of plastic tubing with an interference fit onto the end of the socket; a plastic straw that fits tightly on the injector will work. After pushing the plastic tubing onto the socket about an inch, snip off the rest of the tubing, then use the socket to push the tubing onto the end of the injector (see illustration), sliding it onto the injector until it completely covers the new seal (see illustration). Leave the tubing on for a few hours, then remove it. The seal should now be shrunk back its original outside diameter, or close to it.

Installation

22   Install a new support ring on the top of the injector, then lubricate the new upper O-ring with clean engine oil and install it on the injector. Do NOT oil the new Teflon seal. Note that the O-ring is installed above the support ring (see illustration).
23   Thoroughly clean the injector bores with a small nylon brush.
24   Insert each injector into its bore in the fuel rail. Install the new retaining clips on the injectors.
25   Install the injectors and fuel rail assembly on the cylinder head. Tighten the fuel rail mounting fasteners to the torque listed in this Chapter's Specifications, starting with the center bolt and working outwards.
26   The remainder of installation is the reverse of removal. If removed, install a new fuel rail pressure sensor (see Chapter 6).
27   Reconnect the cable to the negative battery terminal (see Chapter 5), then turn the ignition switch to RUN (but don't operate the starter). This activates the fuel pump for about two seconds, which builds up fuel pressure in the fuel lines and the fuel rail. Repeat this step two or three times, then check the fuel lines, fuel rails and injectors for fuel leaks.

Non-turbocharged models

28   Remove the upper intake manifold (see Chapter 2B).
29   Remove the fuel rail tube support bracket bolt and disconnect the fuel supply line at the fuel rail tube (see illustration).
30   Disconnect the electrical connector from each fuel injector.
31   Detach the three pin-type retainers from the fuel rail, push the wiring harness aside and remove the other three injector electrical connectors.
32   Remove the four fuel rail mounting bolts (see illustration), then remove the fuel rail and injectors as a single assembly.
33   Release the fuel injector retaining clip (see illustration) and remove each injector from its bore in the fuel rail.
34   Remove and discard the upper and lower injector O-rings (see illustration) from each injector.
Note: *Even if you only removed the fuel rail assembly to replace a single injector or a leak-*

Chapter 4 Fuel and exhaust systems

4-11

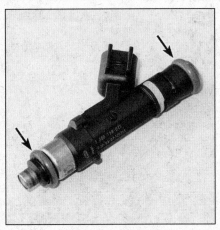

11.34 Injector O-rings

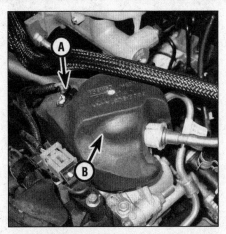

12.5 Disconnect the electrical connector (A), then remove the insulator (B) from the high-pressure fuel pump

12.10 High-pressure fuel pump details
1. Fuel feed line
2. High-pressure fuel line
3. Fuel pump mounting bolts

ing O-ring, it's a good idea to remove all of the injectors from the fuel rail and replace all of the O-rings at the same time.

35 Coat the new upper O-rings with clean engine oil and slide them into place on each of the fuel injectors. Coat the new lower O-rings with clean engine oil and install them on the lower ends of the injectors.

36 Coat the outside surface of each upper O-ring with clean engine oil, then insert each injector into its bore in the fuel rail. Secure the injectors to the fuel rail with the injector retaining clips.

37 Tighten the fuel rail bolts securely. The remainder of installation is the reverse of removal.

38 Reconnect the cable to the negative battery terminal (see Chapter), then turn the ignition switch to RUN (but don't operate the starter). This activates the fuel pump for about two seconds, which builds up fuel pressure in the fuel lines and the fuel rail. Repeat this step two or three times, then check the fuel lines, fuel rails and injectors for fuel leaks.

12 High-pressure fuel pump (turbocharged models) – removal and installation

Note: *This procedure applies to 2.0L, 2.3L and 3.5L turbocharged models with high-pressure direct injection systems.*
Warning: *Gasoline is extremely flammable. See Fuel system warnings in Section 1.*

Removal

1 Remove the engine cover.
2 Relieve the fuel system pressure (see Section 3).
3 Disconnect the cable from the negative battery terminal (see Chapter 5).
4 Remove the air filter housing intake duct.

12.12 Check the high-pressure fuel pump tappet for wear (A). Install a new O-ring (B) on the pump

5 Disconnect the fuel injection pump electrical connector (see illustration).
6 Carefully remove the pump insulator from around the outside of the pump.
7 Disconnect the fuel supply line quick-connect fitting.
8 Loosen then remove the fuel lines using a flare nut wrench.
Note: *Wrap the fittings in a rag to prevent fuel spray or spill. The manufacturer recommends replacing the fuel lines anytime they are removed.*
9 On four-cylinder models, remove the shield bolts and shield. Detach any harness retainers from the shield.
10 On all models, loosen the fuel pump bolts (see illustration), alternating one turn at a time until the bolts and pump can be removed. The bolts must be replaced after they have be removed.
Caution: *The fuel pump is under extreme pressure - if the bolts are not loosened evenly and one complete turn at a time the pump or*

12.14 The cam lobe (and tappet) must be at its lowest point before installing the pump

housing will be damaged.
11 Remove and discard the fuel pump O-ring.

Installation

12 Before installing the fuel injection pump, check the pump tappet in the pump body for wear or damage (see illustration).
13 Install a new O-ring on the pump, then apply clean oil to the pump tappet and the O-ring.
14 Rotate the engine until the cam lobe for the fuel injection pump is on the lowest point or Bottom Dead Center (BDC) (see illustration). The pump tappet will be as low as it can go.
15 Set the pump in the housing, then install the new bolts hand tight.
16 Tighten the bolts in even stages, alternating between the bolts, to the torque listed in this Chapter's Specifications.
17 The remainder of installation is the reverse of removal.

13.3a Disconnect charge air cooler inlet tube…

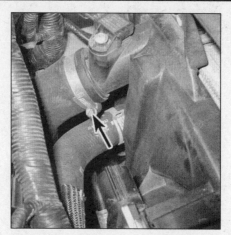

13.3b … and outlet tube clamps

13.4 The charge air cooler is secured by a bolt and retaining tab on each side

14.6 Remove the turbocharger heat shield mounting bolts

14.7a Detach the hoses at each end of the coolant pipe

13 Charge air cooler - removal and installation

Caution: *Whenever the turbocharger, charge air cooler or cooler tubes are removed always cover any opening to prevent debris from falling in. The system is easily damaged, so carefully clean all openings before reassembling.*

1   Remove the radiator (see Chapter 3, Section 7).
2   Remove the Manifold Absolute Pressure Temperature (MAPT) sensor (see Chapter 6).
3   Loosen the clamps securing the inlet and outlet tubes to the charge air cooler (see illustrations), then disconnect the tubes.
4   Remove the charge air cooler bolts (see illustration) and carefully remove the cooler from the vehicle.
5   Installation is the reverse of removal.

14 Turbocharger - removal and installation

Warning: *Wait until the engine is completely cool before beginning this procedure.*
Caution: *Do not disassemble the turbocharger or try to adjust the wastegate actuator. The turbocharger or possibly even the engine could be damaged.*
Caution: *Whenever the turbocharger, charge air cooler or cooler tubes are removed, always cover any opening to prevent debris from falling in. The system is easily damaged, so carefully clean all openings before reassembling.*

Removal - four-cylinder models

1   Raise the vehicle and support it securely on jackstands.
2   Remove the under vehicle cover.
3   Drain the cooling system (see Chapter 1, Section 20).

2.0L engine

4   Remove the catalytic converter (see Chapter 6).
5   Remove the charge air cooler inlet tube (see Section 13) and the turbocharger inlet pipe.
6   Remove the turbocharger heat shield mounting bolts and remove the shield (see illustration).
7   From above, squeeze and slide back the hose clamps on the coolant line (see illustration). From below, remove the coolant outlet line banjo bolt and disconnect the line from the turbocharger (see illustration).
Note: *Always replace the linked sealing washers to the banjo bolts and coolant lines.*
8   Remove the oil supply line banjo bolts and discard the sealing washers (see illustrations), then pull the inline oil supply filter out

Chapter 4 Fuel and exhaust systems

14.7b Disconnect the coolant outlet line from the turbocharger and the engine block

14.8a Remove the oil supply line banjo bolt and detach the line from the turbocharger . . .

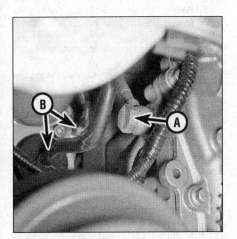

14.8b . . . then remove this banjo bolt (A) from the other end of the oil line. Detach the line from the block, then pull out the filter and install a new one. (B) are coolant lines

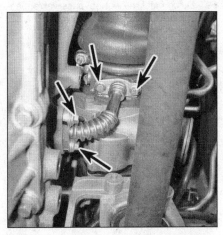

14.10 Turbocharger oil drain tube bolts

14.11 Remove the turbocharger mounting nuts

of the engine block and replace it with a new one.

9   Disconnect the vacuum lines and electrical connectors to the turbocharger.

10   Remove the oil drain tube mounting bolts and drain tube, then discard the gasket and drain tube (see illustration).

Note: *The factory requires the oil drain tube be replaced whenever it is removed.*

11   Remove and discard the turbocharger mounting nuts (see illustration), then carefully maneuver the turbocharger from the vehicle.

2.3L engine

12   Remove the engine cover.

13   Disconnect the turbocharger bypass valve electrical connectors.

14   Loosen the clamps and remove the turbocharger piping connected to the turbocharger, air cleaner assembly, and charge air cooler.

15   Remove the turbocharger piping between the intake hose and the turbocharger inlet. There are two mounting bolts securing the tube to the valve cover.

16   Remove the turbocharger cooling supply line bolt at the turbocharger.

17   Remove the turbocharger cooling supply line bracket bolt and disconnect the line from the hose to remove. Discard the O-rings.

18   Remove the catalytic converter (see Chapter 6, Section 21).

19   Remove the Allen bolt attaching the coolant return line to the engine block.

20   Remove the coolant return line and discard the O-rings.

21   Remove the cowl panel below the windshield (see Chapter 11, Section 12).

22   Remove the four (4) nuts and remove the strut tower brace.

23   Disconnect the turbocharger wastegate vacuum hose at the wastegate.

24   Remove the oil drain tube mounting bolts and drain tube, then discard the gasket and drain tube.

25   Remove the oil supply line banjo bolt at the turbocharger. Discard the sealing washers.

26   Remove the oil supply line bolt at the engine block. Discard the gasket and oil supply line.

27   Carefully remove the oil supply filter from the engine block and discard.

28   Remove the upper turbocharger flange nuts. Remove the lower heat shield bolt and remove the heat shield.

29   Remove the lower turbocharger flange nuts and remove the turbocharger from the vehicle. Discard the flange gasket.

Chapter 4 Fuel and exhaust systems

Removal - V6 models

Front cylinder bank

30 Raise and support the vehicle on jackstands.
31 Remove the engine cover.
32 Drain the cooling system (see Chapter 1, Section 20).
33 Remove the air filter housing assembly (see Section 9).
34 Disconnect the turbocharger wastegate vacuum hose at the wastegate.
35 Remove the bolts and the turbocharger heat shield.
36 Loosen the hose clamp and disconnect the turbocharger inlet pipe.
37 Using an Allen wrench or tool, remove the oil supply line bolt and disconnect the line. Discard the sealing washers.
38 Remove the left side catalytic converter (see Chapter 6, Section 21).
39 Remove the oil return pipe nuts, disconnect the oil return pipe from the turbocharger, then pull out of the engine block. Discard the gasket and O-rings.
40 Remove the banjo bolts from the coolant lines at the turbocharger. Remove the coolant lines and discard the sealing washers.
41 Disconnect the turbocharger electrical connectors.
42 Loosen the hose clamps and disconnect the turbocharger outlet pipe and bypass hose.
43 Remove the turbocharger mounting bolts. Use NEW bolts for installation.
44 Disconnect the turbocharger from the two brackets on the engine and remove the turbocharger.

Rear cylinder bank

45 Raise and support the vehicle on jackstands.
46 Remove the engine cover.
47 Drain the cooling system (see Chapter 1, Section 20).
48 Remove the right side catalytic converter (see Chapter 6, Section 21).
49 Remove the front subframe (see Chapter 10, Section 19).
50 Remove the retainer and disconnect the oil supply line quick-disconnect fitting.
51 Loosen the hose clamp and disconnect the turbocharger inlet pipe.
52 Disconnect the turbocharger wastegate vacuum hose at the wastegate.
53 Disconnect the turbocharger from the lower bracket on the engine.
54 Remove the oil return pipe nuts, disconnect the oil return pipe from the turbocharger, then pull out of the engine block. Discard the gasket and O-rings.
55 Remove the banjo bolts from the coolant lines at the turbocharger. Remove the coolant lines and discard the sealing washers.
56 Disconnect the turbocharger electrical connectors.
57 Loosen the hose clamps and disconnect the turbocharger outlet pipe and bypass hose.
58 Disconnect the wastegate vacuum line from the turbocharger piping.
59 Disconnect the right bypass valve electrical connector and remove the turbocharger outlet pipe from the engine.
60 Remove the bolts and the turbocharger heat shield.
61 Disconnect the turbocharger from the bracket at the exhaust flange.
62 Remove the turbocharger from the engine.
63 Remove the bolt and remove the oil supply line from the turbocharger. Discard the sealing washers.

Installation - all models

64 Install the turbocharger with a new flange gasket and tighten the new bolts or nuts to the torque listed in this Chapter's Specifications.
65 Install the new oil drain tube and new drain tube gaskets and/or O-rings, then install and tighten the bolts to the torque listed in this Chapter's Specifications.
66 Install new oil feed line sealing washers or gasket and coolant line sealing washers and gaskets. Tighten the bolts and banjo bolts to the torque listed in this Chapter's Specifications.
67 The remainder of installation is the reverse of removal.
68 Refill the cooling system and change the engine oil and filter (see Chapter 1).

15.2 The FPCM is mounted behind the right-side quarter trim ("C-pillar") panel

15 Fuel Pump Control Module (FPCM) - replacement

Note: *The FPCM is also referred to as the Fuel Pump Driver Module (FPDM).*

1 Remove the passenger side rear interior quarter panel (see Chapter 11, Section 26).
2 Disconnect the electrical connector from the fuel pump control module (see illustration).
3 Remove the control module mounting nuts and detach the module from the vehicle.
4 Installation is the reverse of removal.

Chapter 5
Engine electrical systems

Contents

	Section		Section
Alternator - removal and installation	7	General information and precautions	1
Battery - disconnection	3	Ignition coil(s) - removal and installation	6
Battery and battery tray - removal and installation	4	Starter motor - removal and installation	8
Battery cables - replacement	5	Troubleshooting	2

1 General information and precautions

General information

Ignition system

1   The electronic ignition system consists of the Crankshaft Position (CKP) sensor, the Camshaft Position (CMP) sensor, the Knock Sensor (KS), the Powertrain Control Module (PCM), the ignition switch, the battery, the individual ignition coils, and the spark plugs. For more information on the CKP, CMP and KS sensors, as well as the PCM, refer to Chapter 6.

Charging system

2   The charging system includes the alternator (with an integral voltage regulator), the Powertrain Control Module (PCM), the Body Control Module (BCM), a charge indicator light on the dash, the battery, a fuse or fusible link and the wiring connecting all of these components. The charging system supplies electrical power for the ignition system, the lights, the radio, etc. The alternator is driven by a drivebelt.

Starting system

3   The starting system consists of the battery, the ignition switch, the starter relay, the Powertrain Control Module (PCM), the Body Control Module (BCM), the Transmission Range (TR) switch, the starter motor and solenoid assembly, and the wiring connecting all of the components.

Precautions

4   Always observe the following precautions when working on the electrical system:
   a) *Be extremely careful when servicing engine electrical components. They are easily damaged if checked, connected or handled improperly.*
   b) *Never leave the ignition switched on for long periods of time when the engine is not running.*
   c) *Never disconnect the battery cables while the engine is running.*
   d) *Maintain correct polarity when connecting battery cables from another vehicle during jump starting - see the "Booster battery (jump) starting" Section at the front of this manual.*
   e) *Always disconnect the cable from the negative battery terminal before working on the electrical system, but read the battery disconnection procedure first (see Section 3).*

5   It's also a good idea to review the safety-related information regarding the engine electrical systems located in the Safety first! Section at the front of this manual before beginning any operation included in this Chapter.

2 Troubleshooting

Ignition system

1 If a malfunction occurs in the ignition system, do not immediately assume that any particular part is causing the problem. First, check the following items:

a) Make sure that the cable clamps at the battery terminals are clean and tight.
b) Test the condition of the battery (see Steps 15 through 19). If it doesn't pass all the tests, replace it.
c) Check the ignition coil or coil pack connections.
d) Check any relevant fuses in the engine compartment fuse and relay box (see Chapter 12). If they're burned, determine the cause and repair the circuit.

Check

Warning: *Because of the high voltage generated by the ignition system, use extreme care when performing a procedure involving ignition components.*

Note: *The ignition system components on these vehicles are difficult to diagnose. In the event of ignition system failure that you can't diagnose, have the vehicle tested at a dealer service department or other qualified auto repair facility.*

Note: *You'll need a spark tester for the following test. Spark testers are available at most auto supply stores.*

2 If the engine turns over but won't start, verify that there is sufficient ignition voltage to fire the spark plugs as follows.

3 Remove a coil and install the tester between the boot at the lower end of the coil and the spark plug (see illustration).

4 Crank the engine and note whether or not the tester flashes.

Caution: *Do NOT crank the engine or allow it to run for more than five seconds; running the engine for more than five seconds may set a Diagnostic Trouble Code (DTC) for a cylinder misfire.*

5 If the tester flashes during cranking, the coil is delivering sufficient voltage to the spark plug to fire it. Repeat this test for each cylinder to verify that the other coils are OK.

6 If the tester doesn't flash, remove a coil from another cylinder and swap it for the one being tested. If the tester now flashes, you know that the original coil is bad. If the tester still doesn't flash, the PCM or wiring harness is probably defective. Have the PCM checked out by a dealer service department or other qualified repair shop (testing the PCM is beyond the scope of the do-it-yourselfer because it requires expensive special tools).

7 If the tester flashes during cranking but a misfire code (related to the cylinder being tested) has been stored, the spark plug could be fouled or defective.

Charging system

8 If a malfunction occurs in the charging system, do not automatically assume the alternator is causing the problem. First check the following items:

a) Check the drivebelt tension and condition, as described in Chapter 1. Replace it if it's worn or deteriorated.
b) Make sure the alternator mounting bolts are tight.
c) Inspect the alternator wiring harness and the connectors at the alternator and voltage regulator. They must be in good condition, tight and have no corrosion.
d) Check the fusible link (if equipped) or main fuse in the underhood fuse/relay box. If it is burned, determine the cause, repair the circuit and replace the link or fuse (the vehicle will not start and/or the accessories will not work if the fusible link or main fuse is blown).
e) Start the engine and check the alternator for abnormal noises (a shrieking or squealing sound indicates a bad bearing).
f) Check the battery. Make sure it's fully charged and in good condition (one bad cell in a battery can cause overcharging by the alternator).
g) Disconnect the battery cables (negative first, then positive). Inspect the battery posts and the cable clamps for corrosion. Clean them thoroughly if necessary (see Chapter 1). Reconnect the cables (positive first, negative last).

Alternator - check

9 Use a voltmeter to check the battery voltage with the engine off. It should be at least 12.6 volts (see illustration 2.16).

10 Start the engine and check the battery voltage again. It should now be approximately 13.5 to 15 volts.

11 If the voltage reading is more or less than the specified charging voltage, the voltage regulator is probably defective, which will require replacement of the alternator (the voltage regulator is not replaceable separately). Remove the alternator and have it bench tested (most auto parts stores will do this for you).

12 The charging system (battery) light on the instrument cluster lights up when the ignition key is turned to ON, but it should go out when the engine starts.

13 If the charging system light stays on after the engine has been started, there is a problem with the charging system. Before replacing the alternator, check the battery condition, alternator belt tension and electrical cable connections.

14 If replacing the alternator doesn't restore voltage to the specified range, have the charging system tested by a dealer service department or other qualified repair shop.

Battery - check

15 Check the battery state of charge. Visually inspect the indicator eye on the top of the battery (if equipped with one); if the indicator eye is black in color, charge the battery as described in Chapter 1. Next perform an open circuit voltage test using a digital voltmeter.

Note: *The battery's surface charge must be removed before accurate voltage measurements can be made. Turn on the high beams for ten seconds, then turn them off and let the vehicle stand for two minutes.*

16 With the engine and all accessories Off, touch the negative probe of the voltmeter to the negative terminal of the battery and the positive probe to the positive terminal of the battery (see illustration). The battery voltage should be 12.6 volts or slightly above. If the battery is less than the specified voltage, charge the battery before proceeding to the next test. Do not proceed with the battery load test unless the battery charge is correct.

17 Disconnect the negative battery cable, then the positive cable from the battery.

18 Perform a battery load test. An accurate check of the battery condition can only be performed with a load tester (see illustration). This test evaluates the ability of the battery to operate the starter and other accessories during periods of high current draw. Connect the load tester to the battery terminals. Load test the battery according to the tool manufacturer's instructions. This tool increases the load

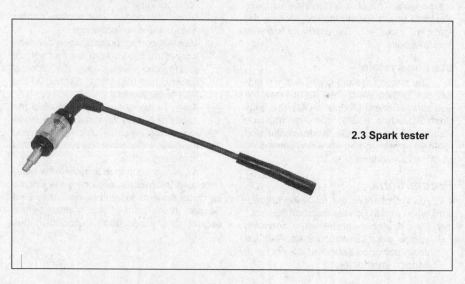

2.3 Spark tester

Chapter 5 Engine electrical systems

2.16 To test the open circuit voltage of the battery, touch the black probe of the voltmeter to the negative terminal and the red probe to the positive terminal of the battery; a fully charged battery should be at least 12.6 volts

2.18 Connect a battery load tester to the battery and check the battery condition under load following the tool manufacturer's instructions

demand (current draw) on the battery.
19 Maintain the load on the battery for 15 seconds and observe that the battery voltage does not drop below 9.6 volts. If the battery condition is weak or defective, the tool will indicate this condition immediately.
Note: *Cold temperatures will cause the minimum voltage reading to drop slightly. Follow the chart given in the manufacturer's instructions to compensate for cold climates. Minimum load voltage for freezing temperatures (32 degrees F) should be approximately 9.1 volts.*

Starting system

The starter rotates, but the engine doesn't
20 Remove the starter (see Section 8). Check the overrunning clutch and bench test the starter to make sure the drive mechanism extends fully for proper engagement with the flywheel ring gear. If it doesn't, replace the starter.
21 Check the flywheel ring gear for missing teeth and other damage. With the ignition turned off, rotate the flywheel so you can check the entire ring gear.

The starter is noisy
22 If the solenoid is making a chattering noise, first check the battery (see Steps 15 through 19). If the battery is okay, check the cables and connections.
23 If you hear a grinding, crashing metallic sound when you turn the key to Start, check for loose starter mounting bolts. If they're tight, remove the starter and inspect the teeth on the starter pinion gear and flywheel ring gear. Look for missing or damaged teeth.
24 If the starter sounds fine when you first turn the key to Start, but then stops rotating the engine and emits a zinging sound, the problem is probably a defective starter drive that's not staying engaged with the ring gear. Replace the starter.

The starter rotates slowly
25 Check the battery (see Steps 15 through 19).
26 If the battery is okay, verify all connections (at the battery, the starter solenoid and motor) are clean, corrosion-free and tight. Make sure the cables aren't frayed or damaged.
27 Check that the starter mounting bolts are tight so it grounds properly. Also check the pinion gear and flywheel ring gear for evidence of a mechanical bind (galling, deformed gear teeth or other damage).

The starter does not rotate at all
28 Check the battery (see Steps 15 through 19).
29 If the battery is okay, verify all connections (at the battery, the starter solenoid and motor) are clean, corrosion-free and tight. Make sure the cables aren't frayed or damaged.
30 Check all of the fuses in the underhood fuse/relay box.
31 Check that the starter mounting bolts are tight so it grounds properly.
32 Check for voltage at the starter solenoid "S" terminal when the ignition key is turned to the start position. If voltage is present, replace the starter/solenoid assembly. If no voltage is present, the problem could be the starter relay, the Transmission Range (TR) switch (see Chapter 6), or with an electrical connector somewhere in the circuit (see the wiring diagrams at the end of this manual). Also, on many modern vehicles, the Powertrain Control Module (PCM) and the Body Control Module (BCM) control the voltage signal to the starter solenoid; on such vehicles a special scan tool is required for diagnosis.

3 Battery - disconnection

Caution: *Always disconnect the cable from the negative battery terminal FIRST and hook it up LAST or the battery may be shorted by the tool being used to loosen the cable clamps.*

1 Some systems on the vehicle require battery power to be available at all times, either to maintain continuous operation (alarm system, power door locks, etc.), or to maintain control unit memory (radio station presets, Powertrain Control Module (PCM) and other control units). When the battery is disconnected, the power that maintains these systems is cut. So, before you disconnect the battery, please note that on a vehicle with power door locks, it's a wise precaution to remove the key from the ignition and to keep it with you, so that it does not get locked inside if the power door locks should engage accidentally when the battery is reconnected!
2 Devices known as "memory-savers" can be used to avoid some of these problems. Precise details vary according to the device used. The typical memory saver is plugged into the cigarette lighter and is connected to a spare battery. Then the vehicle battery can be disconnected from the electrical system. The memory saver will provide sufficient current to maintain audio unit security codes, PCM memory, etc. and will provide power to always hot circuits such as the clock and radio memory circuits.
Warning: *Some memory savers deliver a considerable amount of current in order to keep vehicle systems operational after the main battery is disconnected. If you're using a memory saver, make sure that the circuit concerned is actually open before servicing it.*
Warning: *If you're going to work near any of the airbag system components, the battery MUST be disconnected and a memory saver must NOT be used. If a memory saver is used, power will be supplied to the airbag, which means that it could accidentally deploy and cause serious personal injury.*
3 To disconnect the battery for service procedures requiring power to be cut from the vehicle, loosen the cable end bolt and disconnect the cable from the negative battery terminal. Isolate the cable end to prevent it from coming into accidental contact with the battery terminal.

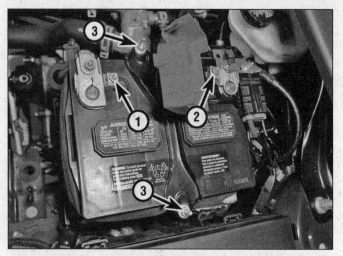

4.1 Battery details - four-cylinder model shown, V6 models similar
1 Negative battery cable clamp nut
2 Positive battery cable clamp nut
3 Hold-down clamp nuts

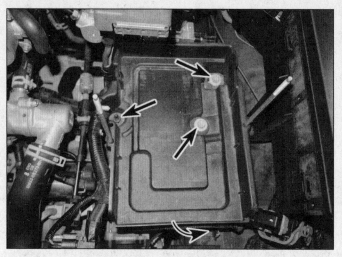

4.5a Battery tray mounting fasteners

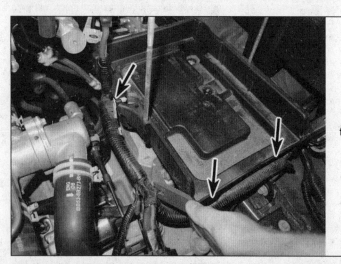

4.5b Pry the wiring harness retainers from the battery tray

4 Battery and battery tray - removal and installation

1 Disconnect the cable from the negative battery terminal first, then disconnect the cable from the positive battery terminal (see illustration).
2 Remove the battery hold-down clamp.
3 Lift out the battery. Be careful - it's heavy.
Note: *Battery straps and handlers are available at most auto parts stores for reasonable prices. They make it easier to remove and carry the battery.*
4 If you are replacing the battery, make sure you get one that's identical, with the same dimensions, amperage rating, cold cranking rating, etc. Also, if the battery was equipped with a heat shield, remove it from the old battery and install it on the new one.
5 If necessary for access to other components, remove the battery tray (see illustrations). The air filter housing will have to be removed first (see Chapter 4).
6 Installation is the reverse of removal. Connect the positive cable first and the negative cable last.

5 Battery cables - replacement

1 When removing the cables, always disconnect the cable from the negative battery terminal first and hook it up last, or you might accidentally short out the battery with the tool you're using to loosen the cable clamps. Even if you're only replacing the cable for the positive terminal, be sure to disconnect the negative cable from the battery first.
2 Disconnect the old cables from the battery, then trace each of them to their opposite ends and disconnect them. Note the routing of each cable before disconnecting it to ensure correct installation.
3 If you are replacing any of the old cables, take them with you when buying new cables. It is vitally important that you replace the cables with identical parts.
4 Clean the threads of the solenoid or ground connection with a wire brush to remove rust and corrosion. Apply a light coat of battery terminal corrosion inhibitor or petroleum jelly to the threads to prevent future corrosion.
5 Attach the cable to the solenoid or ground connection and tighten the mounting nut/bolt securely.
6 Before connecting a new cable to the battery, make sure that it reaches the battery post without having to be stretched.
7 Connect the cable to the positive battery terminal first, then connect the ground cable to the negative battery terminal.

6 Ignition coil(s) - removal and installation

1 Disconnect the cable from the negative battery terminal (see Section 3).
2 Remove the engine cover.
3 On four-cylinder models, remove the air filter outlet duct, then remove the EVAP canister purge valve fastener and move the valve out of the way.
4 On 3.5L turbocharged models, for the front cylinder bank, remove the direct injection fuel pump insulator and disconnect the crankcase vent tube from the front valve cover, and position out of the way. For the rear cylinder bank, remove the turbocharger outlet pipe, bypass valve electrical connector and by-pass valve hose to access the coils.
5 On non-turbocharged V6 models, remove the upper intake manifold to access the rear cylinder bank ignition coils (see Chapter 2B, Section 5).
6 Disconnect the electrical connector from the ignition coil (see illustrations).
7 Remove the mounting fasteners from the ignition coil.
8 Grasp the coil firmly and pull it off the spark plug (see illustration).
9 Installation is the reverse of removal.

Chapter 5 Engine electrical systems

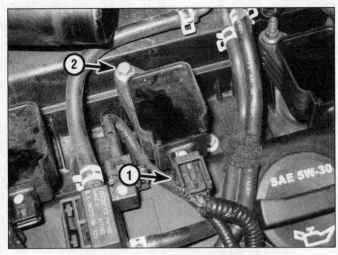

6.6a Ignition coil details (four-cylinder engines)

1 Electrical connector   2 Mounting fasteners

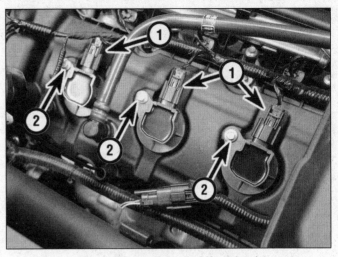

6.6b Ignition coil details (V6 engines)

1 Electrical connector   2 Mounting fasteners

6.8 Grasp the top of the coil and twist the coil while pulling it up and off of the spark plug

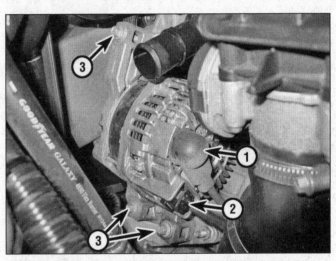

7.6 Alternator mounting details (four-cylinder models)

1 B+ terminal   3 Mounting bolts
2 Electrical connector

7 Alternator - removal and installation

1 Disconnect the cable from the negative battery terminal (see Section 3).
2 Remove the engine cover.
3 Remove the drivebelt (see Chapter 1).

Four-cylinder models

2.0L engine

4 On 2011 through early 2012 models, remove the Charge Air Cooler (CAC) upper pipe.
5 On late 2012 and later models, remove the remove the air conditioning compressor drivebelt (see Chapter 1) and disconnect the air conditioning compressor electrical connectors. Remove the compressor mounting bolts and position the compressor out of the way to allow access to the alternator (see Chapter 3, Section 13).

Warning: *Don't disconnect the refrigerant lines.*

6 Pull back the protective cover from the alternator's battery terminal, remove the nut and disconnect the battery cable from the alternator (see illustration). Set the battery cable aside. Disconnect the electrical connector from the alternator.
7 Remove the alternator mounting fasteners and remove the alternator.
8 On 2011 through early 2012 models, if the alternator is being replaced, remove the idler pulley.

2.3L engine

9 Remove the Charge Air Cooler (CAC) upper pipe.
10 Pull back the protective cover from the alternator's battery terminal, remove the nut and disconnect the battery cable from the alternator (see illustration 7.6). Set the battery cable aside. Disconnect the electrical connector from the alternator.
11 Raise the front of the vehicle and support it securely on jackstands, then remove the engine undercover to access the alternator mounting bolts.
12 Remove the nut from the alternator mounting stud bolt and position bracket aside.

Chapter 5 Engine electrical systems

7.15 Alternator details (V6 models)
1 Protective cover and battery terminal
2 Alternator electrical connector

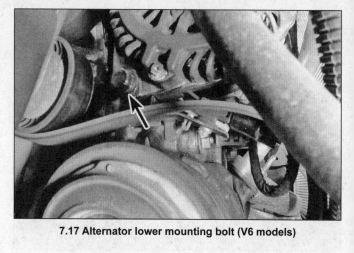

7.17 Alternator lower mounting bolt (V6 models)

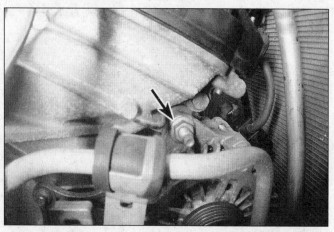

7.18 Alternator upper mounting nut and stud (V6 models)

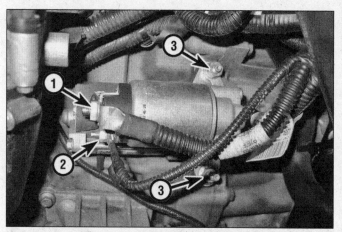

8.3 Starter motor details - 2.0L four-cylinder models
1 Starter motor battery terminal
2 Starter solenoid terminal
3 Mounting stud bolts

13 Remove the alternator mounting fasteners (see illustration 7.6) and remove the alternator.

V6 models

14 Remove the engine cooling fan (see Chapter 3).
15 Push back the protective cover from the alternator's battery terminal, remove the nut and disconnect the battery cable from the alternator. Disconnect the electrical connector from the alternator (see illustration).
Note: *Remove the passenger inner fender splash shield if needed to gain access to the alternator bolt/nut.*
16 On 3.5L turbocharged models, disconnect the O2 sensor connector behind the alternator.
17 On all models, remove the alternator lower mounting bolt (see illustration).
18 Remove the alternator upper mounting stud nut and stud and remove the alternator (see illustration).

All models

19 If you're replacing the alternator, take the old one with you when purchasing the replacement unit. Make sure that the new/rebuilt unit looks identical to the old alternator. Look at the electrical terminals on the backside of the alternator. They should be the same in number, size and location as the terminals on the old alternator. Finally, look at the identification numbers. They will be stamped into the housing or printed on a tag attached to the housing. Make sure that the ID numbers are the same on both alternators.
20 Many new/rebuilt alternators DO NOT have a pulley installed, so you might have to swap the pulley from the old unit to the new/rebuilt one. When buying an alternator, find out the store's policy regarding pulley swaps. Some stores perform this service free of charge. If your local auto parts store doesn't offer this service, you'll have to purchase a puller for removing the pulley and do it yourself.
21 Installation is the reverse of removal. Tighten the alternator mounting bolts securely.
22 Reconnect the cable to the negative terminal of the battery. Check the charging voltage (see Section 2) to verify that the alternator is operating correctly.

8 Starter motor - removal and installation

1 Detach the cable from the negative terminal of the battery (see Section 3).

Four-cylinder models

2.0L engine

2 Raise the vehicle and support it securely on jackstands. Remove the underbody cover.
3 Disconnect the starter motor solenoid wire nut and the starter motor battery cable nut, and disconnect both wires from the starter (see illustration).
Note: *Reposition the coolant hoses as necessary to access the starter bolts.*
4 Remove the nut that secures the ground wire to the starter mounting stud and disconnect the ground wire from the stud. Remove the mounting stud bolts and remove the starter.

Chapter 5 Engine electrical systems

8.13 Starter motor details - V6 models
1 Starter motor battery terminal
2 Starter solenoid terminal

8.14 Starter motor mounting bolts - V6 models

5 Installation is the reverse of removal. Tighten the starter mounting stud bolts securely.

2.3L engine

6 Remove the air filter housing (see Chapter 4, Section 9).
7 Remove the nut that secures the ground wire to the starter mounting stud and disconnect the ground wire from the stud.
8 Remove the mounting stud bolts and reposition the starter to allow access to disconnect the starter motor solenoid wire and the starter motor battery cable.
9 Remove the starter from the vehicle.
10 Installation is the reverse of removal. Tighten the starter mounting stud bolts securely.

V6 models

11 Remove the air filter housing (see Chapter 4).
12 Disconnect the transaxle shift cable from the transaxle shift lever (see Chapter 7A).
13 Remove the starter motor solenoid wire nut and the starter motor battery cable nut, then disconnect both wires from the starter motor (see illustration).
14 Remove the starter motor mounting fasteners (see illustration) and remove the starter motor.
15 Installation is the reverse of removal. Tighten the starter mounting stud bolts securely.

Notes

Chapter 6
Emissions and engine control systems

Contents

	Section
Accelerator Pedal Position (APP) sensor - replacement	4
Camshaft Position (CMP) sensor - replacement	5
Catalytic converter - replacement	21
Crankshaft Position (CKP) sensor - replacement	6
Cylinder Head Temperature (CHT) sensor - replacement	7
Engine Coolant Temperature (ECT) sensor - replacement	8
Engine Oil Pressure (EOP) sensor/oil pressure switch - replacement	9
Evaporative Emissions Control (EVAP) system - component replacement	22
Fuel Rail Pressure (FRP) sensor - replacement	23
General information	1
Intake Air Temperature (IAT) sensor	13
Knock sensors - replacement	10
Manifold Absolute Pressure Temperature (MAPT) sensor - replacement	11
Mass Air Flow/Intake Air Temperature (MAF/IAT) sensor - replacement	12
Obtaining and clearing Diagnostic Trouble Codes (DTCs)	3
On Board Diagnosis (OBD) system	2
Oxygen sensors - replacement	14
Positive Crankcase Ventilation (PCV) valve - replacement	24
Powertrain Control Module (PCM) - removal and installation	20
Throttle Position (TP) sensor - replacement	15
Transmission Range (TR) sensor - removal and installation	16
Turbine Shaft Speed (TSS) sensor - replacement	17
Turbocharger Boost Pressure (TCBP)/Charge Air Cooler Temperature (CACT) sensor - replacement	18
Turbocharger bypass valve - replacement	19
Variable Valve Timing (VVT) system - component replacement	25

Specifications

Torque specifications

Note: *One foot-pound (ft-lb) of torque is equivalent to 12 inch-pounds (in-lbs) of torque. Torque values below approximately 15 ft-lbs are expressed in inch-pounds, since most foot-pound torque wrenches are not accurate at these smaller values.*

	Ft-lbs (unless otherwise indicated)	Nm
Cylinder Head Temperature (CHT) sensor		
Four-cylinder engine	97 in-lbs	11
V6 engines	89 in-lbs	10
Engine Oil Pressure (EOP) sensor		
Step 1	133 in-lbs	15
Step 2	Tighten an additional 180-degrees	
Engine oil pressure switch		
Step 1	124 in-lbs	14
Step 2	Tighten an additional 180-degrees	
Fuel Rail Pressure (FRP) sensor (turbocharged models)		
2.0L engine		
Step 1	53 in-lbs	6
Step 2	Tighten an additional 5-degrees	
Step 3	Loosen 90-degrees	
Step 4	53 in-lbs	6
Step 5	Tighten an additional 21-degrees	
2.3L and 3.5L turbocharged engines	24	33
Knock sensor mounting bolt		
Four-cylinder engines	177 in-lbs	20
V6 engines	97 in-lbs	11
Oil control solenoid bolt		
Four-cylinder engines	89 in-lbs	10
Turbocharged V6 engine	97 in-lbs	11
Non-turbocharged V6 engine		
Step 1	71 in-lbs	8
Step 2	Tighten an additional 20-degrees	
Oxygen sensor/catalyst monitor sensor	35	48
Catalytic converter-to-exhaust manifold nuts		
Step 1	30	40
Step 2	Repeat Step 1	

Chapter 6 Emissions and engine control systems

1 General information

1 To prevent pollution of the atmosphere from incompletely burned and evaporating gases, and to maintain good driveability and fuel economy, a number of emission control systems are incorporated. They include the:

Catalytic converter

2 A catalytic converter is an emission control device in the exhaust system that reduces certain pollutants in the exhaust gas stream. There are two types of converters: oxidation converters and reduction converters.

3 Oxidation converters contain a monolithic substrate (a ceramic honeycomb) coated with the semi-precious metals platinum and palladium. An oxidation catalyst reduces unburned hydrocarbons (HC) and carbon monoxide (CO) by adding oxygen to the exhaust stream as it passes through the substrate, which, in the presence of high temperature and the catalyst materials, converts the HC and CO to water vapor (H_2O) and carbon dioxide (CO_2).

4 Reduction converters contain a monolithic substrate coated with platinum and rhodium. A reduction catalyst reduces oxides of nitrogen (NOx) by removing oxygen, which in the presence of high temperature and the catalyst material produces nitrogen (N) and carbon dioxide (CO_2).

5 Catalytic converters that combine both types of catalysts in one assembly are known as three-way catalysts or TWCs. A TWC can reduce all three pollutants.

Evaporative Emissions Control (EVAP) system

6 The Evaporative Emissions Control (EVAP) system prevents fuel system vapors (which contain unburned hydrocarbons) from escaping into the atmosphere. On warm days, vapors trapped inside the fuel tank expand until the pressure reaches a certain threshold. Then the fuel vapors are routed from the fuel tank through the fuel vapor vent valve and the fuel vapor control valve to the EVAP canister, where they're stored temporarily until the next time the vehicle is operated. When the conditions are right (engine warmed up, vehicle up to speed, moderate or heavy load on the engine, etc.) the PCM opens the canister purge valve, which allows fuel vapors to be drawn from the canister into the intake manifold. Once in the intake manifold, the fuel vapors mix with incoming air before being drawn through the intake ports into the combustion chambers where they're burned up with the rest of the air/fuel mixture. The EVAP system is complex and virtually impossible to troubleshoot without the right tools and training.

Powertrain Control Module (PCM)

7 The Powertrain Control Module (PCM) is the brain of the engine management system. It also controls a wide variety of other vehicle systems. In order to program the new PCM, a technician equipped with the proper scan tool needs the vehicle as well as the new PCM. If you're planning to replace the PCM with a new one, there is no point in trying to do so at home because you won't be able to program it yourself.

Positive Crankcase Ventilation (PCV) system

8 The Positive Crankcase Ventilation (PCV) system reduces hydrocarbon emissions by scavenging crankcase vapors, which are rich in unburned hydrocarbons. A PCV valve or orifice regulates the flow of gases into the intake manifold in proportion to the amount of intake vacuum available.

9 The PCV system generally consists of the fresh air inlet hose, the PCV valve or orifice and the crankcase ventilation hose (or PCV hose). The fresh air inlet hose connects the air intake duct to a pipe on the valve cover. The crankcase ventilation hose (or PCV hose) connects the PCV valve or orifice in the valve cover or crankcase vent oil separator to the intake manifold.

Variable Valve Timing (VVT) system

10 The VVT system controls intake valve timing (turbocharged V6 engines) or intake and exhaust valve timing (four-cylinder and non-turbocharged V6 engines) to increase engine torque in the low and mid-speed range and to increase horsepower in the high-speed range. The VVT system consists of the PCM-controlled Oil Control Valve(s) (OCV), mounted on top of the cylinder head, and the VVT actuator(s), which is mounted on the front end of the camshafts. The PCM-controlled OCV varies the oil pressure in the VVT actuator(s), which continually varies the timing of the camshaft(s).

2 On Board Diagnosis (OBD) system

General description

1 All models are equipped with the second generation OBD-II system. This system consists of an on-board computer known as the Powertrain Control Module (PCM), and information sensors, which monitor various functions of the engine and send data to the PCM. This system incorporates a series of diagnostic monitors that detect and identify fuel injection and emissions control system faults and store the information in the computer memory. This system also tests sensors and output actuators, diagnoses drive cycles, freezes data and clears codes.

2 The PCM is the brain of the electronically controlled fuel and emissions system. It receives data from a number of sensors and other electronic components (switches, relays, etc.). Based on the information it receives, the PCM generates output signals to control various relays, solenoids (fuel injectors) and other actuators. The PCM is specifically calibrated to optimize the emissions, fuel economy and driveability of the vehicle.

3 It isn't a good idea to attempt diagnosis or replacement of the PCM or emission control components at home while the vehicle is under warranty. Because of a Federally mandated warranty which covers the emissions system components and because any owner-induced damage to the PCM, the sensors and/or the control devices may void this warranty, take the vehicle to a dealer service department if the PCM or a system component malfunctions.

Scan tool information

4 Because extracting the Diagnostic Trouble Codes (DTCs) from an engine management system is now the first step in troubleshooting many computer-controlled systems and components, a code reader, at the very least, will be required (see illustration). More powerful scan tools can also perform many of the diagnostics once associated with expensive factory scan tools (see illustration). If

2.4a Simple code readers are an economical way to extract trouble codes when the CHECK ENGINE light comes on

2.4b Hand-held scan tools like these can extract computer codes and also perform diagnostics

Chapter 6 Emissions and engine control systems

Information Sensors

Accelerator Pedal Position (APP) sensor - as you press the accelerator pedal, the APP sensor alters its voltage signal to the PCM in proportion to the angle of the pedal, and the PCM commands a motor inside the throttle body to open or close the throttle plate accordingly

Camshaft Position (CMP) sensor - produces a signal that the PCM uses to identify the number 1 cylinder and to time the firing sequence of the fuel injectors

Crankshaft Position (CKP) sensor - produces a signal that the PCM uses to calculate engine speed and crankshaft position, which enables it to synchronize ignition timing with fuel injector timing, and to detect misfires

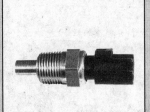

Engine Coolant Temperature (ECT) sensor - a thermistor (temperature-sensitive variable resistor) that sends a voltage signal to the PCM, which uses this data to determine the temperature of the engine coolant

Fuel tank pressure sensor - measures the fuel tank pressure and controls fuel tank pressure by signaling the EVAP system to purge the fuel tank vapors when the pressure becomes excessive

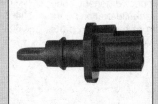

Intake Air Temperature (IAT) sensor - monitors the temperature of the air entering the engine and sends a signal to the PCM to determine injector pulse-width (the duration of each injector's on-time) and to adjust spark timing (to prevent spark knock)

Knock sensor - a piezoelectric crystal that oscillates in proportion to engine vibration which produces a voltage output that is monitored by the PCM. This retards the ignition timing when the oscillation exceeds a certain threshold

Manifold Absolute Pressure (MAP) sensor - monitors the pressure or vacuum inside the intake manifold. The PCM uses this data to determine engine load so that it can alter the ignition advance and fuel enrichment

Mass Air Flow (MAF) sensor - measures the amount of intake air drawn into the engine. It uses a hot-wire sensing element to measure the amount of air entering the engine

Oxygen sensors - generates a small variable voltage signal in proportion to the difference between the oxygen content in the exhaust stream and the oxygen content in the ambient air. The PCM uses this information to maintain the proper air/fuel ratio. A second oxygen sensor monitors the efficiency of the catalytic converter

Throttle Position (TP) sensor - a potentiometer that generates a voltage signal that varies in relation to the opening angle of the throttle plate inside the throttle body. Works with the PCM and other sensors to calculate injector pulse width (the duration of each injector's on-time)

Photos courtesy of Wells Manufacturing, except APP and MAF sensors.

Chapter 6 Emissions and engine control systems

3.3 The Data Link Connector (DLC) is located below the steering column

3 Obtaining and clearing Diagnostic Trouble Codes (DTCs)

1 All models covered by this manual are equipped with on-board diagnostics. When the PCM recognizes a malfunction in a monitored emission or engine control system, component or circuit, it turns on the Malfunction Indicator Light (MIL) on the dash. The PCM will continue to display the MIL until the problem is fixed and the Diagnostic Trouble Code (DTC) is cleared from the PCM's memory. You'll need a scan tool to access any DTCs stored in the PCM.

2 Before outputting any DTCs stored in the PCM, thoroughly inspect ALL electrical connectors and hoses. Make sure that all electrical connections are tight, clean and free of corrosion. And make sure that all hoses are correctly connected, fit tightly and are in good condition (no cracks or tears).

Accessing the DTCs

3 The Diagnostic Trouble Codes (DTCs) can only be accessed with a code reader or scan tool. Professional scan tools are expensive, but relatively inexpensive generic code readers or scan tools (see illustrations 2.4a and 2.4b) are available at most auto parts stores. Simply plug the connector of the scan tool into the diagnostic connector (see illustration). Then follow the instructions included with the scan tool to extract the DTCs.

4 Once you have outputted all of the stored DTCs, look them up on the accompanying DTC chart.

5 After troubleshooting the source of each DTC, make any necessary repairs or replace the defective component(s).

Clearing the DTCs

6 Clear the DTCs with the code reader or scan tool in accordance with the instructions provided by the tool's manufacturer.

Diagnostic Trouble Codes

7 The accompanying tables are a list of the Diagnostic Trouble Codes (DTCs) that can be accessed by a do-it-yourselfer working at home (there are many, many more DTCs available to professional mechanics with proprietary scan tools and software, but those codes cannot be accessed by a generic scan tool). This list of DTCs may not be complete. If, after you have checked and repaired the connectors, wire harness and vacuum hoses (if applicable) for an emission-related system, component or circuit, the problem persists, have the vehicle checked by a dealer service department or other qualified repair shop.

you're planning to obtain a generic scan tool for your vehicle, make sure that it's compatible with OBD-II systems. If you don't plan to purchase a code reader or scan tool and don't have access to one, you can have the codes extracted by a dealer service department or an independent repair shop.
Note: *Some auto parts stores even provide this service.*

Diagnostic Trouble Codes

Code	Probable cause
P0001	Fuel volume regulator control circuit open
P0003	Fuel volume regulator control circuit low
P0004	Fuel volume regulator control circuit high
P000A	Intake camshaft position slow response (Bank 1)
P000B	Exhaust camshaft position slow response (Bank 1)
P0010	Intake camshaft position actuator, open circuit (Bank 1)
P0011	Intake camshaft position timing over-advanced (Bank 1)
P0012	Intake camshaft position timing, over-retarded (Bank 1)
P0013	Exhaust camshaft position actuator, open circuit (Bank 1)
P0014	Exhaust camshaft position timing over-advanced (Bank 1)
P0015	Exhaust camshaft position timing, over-retarded (Bank 1)
P0016	Crankshaft position-to-camshaft position correlation (Bank 1 Sensor A)
P0017	Crankshaft position-to-camshaft position correlation (Bank 1 Sensor B)

Chapter 6 Emissions and engine control systems

Code	Probable cause
P0018	Crankshaft position-to-camshaft position correlation (Bank 2 Sensor A)
P0019	Crankshaft position-to-camshaft position correlation (Bank 2 Sensor B)
P0020	Intake camshaft position actuator, open circuit (Bank 2)
P0021	Intake camshaft position timing over-advanced (Bank 2)
P0022	Intake camshaft position timing over-retarded (Bank 2)
P0023	Exhaust camshaft position actuator, open circuit (Bank 2)
P0024	Exhaust camshaft position timing over-advanced (Bank 2)
P0025	Exhaust camshaft position timing over-retarded (Bank 2)
P0030	Oxygen sensor heater control circuit (Bank 1, Sensor 1)
P0040	Oxygen sensor signals swapped (Bank 1, Sensor 1/Bank 2, Sensor 1)
P0041	Oxygen sensor signals swapped (Bank 1, Sensor 2/Bank 2, Sensor 2)
P0050	Oxygen sensor heater control circuit (Bank 2, Sensor 1)
P0053	Oxygen sensor heater resistance (Bank 1, Sensor 1)
P0054	Oxygen sensor heater resistance (Bank 1, Sensor 2)
P0055	Oxygen sensor heater resistance (Bank 1, Sensor 3)
P0059	Oxygen sensor heater resistance (Bank 2, Sensor 1)
P0060	Oxygen sensor heater resistance (Bank 2, Sensor 2)
P0068	Manifold Absolute Pressure (MAP) sensor/Mass Air Flow (MAF) sensor-to-throttle position correlation
P0097	Intake Air Temperature (IAT) sensor 2 circuit, low voltage
P0098	Intake Air Temperature (IAT) sensor 2 circuit, high voltage
P00C1	Turbocharger bypass valve B control circuit low
P00C2	Turbocharger bypass valve B control circuit high
P00DF	Charge air cooler coolant temperature sensor A circuit range/performance
P00E0	Charge air cooler coolant temperature sensor A circuit low
P00E1	Charge air cooler coolant temperature sensor A circuit high
P00E2	Charge air cooler coolant temperature sensor A intermittent
P0102	Mass or volume air flow A circuit, low voltage
P0103	Mass or volume air flow A circuit, high voltage
P0104	Mass Air Flow (MAF) sensor A circuit, intermittent or erratic signal
P0106	Manifold Absolute Pressure (MAP) sensor circuit, range or performance problem
P0107	Manifold Absolute Pressure (MAP) sensor circuit, low voltage

Diagnostic Trouble Codes (continued)

Code	Probable cause
P0108	Manifold Absolute Pressure (MAP) sensor circuit, high voltage
P0109	Manifold Absolute Pressure (MAP) sensor circuit, intermittent signal
P0111	Intake Air Temperature (IAT) sensor circuit, range or performance problem
P0112	Intake Air Temperature (IAT) sensor circuit, low voltage
P0113	Intake Air Temperature (IAT) sensor circuit, high voltage
P0114	Intake Air Temperature (IAT) sensor circuit, intermittent or erratic signal
P0116	Engine Coolant Temperature (ECT) sensor circuit, range or performance problem
P0117	Engine Coolant Temperature (ECT) sensor circuit, low voltage
P0118	Engine Coolant Temperature (ECT) sensor circuit, high voltage
P0119	Engine Coolant Temperature (ECT) sensor circuit, intermittent or erratic signal
P0121	Throttle Position (TP) sensor A circuit, range or performance problem
P0122	Throttle Position (TP) sensor A circuit, low voltage
P0123	Throttle Position (TP) sensor A circuit, high voltage
P0124	Throttle Position (TP) sensor A intermittent
P0125	Insufficient coolant temperature for closed loop fuel control
P0128	Coolant temperature below coolant thermostat's regulating temperature
P012B	Turbocharger inlet pressure sensor A circuit, range/performance
P012C	Turbocharger inlet pressure sensor A circuit, low
P012D	Turbocharger inlet pressure sensor A circuit, high
P012E	Turbocharger inlet pressure sensor A intermittent
P0130	Oxygen sensor circuit malfunction (Bank 1, Sensor 1)
P0132	Oxygen sensor circuit, high voltage (Bank 1, Sensor 1)
P0133	Oxygen sensor circuit, slow response (Bank 1, Sensor 1)
P0134	Oxygen sensor circuit, no activity detected (Bank 1, Sensor 1)
P0135	Oxygen sensor heater circuit malfunction (Bank 1, Sensor 1)
P0138	Oxygen sensor circuit, high voltage (Bank 1, Sensor 2)
P0139	Oxygen sensor circuit, slow response (Bank 1, Sensor 2)
P013A	Oxygen sensor slow response, rich to lean (Bank 1, Sensor 2)
P013C	Oxygen sensor slow response, rich to lean (Bank 2, Sensor 2)
P013E	Oxygen sensor delayed response, rich to lean (Bank 1, Sensor 2)

Chapter 6 Emissions and engine control systems

Code	Probable cause
P0144	Oxygen sensor circuit, high voltage (Bank 1, Sensor 3)
P0147	Oxygen sensor heater circuit malfunction (Bank 1, Sensor 3)
P0148	Fuel delivery error
P014A	Oxygen sensor delayed response, rich to lean (Bank 2, Sensor 2)
P0150	Oxygen sensor circuit malfunction (Bank 2, Sensor 1)
P0152	Oxygen sensor circuit, high voltage (Bank 2, Sensor 1)
P0153	Oxygen sensor circuit, slow response (Bank 2, Sensor 1)
P0154	Oxygen sensor circuit, no activity detected (Bank 2, Sensor 1)
P0155	Oxygen sensor heater circuit malfunction (Bank 2, Sensor 1)
P0158	Oxygen sensor circuit, high voltage (Bank 2, Sensor 2)
P0159	Oxygen sensor circuit, slow response (Bank 2, Sensor 2)
P0161	Oxygen sensor heater circuit malfunction (Bank 2, Sensor 2)
P0171	System too lean (Bank 1)
P0172	System too rich (Bank 1)
P0174	System too lean (Bank 2)
P0175	System too rich (Bank 2)
P017C	Cylinder head temperature sensor circuit, low
P017D	Cylinder head temperature sensor circuit, high
P017E	Cylinder head temperature sensor intermittent
P0180	Fuel temperature sensor circuit malfunction
P0181	Fuel temperature sensor circuit, range or performance problem
P0182	Fuel temperature sensor circuit, low voltage
P0183	Fuel temperature sensor circuit, high voltage
P0191	Fuel rail pressure sensor circuit, range or performance problem
P0192	Fuel rail pressure sensor circuit, low voltage
P0193	Fuel rail pressure sensor circuit, high voltage
P0196	Engine Oil Temperature (EOT) sensor circuit, range or performance problem
P0197	Engine Oil Temperature (EOT) sensor circuit, low voltage
P0198	Engine Oil Temperature (EOT) sensor circuit, high voltage
P0201	Injector open circuit, cylinder 1
P0202	Injector open circuit, cylinder 2

Diagnostic Trouble Codes (continued)

Code	Probable cause
P0203	Injector open circuit, cylinder 3
P0204	Injector open circuit, cylinder 4
P0205	Injector open circuit, cylinder 5
P0206	Injector open circuit, cylinder 6
P0217	Engine coolant over-temperature condition
P0218	Transaxle fluid temperature over-temperature condition
P0219	Engine over-speed condition
P0221	Throttle Position (TP) sensor circuit, range or performance problem
P0222	Throttle Position (TP) sensor circuit, low voltage
P0223	Throttle Position (TP) sensor circuit, high voltage
P0230	Fuel pump primary circuit malfunction
P0231	Fuel pump secondary circuit, low voltage
P0232	Fuel pump secondary circuit, high voltage
P0234	Turbocharger A overboost condition
P0236	Turbocharger boost sensor A circuit, range/performance
P0237	Turbocharger boost sensor A circuit, low
P0238	Turbocharger boost sensor A circuit, high
P0243	Turbocharger wastegate actuator A
P0244	Turbocharger wastegate actuator A circuit, range/performance
P0245	Turbocharger wastegate actuator A circuit, low
P0246	Turbocharger wastegate actuator A circuit, high
P0247	Turbocharger wastegate actuator B
P0248	Turbocharger wastegate actuator B circuit, range/performance
P0249	Turbocharger wastegate actuator B circuit, low
P0250	Turbocharger wastegate actuator B circuit, high
P025A	Fuel pump module control circuit open
P025B	Fuel pump module control circuit range or performance problem
P0297	Vehicle over-speed condition
P0298	Engine oil over-temperature condition
P0300	Random misfire detected

Chapter 6 Emissions and engine control systems

Code	Probable cause
P0301	Cylinder 1 misfire
P0302	Cylinder 2 misfire
P0303	Cylinder 3 misfire
P0304	Cylinder 4 misfire
P0305	Cylinder 5 misfire
P0306	Cylinder 6 misfire
P0315	Crankshaft position system variation not learned
P0316	Misfire detected on start-up (first 1000 revolutions)
P0320	Ignition/distributor engine speed input circuit
P0325	Knock sensor 1 circuit malfunction (Bank 1)
P0326	Knock sensor 1 circuit, range or performance problem (Bank 1)
P0330	Knock sensor 2 circuit malfunction (Bank 2)
P0331	Knock sensor 2 circuit, range or performance problem (Bank 2)
P0340	Camshaft Position (CMP) sensor circuit malfunction (Bank 1 or single sensor)
P0341	Camshaft Position (CMP) sensor circuit, range or performance problem (Bank 1 or single sensor)
P0344	Camshaft Position (CMP) sensor circuit, intermittent signal (Bank 1 or single sensor)
P0345	Camshaft Position (CMP) sensor circuit malfunction (Bank 2)
P0346	Camshaft Position (CMP) sensor circuit, range or performance problem (Bank 2)
P0349	Camshaft Position (CMP) sensor circuit, intermittent signal (Bank 2)
P0350	Ignition coil primary/secondary circuit malfunction
P0351	Ignition coil A primary/secondary circuit malfunction
P0352	Ignition coil B primary/secondary circuit malfunction
P0353	Ignition coil C primary/secondary circuit malfunction
P0354	Ignition coil D primary/secondary circuit malfunction
P0355	Ignition coil E primary/secondary circuit malfunction
P0356	Ignition coil F primary/secondary circuit malfunction
P0420	Catalyst system efficiency below threshold (Bank 1)
P0430	Catalyst system efficiency below threshold (Bank 2)
P0442	Evaporative Emission (EVAP) system, small leak detected
P0443	Evaporative Emission (EVAP) system, purge control valve circuit malfunction
P0446	Evaporative Emission (EVAP) system, vent control circuit malfunction

Diagnostic Trouble Codes (continued)

Code	Probable cause
P0451	Evaporative Emission (EVAP) system, pressure sensor range or performance problem
P0452	Evaporative Emission (EVAP) system, pressure sensor, low voltage
P0453	Evaporative Emission (EVAP) system, pressure sensor, high voltage
P0454	Evaporative Emission (EVAP) system, pressure sensor, intermittent signal
P0455	Evaporative Emission (EVAP) system, gross leak detected/no flow
P0456	Evaporative Emission (EVAP) system, very small leak detected
P0457	Evaporative Emission (EVAP) system, leak detected (fuel cap loose or off)
P0460	Fuel level sensor circuit malfunction
P0461	Fuel level sensor circuit, range or performance problem
P0462	Fuel level sensor circuit, low voltage
P0463	Fuel level sensor circuit, high voltage
P0480	Fan 1 control circuit malfunction
P0481	Fan 2 control circuit malfunction
P0483	Fan performance
P0491	Secondary Air Injection (AIR) system, insufficient flow (Bank 1)
P0500	Vehicle Speed Sensor (VSS)
P0503	Vehicle Speed Sensor (VSS), intermittent, erratic or high signal
P0505	Idle Air Control (IAC) system
P0506	Idle Air Control (IAC) system, rpm lower than expected
P0507	Idle Air Control (IAC) system, rpm higher than expected
P050A	Cold start idle air control performance
P050B	Cold start ignition timing performance
P050E	Cold start engine exhaust temperature out of range
P0511	Idle Air Control (IAC) system circuit malfunction
P0512	Starter request circuit malfunction
P0528	Fan speed sensor circuit, no signal
P052A	Cold start camshaft position timing over-advanced (Bank 1)
P052B	Cold start camshaft position timing over-retarded (Bank 1)
P052C	Cold start camshaft position timing over-advanced (Bank 2)
P052D	Cold start camshaft position timing over-retarded (Bank 2)

Code	Probable cause
P0532	Air conditioning refrigerant pressure sensor circuit, low voltage
P0533	Air conditioning refrigerant pressure sensor circuit, high voltage
P0534	Air conditioning refrigerant charge loss
P0537	Air conditioning evaporator temperature sensor circuit, low voltage
P0538	A/C evaporator temperature sensor circuit, high voltage
P053A	Positive Crankcase Ventilation (PCV) heater control circuit open
P0552	Power Steering Pressure (PSP) sensor circuit, low voltage
P0553	Power Steering Pressure (PSP) sensor circuit, high voltage
P0562	System voltage low
P0563	System voltage high
P0571	Brake switch circuit malfunction
P0572	Brake switch circuit, low voltage
P0573	Brake switch circuit, high voltage
P0579	Cruise control multifunction input circuit, ranger or performance problem
P0581	Cruise control multifunction input circuit, high voltage
P0600	Serial communication link
P0601	Powertrain Control Module (PCM), memory checksum error
P0602	Powertrain Control Module (PCM) programming error
P0603	Powertrain Control Module (PCM), Keep Alive Memory (KAM) error
P0604	Powertrain Control Module (PCM), Random Access Memory (RAM) error
P0605	Powertrain Control Module (PCM), Read Only Memory (ROM) error
P0606	Powertrain Control Module (PCM) processor
P0607	Powertrain Control Module (PCM) performance
P060A	Internal control module monitoring processor performance
P060B	Internal control module analog/digital processing performance
P060C	Internal control module main processor performance
P060D	Internal control module accelerator pedal position performance
P0610	Powertrain Control Module (PCM) options error
P061B	Internal control module torque calculation performance
P061C	Internal control module engine rpm performance
P061D	Internal control module engine air mass performance

Diagnostic Trouble Codes (continued)

Code	Probable cause
P061F	Internal control module throttle actuator controller performance
P0620	Alternator control circuit malfunction
P0622	Alternator field terminal, circuit malfunction
P0625	Alternator field terminal, low circuit voltage
P0626	Alternator field terminal, high circuit voltage
P0627	Fuel pump, open control circuit
P062C	Internal control module vehicle speed performance
P062F	Internal control module EEPROM error
P0642	Sensor reference voltage (VREF) circuit below VREF minimum voltage
P0643	Sensor reference voltage (VREF) circuit, high voltage
P0645	Air conditioning clutch relay control circuit malfunction
P064D	Internal control module oxygen sensor processor performance (Bank 1)
P064E	Internal control module oxygen sensor processor performance (Bank 2)
P0657	Actuator supply voltage, open circuit
P065B	Alternator control circuit range or performance problem
P065B	Alternator control circuit range or performance problem
P0660	Intake Manifold Tuning Valve (IMTV) control circuit, open circuit (Bank 1)
P0663	Intake Manifold Tuning Valve (IMTV) control circuit, open circuit (Bank 2)
P0685	Powertrain Control Module (PCM) power relay control circuit open
P0689	Powertrain Control Module (PCM) power relay sense circuit, low voltage
P0690	Powertrain Control Module (PCM) power relay sense circuit, high voltage
P06B8	Internal control module Non-volatile random access memory (NVRAM) error
P0703	Brake switch input circuit malfunction
P0704	Clutch switch input circuit malfunction
P0705	Transmission Range (TR) sensor circuit (PRNDL) input problem
P0706	Transmission Range (TR) sensor circuit, range or performance problem
P0707	Transmission Range (TR) sensor circuit, low voltage
P0708	Transmission range sensor circuit, high voltage
P0711	Transmission fluid temperature sensor circuit, range or performance problem
P0712	Transmission fluid temperature sensor circuit, low input

Chapter 6 Emissions and engine control systems

Code	Probable cause
P0713	Transmission fluid temperature sensor circuit, high input
P0715	Input/turbine speed sensor circuit malfunction
P0716	Input/turbine speed sensor circuit, range or performance problem
P0717	Input/turbine speed sensor circuit, no signal
P0720	Output Shaft Speed (OSS) sensor circuit malfunction
P0721	Output Shaft Speed (OSS) sensor circuit, range or performance problem
P0722	No signal from Output Shaft Speed (OSS) sensor
P0723	Output Shaft Speed (OSS) sensor circuit, intermittent signal
P0729	Gear 6 incorrect ratio
P072C	Stuck in Gear 1
P072E	Stuck in Gear 3
P072F	Stuck in Gear 4
P0730	Incorrect gear ratio
P0731	Incorrect gear ratio, first gear
P0732	Incorrect gear ratio, second gear
P0733	Incorrect gear ratio, third gear
P0734	Incorrect gear ratio, fourth gear
P0735	Incorrect gear ratio, fifth gear
P0736	Incorrect gear ratio, reverse gear
P0740	Torque converter clutch solenoid circuit, open
P0741	Torque converter clutch, circuit performance problem or stuck in Off position
P0742	Torque converter clutch circuit, stuck in On position
P0743	Torque converter clutch solenoid circuit, electrical malfunction
P0744	Torque converter clutch circuit, intermittent
P0745	Pressure control solenoid A electrical malfunction
P0748	Pressure control solenoid malfunction
P0750	Shift solenoid A, performance problem
P0751	Shift solenoid A, performance problem or stuck in Off position
P0752	Shift solenoid A, stuck in On position
P0753	Shift solenoid A, electrical problem
P0755	Shift solenoid B, performance problem

Diagnostic Trouble Codes (continued)

Code	Probable cause
P0756	Shift solenoid B, performance problem or stuck in Off position
P0757	Shift solenoid B, stuck in On position
P0758	Shift solenoid B, electrical problem
P0760	Shift solenoid C, performance problem
P0761	Shift solenoid C, performance problem or stuck in Off position
P0762	Shift solenoid C, stuck in On position
P0763	Shift solenoid C, electrical problem
P0765	Shift solenoid D, performance problem
P0766	Shift solenoid D, performance problem or stuck in Off position
P0767	Shift solenoid D, stuck in On position
P0768	Shift solenoid D, electrical problem
P0770	Shift solenoid E, performance problem
P0771	Shift solenoid E, performance problem or stuck in Off position
P0772	Shift solenoid E, stuck in On position
P0773	Shift solenoid E, electrical problem
P0774	Shift solenoid E, intermittent electrical problem
P0777	Pressure control solenoid "B" stuck On
P0778	Pressure control solenoid "B" electrical
P0780	Shift malfunction
P0791	Intermediate shaft speed sensor circuit malfunction
P0812	Reverse input circuit malfunction
P0815	Upshift switch circuit malfunction
P0816	Downshift switch circuit malfunction
P0817	Starter disable circuit malfunction
P0830	Clutch pedal switch circuit malfunction
P0840	Transmission fluid pressure sensor circuit malfunction
P0841	Transmission fluid pressure sensor/switch "A" circuit range/performance problem
P0850	Park/neutral switch input circuit malfunction
P0882	Transmission control module (TCM) power input signal low
P0894	Transmission component slipping

Chapter 6 Emissions and engine control systems

Code	Probable cause
P0960	Pressure control (PC) solenoid A - control circuit open
P0961	Pressure control (PC) solenoid A - control circuit range/performance problem
P0962	Pressure control (PC) solenoid A - control circuit low
P0963	Pressure control (PC) solenoid A - control circuit high
P0973	Shift solenoid (SS) A - control circuit low
P0974	Shift solenoid (SS) A - control circuit high
P0976	Shift solenoid (SS) B - control circuit low
P0977	Shift solenoid (SS) B - control circuit high
P0978	Shift solenoid (SS) C - control circuit range/performance problem
P0979	Shift solenoid (SS) C - control circuit low
P0980	Shift solenoid (SS) C - control circuit high
P0981	Shift solenoid (SS) D - control circuit range/performance problem
P0982	Shift solenoid (SS) D - control circuit low
P0983	Shift solenoid (SS) D - control circuit high
P0984	Shift solenoid (SS) E - control circuit range/performance problem
P0985	Shift solenoid (SS) E - control circuit low
P0986	Shift solenoid (SS) E - control circuit high
P0997	Shift solenoid (SS) F - control circuit range/performance problem
P0998	Shift solenoid (SS) F - control circuit low
P0999	Shift solenoid (SS) F - control circuit high
UXXXX	Network connection error DTC

4 Accelerator Pedal Position (APP) sensor - replacement

1 Disconnect the electrical connector from the upper end of the APP sensor assembly (see illustration).
2 Remove the accelerator pedal/APP sensor assembly mounting bolts and remove the assembly.
3 Installation is the reverse of removal.

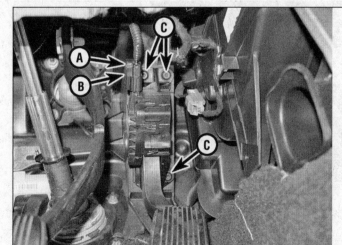

4.1 Slide the connector lock (A) up and depress the release tab (B) to disconnect the electrical connector, then remove the APP sensor mounting bolts (C)

Chapter 6 Emissions and engine control systems

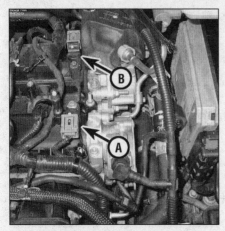

5.3 The intake CMP sensor (A) and the exhaust CMP sensor (B) are located on the top of the valve cover (four-cylinder engine)

5.9a Identifying the intake (A) and exhaust (B) CMP sensors on V6 engines - front cylinder bank

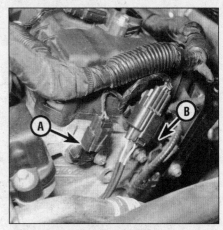

5.9b Identifying the intake (A) and exhaust (B) CMP sensors on V6 engines - rear cylinder bank

5 Camshaft Position (CMP) sensor – replacement

1 Disconnect the cable from the negative battery terminal (see Chapter 5). Remove the engine cover.

Four-cylinder models

Note: *The CMP sensors are located at the top corners of the valve cover, above the left end of the intake and exhaust camshaft.*

2 Remove the intake air duct running across the top of the engine to access the exhaust CMP sensor.
3 Disconnect the CMP sensor electrical connector(s) (see illustration).
4 Remove the CMP sensor mounting bolt and remove the sensor from the valve cover.
5 Inspect the condition of the CMP sensor O-ring. If it's cracked, torn or deteriorated, replace it.
6 Installation is the reverse of removal.

V6 models

Note: *There are four CMP sensors, two on the end of each cylinder head.*

7 Remove the air filter housing and air intake duct (see Chapter 4).
8 Disconnect the CMP sensor electrical connector.
9 Remove the CMP sensor mounting bolt and remove the CMP sensor from the cylinder head (see illustrations). The intake CMP sensor is near the intake side of the engine and the exhaust CMP sensor is near the exhaust side of the engine.
10 Remove the CMP sensor O-ring and inspect its condition. If it's cracked, torn or otherwise deteriorated, replace it.
11 Lubricate the sensor O-ring with clean engine oil. Installation is otherwise the reverse of removal.

6 Crankshaft Position (CKP) sensor - replacement

Note: *Replacing the CKP sensor might set a Diagnostic Trouble Code (DTC). If it does, and you have a generic scan tool that can clear codes, erase the code and see if it reappears. If it does, drive the vehicle to a dealer service department to perform a Misfire Monitor Neutral Profile Correction procedure with a factory scan tool.*

1 Disconnect the cable from the negative battery terminal (see Chapter 5).
2 If you're working on a four-cylinder model, loosen the right front wheel lug nuts. Raise the front of the vehicle and place it securely on jackstands. Remove the right front wheel, then remove the inner fender splash shield (see Chapter 11). On all models, remove the engine under-cover.

Four-cylinder models

Removal

Note: *The CKP sensor is located at the lower front corner of the timing chain cover.*

3 Put the No. 1 cylinder at Top Dead Center (TDC) (see Chapter 2A).
4 Disconnect the electrical connector from the CKP sensor (see illustration).
5 Unscrew the CKP sensor mounting bolts and remove the sensor.

Installation

Note: *If you're installing a new CKP sensor, the new sensor comes with a special alignment jig (303-1521) that, according to the manufacturer, is available only with a new sensor. However, you might be able to find an aftermarket tool that does the same thing. If you removed the CKP sensor simply to access some other component, like the timing chain, use the alternate method of aligning the CKP sensor included here.*

6 Install the CKP sensor but don't tighten the sensor mounting bolts.
7 If you're installing a new CKP sensor, align the sensor with the special alignment jig (included with the new sensor) in accordance with the manufacturer's instructions. With the alignment jig in place, tighten the CKP sensor bolts securely, then remove the alignment jig.
8 If you're installing the old CKP sensor, look at the sensor trigger wheel, or timing plate, that's mounted on the backside of the crankshaft pulley. Note the small teeth that

6.4 CKP sensor details (four-cylinder models):

1 Electrical connector
2 Mounting bolts
3 CKP sensor

Chapter 6 Emissions and engine control systems

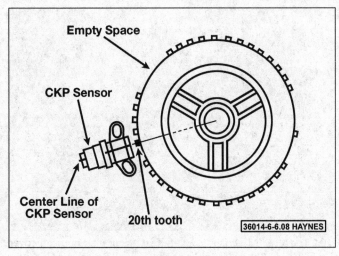

6.8 CKP sensor alignment details (four-cylinder models)

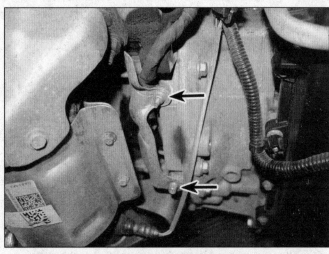

6.12 Remove the CKP heat shield mounting bolt, nut and heat shield (non-turbo model shown)

stick out from the circumference of the timing plate. There is a blank spot on the edge of the timing plate, where there are no teeth. From this blank area, count 20 teeth in a counter-clockwise direction, then use a straightedge to draw a straight line from the center of the 20th tooth through the center of the crankshaft pulley (see illustration). Position the centerline of the CKP sensor with the line that you made, then tighten the sensor bolts securely.

9 Installation is otherwise the reverse of removal.

10 After replacement, a scan tool must be used to perform the Misfire Monitor Neutral Profile Correction procedure.

V6 models

Note: *The CKP sensor is located on the left front side of the engine block, near the transaxle.*

11 On turbocharged models, remove the front cylinder bank turbocharger assembly (see Chapter 4, Section 14).

12 On all models, remove the CKP sensor heat shield mounting bolt and nut, then remove the heat shield (see illustration).

13 Remove the rubber grommet cover (see illustration).

14 Disconnect the CKP sensor electrical connector.

15 Remove the CKP sensor mounting bolt and remove the sensor.

16 Installation is the reverse of removal.

17 After replacement, a scan tool must be used to perform the Misfire Monitor Neutral Profile Correction procedure.

7 Cylinder Head Temperature (CHT) sensor - replacement

Warning: *Wait until the engine has cooled completely before beginning this procedure.*

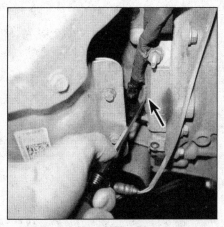

6.13 Remove the CKP rubber grommet cover (non-turbo model shown)

Four-cylinder models

1 Disconnect the cable from the negative battery terminal (see Chapter 5).

2 Remove the air filter outlet duct, then disconnect the charge air cooler inlet pipe (see Chapter 4).

3 Pull back the rubber weather cover from the CHT sensor electrical connector and disconnect the connector (see illustration).

4 Unscrew the CHT sensor from the cylinder head.

5 When installing the CHT sensor, tighten it to the torque listed in this Chapter's Specifications. Installation is the reverse of removal.

V6 models

Non-turbocharged models

Note: *The CHT sensor is screwed into the rear cylinder head below the lower intake manifold.*

6 Disconnect the cable from the negative battery terminal (see Chapter 5).

7 Remove the lower intake manifold (see

7.3 On four-cylinder models, the CHT sensor is located on the back side of the cylinder head

Chapter 2B, Section 5).

8 Disconnect the CHT sensor electrical connector.

9 Unscrew and remove the CHT sensor.

10 When installing the CHT sensor, tighten it to the torque listed in this Chapter's Specifications. Installation is otherwise the reverse of removal.

Turbocharged models

Note: *The CHT sensor is screwed into the end of the rear cylinder head near the thermostat.*

11 Disconnect the cable from the negative battery terminal (see Chapter 5, Section 3).

12 Remove the air filter outlet duct, then disconnect the charge air cooler inlet pipe.

13 Disconnect the CHT sensor electrical connector.

14 Unscrew and remove the CHT sensor.

15 When installing the CHT sensor, tighten it to the torque listed in this Chapter's Specifications. Installation is otherwise the reverse of removal.

Chapter 6 Emissions and engine control systems

10.4 The right hand knock sensor (A), must be installed with the wire end of the sensor pointing towards the 6 o'clock position and the left side knock sensor (B - not visible) must be installed with the wire end pointing towards the 3 o'clock position (2.0L four-cylinder model shown)

12.2 The MAF/IAT sensor is located on the air filter housing cover

8 Engine Coolant Temperature (ECT) sensor - replacement

Warning: *Wait until the engine is completely cool before beginning this procedure.*
Note: *The ECT sensor is used only on four-cylinder engines and is located near the driver's side of the cylinder head on the water outlet housing.*

1 Remove the engine cover.
2 Remove the air intake duct for access to the ECT (see Chapter 4, Section 9).
3 Locate and disconnect the ECT sensor connector.
Note: *Coolant will come out when you remove the ECT from the cylinder head. Have the new sensor ready to install to minimize the loss of coolant.*
4 Pull the retaining clip under the sensor and pull the sensor out of the cylinder head.
5 Installation is reverse of removal. Top off the coolant as necessary.

9 Engine Oil Pressure (EOP) sensor/oil pressure switch - replacement

1 Raise and support the front of the vehicle on jackstands.
2 Remove the vehicle under cover.
3 Remove the under-vehicle air duct (if equipped).
4 On four-cylinder engines, locate the EOP sensor on the oil filter housing adapter and disconnect the sensor electrical connector.
5 On V6 engines, locate the oil pressure switch on the engine block, near the oil filter housing adapter and disconnect the sensor electrical connector.
Note: *Place a drain pan under the work area to catch any oil that may drip when replacing the EOP sensor.*
6 On all engines, unscrew the EOP sensor/switch to remove from the oil filter housing adapter.
7 Installation is reverse of removal. Apply PTFE thread sealant to the threads before installation.
8 Tighten to the torque listed in this Chapter's Specifications.

10 Knock sensors - replacement

1 Disconnect the cable from the negative battery terminal (see Chapter 5).

Four-cylinder models
Note: *The knock sensors are located on the front side of the block, behind the intake manifold.*
2 On 2.0L models, remove the intake manifold (see Chapter 2A).
3 On all models, disconnect the electrical connectors from the knock sensors.
4 Remove the knock sensor bolt and remove the knock sensor (see illustration).
Note: *The right hand knock sensor must be installed with the wire end of the sensor pointing towards the 6 o'clock position and the left side knock sensor must be installed with the wire end pointing towards the 3 o'clock position.*
5 Installation is the reverse of removal. Tighten the knock sensor bolt to the torque listed in this Chapter's Specifications.

V6 models
Note: *The knock sensors are located in the valley between the cylinder heads.*
6 Remove the upper and lower intake manifolds (see Chapter 2B, Section 5).
7 Remove the thermostat housing and the coolant pipe in the engine block valley to access the knock sensors.
8 Disconnect the knock sensor harness electrical connector.
9 Remove the knock sensor bolts and remove the knock sensors with the knock sensor harness.
10 Installation is the reverse of removal.
11 Tighten the knock sensor bolts to the torque listed in this Chapter's Specifications.

11 Manifold Absolute Pressure Temperature (MAPT) sensor - replacement

1 Disconnect the cable from the negative battery terminal (see Chapter 5).
2 Disconnect the electrical connector from the MAPT sensor.
3 Remove the MAPT sensor retaining screw and remove the sensor from the intake manifold.
4 Check the condition of the sensor O-ring and replace as necessary.
5 Installation is the reverse of removal. Lubricate the O-ring with clean motor oil prior to installation.

12 Mass Air Flow/Intake Air Temperature (MAF/IAT) sensor - replacement

Note: *2.0L four-cylinder and 2015 and earlier non-turbocharged V6 engines are equipped with a combination MAF/IAT sensor installed in the air filter housing lid.*

1 Remove the air intake duct (see Chapter 4).
2 Disconnect the electrical connector from the MAF/IAT sensor (see illustration).
3 Remove the MAF/IAT sensor mounting fasteners and remove the sensor from the air filter housing.
4 Inspect the seal and replace as necessary.
5 Installation is the reverse of removal.

Chapter 6 Emissions and engine control systems

14.4a Disconnect the electrical connector and remove the upstream oxygen sensor (2.0L four-cylinder engine shown)

14.4b Upstream oxygen sensor - V6 engine, front cylinder bank (non-turbocharged model shown)

13 Intake Air Temperature (IAT) sensor

Note: *Only 2016 and later turbocharged engines are equipped with an IAT sensor installed in the air filter housing lid.*

1 Disconnect the electrical connector from the IAT sensor.
2 Rotate the IAT sensor 45 degrees to unlock from the air filter housing lid, and remove the sensor.
3 Inspect the O-ring and replace as necessary.
4 Installation is reverse of removal. Rotate the IAT sensor until a click is heard and it is locked in position.

14 Oxygen sensors - replacement

Note: *Because it is installed in the exhaust manifold or pipe, both of which contract when cool, an oxygen sensor might be very difficult to loosen when the engine is cold. Rather than risk damage to the sensor or its mounting threads, start and run the engine for a minute or two, then shut it off. Be careful not to burn yourself during the following procedure.*

1 Be particularly careful when servicing an oxygen sensor:

a) Oxygen sensors have a permanently attached pigtail and an electrical connector that cannot be removed. Damaging or removing the pigtail or electrical connector will render the sensor useless.
b) Keep grease, dirt and other contaminants away from the electrical connector and the louvered end of the sensor.
c) Do not use cleaning solvents of any kind on an oxygen sensor.
d) Oxygen sensors are extremely delicate. Do not drop a sensor or handle it roughly.

14.4c Upstream oxygen sensor - V6 engine, rear cylinder bank (non-turbocharged model shown)

e) Make sure that the silicone boot on the sensor is installed in the correct position. Otherwise, the boot might melt and it might prevent the sensor from operating correctly.

Replacement

2 Disconnect the cable from the negative battery terminal (see Chapter 5).
3 To access downstream sensors on all models, and the upstream sensor on the exhaust manifold for the rear cylinder head on V6 models, raise the front of the vehicle and place it securely on jackstands. Remove the engine under-cover.

Upstream oxygen sensor

Note: *There is an upstream oxygen sensor in each catalytic converter on V6 models.*

4 Locate the upstream oxygen sensor (see illustrations), then trace the sensor's electrical lead to its electrical connector and disconnect the connector. Disengage any harness clips.

5 Using a wrench or an oxygen sensor socket, unscrew the upstream oxygen sensor.
6 If you're going to install the old sensor, apply anti-seize compound to the threads of the sensor to facilitate future removal. If you're going to install a new oxygen sensor, it's not necessary to apply anti-seize compound to the threads; the threads on new sensors already have anti-seize compound on them.
7 Installation is the reverse of removal. Tighten the oxygen sensor to the torque listed in this Chapter's Specifications.

Downstream oxygen sensor (catalyst monitor sensor)

Note: *The downstream oxygen sensor is located in the lower end of the exhaust manifold/catalytic converter assembly or on the short section of exhaust pipe, just below the catalyst. There is a downstream oxygen sensor in each exhaust pipe on V6 models.*

14.8a Disconnect the electrical connector and remove the downstream oxygen sensor (2.0L four-cylinder engine shown)

14.8b Downstream oxygen sensor - V6 engine, front cylinder bank (non-turbocharged model shown)

14.8c Downstream oxygen sensor - V6 engine, rear cylinder bank (non-turbocharged model shown)

8 Locate the downstream oxygen sensor (see illustrations), then trace the lead up to the electrical connector and disconnect the connector. Disengage any harness clips.
9 Using a wrench or an oxygen sensor socket, unscrew the downstream oxygen sensor.
10 If you're going to install the old sensor, apply anti-seize compound to the threads of the sensor to facilitate future removal. If you're going to install a new oxygen sensor, it's not necessary to apply anti-seize compound to the threads. The threads on new sensors already have anti-seize compound on them.
11 Installation is the reverse of removal. Tighten the oxygen sensor to the torque listed in this Chapter's Specifications.

15 Throttle Position (TP) sensor - replacement

Note: *Only the 3.5L turbocharged engine is equipped with a separate TP sensor. On all other models the TP sensor is an integral component of the electronic throttle body, and is not separately serviceable. If you need to replace the TP sensor, you must replace the throttle body (see Chapter 4).*
1 Remove the engine cover.
2 Remove the charge air cooler pipe from the throttle body.
3 Locate the TP sensor on the throttle body, opposite of the drive-by-wire components. Disconnect the TP sensor electrical connector.
4 Using a heat gun, bring the temperature of the top TP sensor screw mounting ear to approximately 130 degrees F (33 degrees C).
Note: *DO NOT overheat the TP sensor or damage to the plastic parts and internal components may occur.*
5 Remove the TP sensor screw OPPO-SITE of the screw that was heated up.
6 Remove the remaining TP sensor screw and remove the TP sensor from the throttle body.
7 Installation is reverse of removal.

16 Transmission Range (TR) sensor - removal and installation

1 The TR sensor is located in the valve body and repair is beyond the scope of the home mechanic. If you need to replace the TR sensor, we recommend taking the vehicle to your local dealer or other qualified repair shop for repair.

17 Turbine Shaft Speed (TSS) sensor - replacement

Note: *On 6F35 transaxles in 2.0L and 2.3L equipped models, the Turbine Shaft Speed (TSS) sensor is located on top of the transaxle. The TSS sensor for V6 models is located in the valve body of the transaxle and should be taken to your local dealer or other qualified repair shop for repair.*
1 On 2015 and earlier models, raise and support the front of the vehicle on jack stands and remove the driver front wheel and inner fender liner.
2 On 2016 and later models, remove the air filter housing (see Chapter 4, Section 9).
3 Disconnect the electrical connector from the sensor.
4 Remove the sensor mounting bolt and remove the sensor.
5 Inspect and discard the sensor O-ring if damaged.
6 When installing the sensor, use a new O-ring. Installation is otherwise the reverse of removal.

18 Turbocharger Boost Pressure (TCBP)/Charge Air Cooler Temperature (CACT) sensor - replacement

Note: *The sensor may also be referred to as the Charge Air Cooler Manifold Absolute Pressure and Temperature (CAC MAPT) sensor.*
Note: *On turbocharged models are equipped with a TCBP/CACT sensor. On four-cylinder models, the sensor is mounting in the charge air cooler piping below the air filter housing. On V6 models, the sensor is installed in the charge air cooler piping near the throttle body.*
1 On four-cylinder models, remove the air filter housing as necessary to gain access.
2 On V6 models, remove the engine cover.
3 Locate the TCBP/CACT sensor and disconnect the electrical connector.
4 Remove the screw(s) attaching the sensor to the piping. V6 models are attached with two Torx screws.
5 Remove the sensor and inspect the O-ring. If it is damaged, replace it.
6 Installation is reverse of removal.

19 Turbocharger bypass valve - replacement

Note: *The turbocharger bypass valve may also be known as a Blow Off Valve (BOV).*
Note: *Four-cylinder models are equipped with a single bypass valve - located on the rear of the engine in the charge air cooler piping. V6 models are equipped with two; a bypass valve for each turbocharger, located above their respective valve cover in the charge air cooler piping.*
1 Remove the engine cover.

2 On four-cylinder models, remove the turbocharger inlet duct and intake piping that goes over the top of the engine to gain access to the bypass valve. Use care when disconnecting the crankcase vent tube to prevent damage.
3 On all models, identify the bypass valve(s) and disconnect the electrical connector(s).
Note: On V6 models, the rear bypass valve is located under the strut tower, near the cowl.
4 Disconnect the hose clamp and hose attached to the bypass valve (if equipped).
5 Remove the bolts/screws attaching the bypass valve(s) to the piping and remove the valve(s).
6 Inspect the bypass valve O-ring for damage. if the O-ring is damaged, replace it.
7 Installation is reverse of removal.

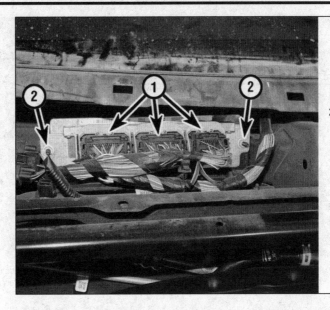

20.3 Powertrain Control Module (PCM) details: (cowl panel removed for clarity)

1 Electrical connectors
2 PCM mounting nuts

20 Powertrain Control Module (PCM) - removal and installation

Caution: *To avoid electrostatic discharge damage to the PCM, handle the PCM only by its case. Do not touch the electrical terminals during removal and installation. If available, ground yourself to the vehicle with an anti-static ground strap, available at computer supply stores.*

Note: *This procedure applies only to disconnecting, removing and installing the PCM that is already installed in your vehicle. If the PCM is defective and has to be replaced, it must be programmed with new software and calibrations. This procedure requires the use of Vehicle Communication Module (VCM) and Integrated Diagnostic System (IDS) software with appropriate hardware, or equivalent scan tool, so you WILL NOT BE ABLE TO REPLACE THE PCM AT HOME.*

1 Disconnect the cable from the negative battery terminal (see Chapter 5).
2 Remove the cowl panel (see Chapter 11, Section 12). The PCM is located on the passenger side, under the cowl panel.
3 Disconnect the electrical connectors from the PCM (see illustration).
4 Remove the PCM mounting nuts and remove the PCM.
5 Installation is reverse of removal.

21 Catalytic converter - replacement

Warning: *Wait until the engine has cooled completely before beginning this procedure.*

1 Raise the front of the vehicle and support it securely on jackstands. Remove the engine under-cover and under vehicle air duct (if equipped).

Four-cylinder engines
2.0L models

2 Locate the downstream oxygen sensor on the upper central part of the converter, trace the sensor to its electrical connector and disconnect it. Unscrew and remove the downstream oxygen sensor/catalyst sensor (see Section 14).
3 Remove the nuts and bolts that secure the flange at the rear end of the converter to the exhaust pipe. If the threads are severely damaged or rusted, apply some penetrant to the threads and wait awhile before loosening them.
4 Remove the nuts securing the catalytic converter to the bracket.
5 Loosen the clamp that secures the front end of the catalyst to the turbocharger. If the threads are severely damaged or rusted, apply some penetrant to the threads and wait awhile before loosening them.
6 Remove the catalytic converter.
7 Remove and discard the old gasket between the exhaust down pipe and the catalyst mounting flange.
8 Installation is the reverse of removal. Use a new gasket and, if necessary, new fasteners.

2.3L models

9 On AWD models, remove the battery and battery tray (see Chapter 5, Section 4).
10 Locate the oxygen sensor electrical connectors and disconnect them. Disconnect the harness retainers.
11 Remove the catalytic converter-to-turbocharger nuts. Discard them.
12 Remove the vehicle undercover and under body air duct.
13 Locate the catalyst monitor sensor electrical connector and disconnect it. Disconnect the harness retainers.
14 Remove the two (2) nuts at each end of the flexible exhaust pipe and remove the pipe from the vehicle.
15 Remove the bolt at the catalytic converter support bracket.
16 Remove the catalytic converter.
17 Remove and discard the old gasket between the turbocharger and the catalyst mounting flange.
18 Installation is the reverse of removal. Use a new gasket and, if necessary, new fasteners.

V6 engines
Non-turbocharged models

Note: On 2012 and earlier models, the right catalyst can be serviced separately. On 2013 and later models, both catalytic converters on non-turbocharged models are an integral component of the exhaust manifold and cannot be serviced separately.

2012 and earlier models RH

19 Locate the upstream oxygen sensor and downstream catalyst monitor sensor, trace their electrical leads to their respective connectors and disconnect them.
20 Remove the fasteners that secure the lower mounting flange to the exhaust Y-pipe. If the threads are severely damaged or rusted, apply some penetrant to the threads and wait awhile before loosening them.
21 Pull the exhaust Y-pipe down far enough to clear the lower end of the catalyst.
22 On AWD models, remove the anti-roll mount (see Chapter 7A, Section 9).
23 Rotate the engine forward and remove the four nuts securing the catalyst to the manifold.
24 Remove the catalyst from the vehicle.
25 Remove and discard the old gaskets between the exhaust manifold and the catalyst and, if equipped, between the lower catalyst flange and the exhaust pipe flange.
26 Installation is the reverse of removal. Use new flange gaskets. Tighten the fasteners to the torque listed in this Chapter's Specifications.

2012 and earlier models LH converter and 2013 and later models both converters

27 Remove the engine cover.

Chapter 6 Emissions and engine control systems

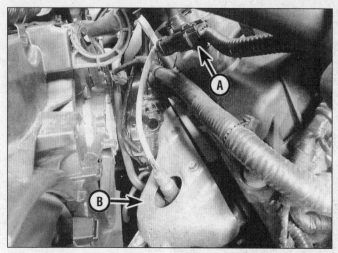

21.28 Disconnect the upper O2 sensor connector (A) and remove the heat shield (B)

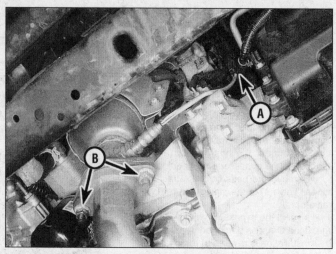

21.33 Disconnect the downstream catalyst monitor sensor connector (A) and remove the Y-pipe nuts (B)

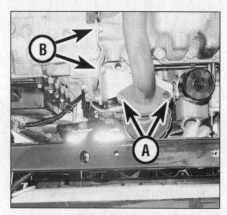

21.35 Remove the nuts for the Y-pipe (A) and bolts for the bracket (B)

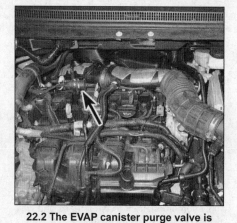

22.2 The EVAP canister purge valve is attached to the valve cover under the engine cover mount on four-cylinder models

28 Locate the upstream oxygen sensor and downstream catalyst monitor sensor, trace their electrical leads to their respective connectors and disconnect them (see illustration).
29 Remove the heat shield bolts and remove the heat shield from the exhaust manifold (see illustration). There is another heat shield protecting the steering gear assembly, remove this heat shield as well.
30 On AWD models, disconnect the driveshaft at the transfer case and position out of the way (see Chapter 8, Section 3).
31 On AWD models, remove the anti-roll mount (see Chapter 7A, Section 9).
32 On all models, remove the nuts that secure the exhaust manifold to the cylinder head. If the threads are severely damaged or rusted, apply some penetrant to the threads and wait awhile before loosening them.
33 Remove the fasteners that secure the lower mounting flange to the exhaust Y-pipe (see illustration). If the threads are severely damaged or rusted, apply some penetrant to the threads and wait awhile before loosening them.

34 Pull the exhaust Y-pipe down far enough to clear the lower end of the catalyst.
35 Remove the catalyst bracket bolts at the transaxle (see illustration).
36 Remove the catalyst and exhaust manifold from the vehicle.
37 Remove and discard the old gaskets between the exhaust manifold flange and the cylinder head and, if equipped, between the lower catalyst flange and the exhaust pipe flange.
38 Installation is the reverse of removal. Use new flange gaskets. Tighten the fasteners to the torque listed in this Chapter's Specifications.
Note: *It is recommended to use NEW nuts and studs at the manifold-to-cylinder head.*

Turbocharged models

39 For the rear converter, remove the passenger front wheel.
40 For both sides, locate the upstream oxygen sensor and downstream catalyst monitor sensor, trace their electrical leads to their respective connectors and disconnect them.

Disconnect the harness retainers.
41 Remove the two nuts at each end of the flexible exhaust pipe and remove the pipe from the vehicle.
42 Remove the ten retainers and the skid plate attached to the subframe.
43 Remove the three catalytic converter-to-turbocharger nuts. Discard them.
44 Remove the catalyst from the vehicle.
45 Remove and discard the old gaskets between the turbocharger and catalyst and, if equipped, between the lower catalyst flange and the exhaust pipe flange.
46 Installation is the reverse of removal. Use new flange gaskets. Tighten the fasteners to the torque listed in this Chapter's Specifications.

22 Evaporative Emissions Control (EVAP) system - component replacement

EVAP canister purge valve

Four-cylinder engines

1 Disconnect the cable from the negative battery terminal (see Chapter 5).
2 Disconnect the electrical connector from the canister purge valve (see illustration).
3 Disconnect the EVAP line quick-connect fittings (see Chapter 4 for information on quick-connect fittings).
4 Remove the purge valve mounting stud-bolt and remove the assembly.
5 Installation is the reverse of removal.

V6 engines

Note: *On turbocharged models, the valve is located on the rear of the intake manifold, above the valve cover. On non-turbocharged models, the valve is mounted on the intake manifold, next to the throttle body.*

6 Disconnect the cable from the negative battery terminal (see Chapter 5).

Chapter 6 Emissions and engine control systems 6-23

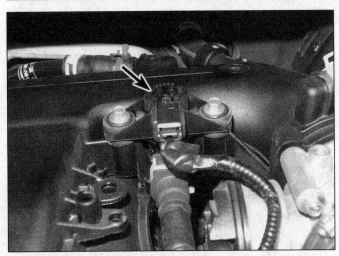

22.7 The canister purge valve is located on the upper intake manifold, near the throttle body (non-turbocharged V6)

22.14 Disconnect the two EVAP line quick-connect fittings (A), the filter fresh air hose (B), and the canister vent solenoid connector (C)

7 Disconnect the electrical connector from the canister purge valve (see illustration).
8 Disconnect the EVAP line quick-connect fitting (see Chapter 4 for information on quick-connect fittings).
9 On turbocharged models, slide the retainer off of the valve cover and remove the purge valve.
10 On non-turbocharged models, remove the purge valve mounting bolts and remove the purge valve.
11 Installation is the reverse of removal. On non-turbocharged models, use a new O-ring if the O-ring is damaged.

EVAP canister

Note: *The EVAP canister is located under the vehicle, towards the rear.*

12 Disconnect the cable from the negative battery terminal (see Chapter 5).
13 Raise the vehicle and support it securely on jackstands.
14 Disconnect the two EVAP line quick-connect fittings (see illustration) from the EVAP canister (see Chapter 4 for information on quick-connect fittings).
15 Disconnect the electrical connector and the fresh air hose from the EVAP canister vent solenoid (see illustration).
16 Remove the mounting fasteners that secure the canister and the shield and remove the canister with the shield from the vehicle.
17 Installation is the reverse of removal.

EVAP canister vent solenoid and filter assembly

Note: *The EVAP canister vent solenoid and filter assembly is mounted on the EVAP canister. The EVAP canister vent solenoid is not serviceable separately. The filter may also be referred to as the dust separator.*

18 Remove the EVAP canister.
19 Release the three (3) lock tabs and remove the EVAP canister vent solenoid and filter assembly from the canister.
20 Installation is the reverse of removal.

Fuel tank pressure sensor

Note: *The fuel tank pressure sensor is located on top of the fuel tank and is replaced with the connected tubing. The fuel tank pressure sensor and tubing may also be referred to as the fuel vapor tube assembly.*

21 Remove the fuel tank (see Chapter 4, Section 8).
22 Disconnect the fuel tank pressure sensor electrical connector.
23 Disconnect the quick-connect fitting at the fuel pump module.
24 Detach the tubing from the fuel tank retainer and remove the adhesive pad (if equipped) securing the tubing in place and remove the tubing and sensor from the fuel tank.
25 Installation is the reverse of removal.

23.5 Location of the Fuel Rail Pressure (FRP) sensor - (four-cylinder shown, manifold removed for clarity)

23 Fuel Rail Pressure (FRP) sensor - replacement

Note: *Only turbocharged models are equipped with a FRP sensor. The FRP sensor is located on the fuel rail.*

1 Relieve the fuel system pressure (see Chapter 4), then disconnect the cable from the negative terminal of the battery (see Chapter 5).

Four-cylinder models

2 Relieve the fuel system pressure (see Chapter 4), then disconnect the cable from the negative terminal of the battery (see Chapter 5).
3 Remove the engine cover.
4 Remove the charge air cooler inlet and middle tubes (see Chapter 4).
5 Locate the FRP sensor at the right end of the fuel rail. Disconnect the FRP sensor electrical connector (see illustration).

24.2 Location of the vent oil separator (four-cylinder engine)

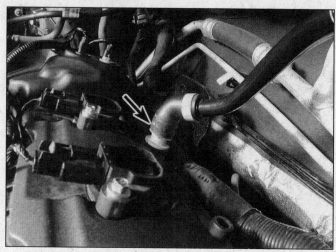

24.7 Location of the PCV valve (non-turbocharged V6 shown)

6 Unscrew and remove the FRP sensor.
7 Installation is the reverse of removal, tightening the sensor to the torque listed in this Chapter's Specifications.

V6 models

8 Relieve the fuel system pressure (see Chapter 4, Section 3).
9 Disconnect the negative battery cable (see Chapter 5, Section 3).
10 Remove the upper and lower intake manifolds (see Chapter 2B, Section 5).
11 Disconnect the FRP sensor connector.
12 Unscrew the FRP from the fuel rail to remove.
13 Installation is the reverse of removal, tightening the sensor to the torque listed in this Chapter's Specifications.

24 Positive Crankcase Ventilation (PCV) valve - replacement

Four-cylinder models

Note: The PCV valve is an integral part of the crankcase vent oil separator, which is located on the front side of the engine block and is replaced as one unit.

1 Remove the intake manifold (see Chapter 2A, Section 6).
2 Remove the oil vent separator mounting bolts (see illustration) and gasket.
3 Clean the engine block and install a new gasket to the separator.
4 Installation is the reverse of removal, tightening the oil vent separator bolts securely.

V6 models

Note: The PCV valve is located on the rear valve cover. According to the manufacturer, the PCV valve must be replaced if it's removed.

5 Remove the engine cover.
6 On non-turbocharged models, remove the upper intake manifold (see Chapter 2B, Section 5).
7 Disconnect the crankcase vent tube from the PCV valve (see illustration).
8 Turn the PCV valve counterclockwise and remove it from the valve cover.
Note: You must discard the PCV valve every time that you remove it and replace it with a new unit, because the locking mechanism of the PCV valve is damaged by removal.
9 Installation is the reverse of removal.

Ensure the PCV valve locks into position once installed.

25 Variable Valve Timing (VVT) system - component replacement

Oil Control Valve (OCV)

Note: On four-cylinder models, two OCVs are attached to the cylinder head, under the valve cover. On 3.5L turbocharged models, one OCV is attached to each cylinder head, under the valve cover. On 3.5L non-turbocharged models, two OCVs are attached to each cylinder head, under the valve cover, for a total of four OCVs.

1 Remove the valve cover for the OCV to be serviced (see Chapter 2A, Section 4 or Chapter 2B, Section 4).
2 Remove the bolt and pull the OCV from the cylinder head.
3 Inspect the OCV passages for debris and clean as necessary.
4 Inspect the O-ring and replace if damaged.
5 Installation is reverse of removal. Tighten the bolt to specifications in this Chapter's Specifications.

Chapter 7 Part A
Automatic transaxle

Contents

	Section
Automatic transaxle - removal and installation	6
Automatic transaxle overhaul - general information	7
Diagnosis - general	2
Driveaxle oil seals - replacement	8
General information	1

	Section
Shift cable - replacement and adjustment	4
Shift lever - replacement	3
Tow/haul switch - replacement	5
Transaxle mount - replacement	9

Specifications

General

Fluid type and capacity .... See Chapter 1

Torque specifications

Ft-lbs (unless otherwise indicated)   Nm

Note: *One foot-pound (ft-lb) of torque is equivalent to 12 inch-pounds (in-lbs) of torque. Torque values below approximately 15 foot-pounds are expressed in inch-pounds, because most foot-pound torque wrenches are not accurate at these smaller values.*

	Ft-lbs	Nm
Anti-roll mount		
Bracket-to-transaxle bolts		
Left mount (turbocharged V6 models)	66	90
Right mount		
Four-cylinder models	66	90
V6 models	76	103
Through-bolt		
Four-cylinder models	66	90
V6 models	76	103
Crossmember mounting nuts	115	156
Crossmember bracket bolts	76	103
Driveshaft bolts (AWD)	See Chapter 8	
Fluid pan bolts	80 in-lbs	9
Crossmember brace bolts (3.5L turbocharged models)	52	70
Oil cooler adapter bolts	18	25
Torque converter-to-driveplate		
Bolts (V6 models)	41	55
Nuts (four-cylinder models)	30	40
Transaxle-to-engine mounting bolts	35	48
Transaxle mount		
Four-cylinder models		
Bolts	111	150
Nuts	46	63
V6 models		
Upper bolts	52	70
Lower bolts	41	55
Bracket bolt	59	80
Bracket nuts	46	63

Chapter 7 Part A Automatic transaxle

1 General information

1 Information on the automatic transaxle is included in this Part of Chapter 7A. The transfer case portion of the transaxle is covered in Chapter 7B.
2 The transaxle electronic operation is controlled by the Powertrain Control Module (PCM) on V6 models, or the Transmission Control Module (TCM) on four-cylinder models.
3 Because of the complexity of the automatic transaxles and the specialized equipment necessary to perform most service operations, this Chapter contains only those procedures related to general diagnosis, routine maintenance, adjustment and removal and installation.
4 If the transaxle requires major repair work, it should be left to a dealer service department or an automotive or transmission repair shop. Once properly diagnosed you can, however, remove and install the transaxle yourself and save the expense, even if the repair work is done by a transmission shop. After transaxle replacement, the Misfire Monitor Neutral Profile Correction procedure will need to be performed using a scan tool. If the valve body is replaced, the valve (solenoid) body strategy will need to be updated using a scan tool. These procedures will need to be performed by a qualified technician or repair facility using the proper diagnostic and repair equipment.

2 Diagnosis - general

1 Automatic transaxle malfunctions may be caused by five general conditions:
   a) *Poor engine performance*
   b) *Improper adjustments*
   c) *Hydraulic malfunctions*
   d) *Mechanical malfunctions*
   e) *Malfunctions in the computer or its signal network*

2 Diagnosis of these problems should always begin with a check of the easily repaired items: fluid level and condition (see Chapter 1), shift cable adjustment and shift lever installation. Next, perform a road test to determine if the problem has been corrected or if more diagnosis is necessary. If the problem persists after the preliminary tests and corrections are completed, additional diagnosis should be performed by a dealer service department or other qualified transmission repair shop. On modern electronically-controlled automatic transaxles, a scan tool is helpful in retrieving trouble codes relating to the transaxle. Refer to the "Troubleshooting" Section at the front of this manual for information on symptoms of transaxle problems.

Preliminary checks

3 Drive the vehicle to warm the transaxle to normal operating temperature.
4 Check the fluid level as described in Chapter 1:
   a) *If the fluid level is unusually low, add enough fluid to bring the level within the designated area of the dipstick, then check for external leaks (see following).*
   b) *If the fluid level is abnormally high, drain off the excess, then check the drained fluid for contamination by coolant. The presence of engine coolant in the automatic transmission fluid indicates that a failure has occurred in the internal radiator oil cooler walls that separate the coolant from the transmission fluid (see Chapter 3).*
   c) *If the fluid is foaming, drain it and refill the transaxle, then check for coolant in the fluid, or a high fluid level.*

5 Check the engine idle speed.
Note: *If the engine is malfunctioning, do not proceed with the preliminary checks until it has been repaired and runs normally.*
6 Check and adjust the shift cable, if necessary (see Section 4).
7 If hard shifting is experienced, inspect the shift cable under the steering column and at the manual lever on the transaxle (see Section 4).

Fluid leak diagnosis

8 Most fluid leaks are easy to locate visually. Repair usually consists of replacing a seal or gasket. If a leak is difficult to find, the following procedure may help.
9 Identify the fluid. Make sure it's transmission fluid and not engine oil or brake fluid (automatic transmission fluid is a deep red color).
10 Try to pinpoint the source of the leak. Drive the vehicle several miles, then park it over a large sheet of cardboard. After a minute or two, you should be able to locate the leak by determining the source of the fluid dripping onto the cardboard.
11 Make a careful visual inspection of the suspected component and the area immediately around it. Pay particular attention to gasket mating surfaces. A mirror is often helpful for finding leaks in areas that are hard to see.
12 If the leak still cannot be found, clean the suspected area thoroughly with a degreaser or solvent, then dry it thoroughly.
13 Drive the vehicle for several miles at normal operating temperature and varying speeds. After driving the vehicle, visually inspect the suspected component again.
14 Once the leak has been located, the cause must be determined before it can be properly repaired. If a gasket is replaced but the sealing flange is bent, the new gasket will not stop the leak. The bent flange must be straightened.
15 Before attempting to repair a leak, check to make sure that the following conditions are corrected or they may cause another leak.
Note: *Some of the following conditions cannot be fixed without highly specialized tools and expertise. Such problems must be referred to a qualified transmission shop or a dealer service department.*

Gasket leaks

16 Check the pan periodically. Make sure the bolts are tight, no bolts are missing, the gasket is in good condition and the pan is flat (dents in the pan may indicate damage to the valve body inside).
17 If the pan gasket is leaking, the fluid level or the fluid pressure may be too high, the vent may be plugged, the pan bolts may be too tight, the pan sealing flange may be warped, the sealing surface of the transaxle housing may be damaged, the gasket may be damaged or the transaxle casting may be cracked or porous. If sealant instead of gasket material has been used to form a seal between the pan and the transaxle housing, it may be the wrong type of sealant.

Seal leaks

18 If a transaxle seal is leaking, the fluid level or pressure may be too high, the vent may be plugged, the seal bore may be damaged, the seal itself may be damaged or improperly installed, the surface of the shaft protruding through the seal may be damaged or a loose bearing may be causing excessive shaft movement.
19 Make sure the dipstick tube seal is in good condition and the tube is properly seated. Periodically check the area around the sensors for leakage. If transmission fluid is evident, check the seals for damage.

Case leaks

20 If the case itself appears to be leaking, the casting is porous and will have to be repaired or replaced.
21 Make sure the oil cooler hose fittings are tight and in good condition.

Fluid comes out vent pipe or fill tube

22 If this condition occurs the possible causes are: the transaxle is overfilled; there is coolant in the fluid; the case is porous; the dipstick is incorrect; the vent is plugged or the drain-back holes are plugged.

3 Shift lever - replacement

Warning: *These models are equipped with a Supplemental Restraint System (SRS), more commonly known as airbags. Always disable the airbag system before working in the vicinity of any airbag system component to avoid the possibility of accidental deployment of the airbag(s), which could cause personal injury (see Chapter 12).*
Warning: *Do not use a memory saving device to preserve the PCM or radio memory when working on or near airbag system components.*

Shift knob

1 Remove the upper bezel of the floor console around the shifter (see Chapter 11).
2 Pull the shift knob boot up and disconnect the electrical connector. Remove the two

Chapter 7 Part A Automatic transaxle

3.2 Lift the boot up, disconnect the electrical connector (A) and remove the shift lever mounting screws (B)

3.6 Pry the cable end from the ballstud on the shifter

3.7 Pry the tabs forward to release the cable retainer from the console

screws and pull the knob off the shaft (see illustration).

3 Installation is the reverse of the removal procedure.

Shift lever assembly

4 Disconnect the cable from the negative battery terminal (see Chapter 5). Wait at least two minutes before proceeding.
5 Remove the floor console center trim cover (see Chapter 11) and side panels.
6 Use a trim tool to pry the cable eye from the ballstud on the shifter (see illustration).
7 Release the tabs on the sides of the shift cable and detach it from the console (see illustration).
8 Disconnect any electrical connectors at the shifter assembly.
9 Remove the mounting bolts and remove the shifter assembly from the floor (see illustration).
10 Installation is the reverse of removal. The shift cable should be adjusted any time it has been disconnected (see Section 4).
Note: *It is recommended to use a NEW plastic cable retaining clip.*

Park/neutral position switch

Note: *The following procedure applies to models with push-button start. On vehicles without push-button start, the park/neutral switch is part of the shift lever assembly and must be replaced as a unit.*

11 Remove the shift lever assembly from the vehicle.
12 Place the shift lever in the Neutral (N) position.
13 On the left rear of the shift lever assembly, just below the shift plate, remove the screw and the park/neutral switch cover.
14 Disconnect the Brake Shift Interlock connector at the bottom left side of the shift lever assembly.
15 Disconnect the park/neutral switch harness from the shift lever assembly.

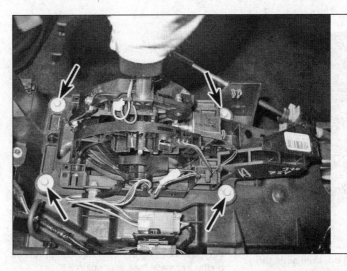

3.9 To remove the shifter assembly, remove the four mounting bolts

16 Note the position and location of the park/neutral harness terminal pins in the shift lever assembly connector.
17 Remove the terminals for the park/neutral position switch from the shift lever assembly connector.
18 Carefully pry the park/neutral position switch from the shift lever assembly.
19 Installation is reverse of removal. Ensure the harness terminals are inserted into the correct positions in the shift lever assembly connector.

Paddle shift levers

20 Disconnect the cable from the negative battery terminal (see Chapter 5). Wait at least two minutes before proceeding.
Note: *On 2015 and earlier models, the paddle shift levers can be replaced by removing the airbag (see Chapter 12, Section 27) and front trim only, without steering wheel removal.*
21 Remove the steering wheel (see Chapter 10, Section 14).
22 Remove the rear steering wheel trim panel screws and trim panel.
23 Using a trim tool, carefully remove the front steering wheel trim by prying away from the steering wheel, starting at the top.
24 Disconnect the electrical connectors from the switches.
25 Remove the front multi-function switches to access the screws for the paddle shifters.
26 Remove the mounting screws and remove the paddle shifter(s).
27 Installation is the reverse of removal.

4 Shift cable - replacement and adjustment

Warning: *The models covered by this manual are equipped with a Supplemental Restraint System (SRS), more commonly known as airbags. Always disarm the airbag system before working in the vicinity of any airbag system component to avoid the possibility of accidental deployment of the airbag, which could cause personal injury (see Chapter 12). Do not use a memory saving device to preserve*

Chapter 7 Part A Automatic transaxle

4.3 Pry the cable end from the ballstud on the shift control lever (A), then squeeze the tabs (B) and slide the cable housing out of the bracket

4.11 Shift cable grommet plate nuts

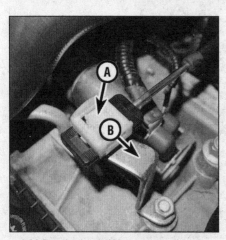

4.21 Pry the latch (A) up, then slide the lock to the side (B)

the PCM or radio memory when working on or near airbag system components.

Note: *This is a difficult procedure for the home mechanic, since replacement of the cables requires removal of the instrument panel and the HVAC housing under the instrument panel.*

Replacement

1 If the vehicle has just been driven, wait several hours to allow the engine to cool down before beginning this procedure. Disconnect both the negative and positive cables from the battery (see Chapter 5). Wait at least two minutes before proceeding.
2 Remove the air filter housing (see Chapter 4).
3 To disconnect the shift cable at the transaxle control lever, pop the cable end off of the lever ballstud (see illustration).
4 Squeeze the retaining tabs and detach the cable housing from the mounting bracket on the transaxle.
5 On console shift models, move the seat to the furthest away position from the instrument panel, position the seat as far down as possible (closest to the floor), and fully recline the seatback.
6 Remove the driver's side trim panel of the center console (see Chapter 11, Section 21).
7 On column shift models, remove the driver side trim panel above the accelerator pedal and near the center of the instrument panel.
8 On all models, to disconnect the shift cable at the shift lever assembly, pop the cable end off of the lever ballstud (see illustration 3.6).
9 Squeeze the retaining tabs and detach the cable housing from the mounting bracket on the transaxle (see illustration 3.7).
10 Pull back the carpet near the accelerator pedal to expose the shift cable grommet at the floor.
11 On console shift models, remove the grommet plate nuts from the floorpan studs below the HVAC housing (see illustration).
12 Carefully pull the grommet out of the floorpan.
13 Disconnect any cable retainers along the length of the cable.
14 Pull the shift cable through the floor from the inside to remove.

Note: *The grommet is integral to the cable assembly.*

15 Installation is the reverse of removal. Adjust the new cable after installation.

Adjustment

Note: *The transaxle end of the shift cable is equipped with an adjuster mechanism.*

16 Set the parking brake, then disconnect the cable from the negative battery terminal (see Chapter 5).
17 Remove the air filter housing (see Chapter 4).
18 Place the shifter in Drive.
19 Disconnect the cable from the transmission shift lever, pop the cable end off of the lever ballstud (see illustration 4.4).
20 Manually move the shift arm at the transaxle clockwise until it stops, then rotate the lever counterclockwise one click.
21 Pry the latch up, then slide the lock tab to the side (see illustration).
22 Slide the shift lever cable end until the cable is aligned with the control lever.
23 Connect the cable end back onto the manual lever ballstud, with the lock still released.
24 Lock the adjuster, then slide the lock tab back on the adjuster.

Note: *Make sure the adjuster is locked and the cable is securely seated onto the ball stud.*

25 After installing any remaining components, apply the parking brake, start the engine and shift into each range to verify the adjustment is correct.

5 Tow/haul switch - replacement

1 Using a trim tool, carefully pry and remove the trim panel at the driver's end of the instrument panel (see Chapter 11, Section 22).
2 Remove the trim panel against the door jamb, and in front of the instrument panel opening.
3 Pull firmly outwards to remove the trim panel containing the headlight switch and tow/haul switch (see Chapter 11, Section 22).
4 Disconnect the electrical connectors to remove the trim panel and switches.
5 Depress the tabs at the end of the tow/haul switch bank and push the switch out through the front of the trim panel.
6 Installation is reverse of removal.

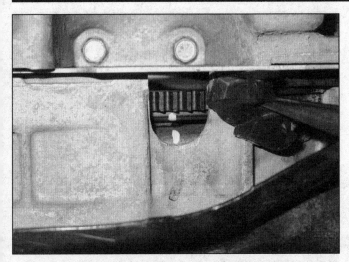

6.9 Mark the relationship of the torque converter to the driveplate through the starter opening (four-cylinder models)

6.11 Use an overhead support fixture to secure the engine while the transaxle is removed

6 Automatic transaxle - removal and installation

Removal

1 Raise and support the vehicle on a lift or equivalent and place the vehicle in Neutral (N).
2 Disconnect the cables from the battery and remove the battery and battery tray (see Chapter 5).
3 Remove the air filter housing and duct (see Chapter 4, Section 9), then remove the housing bracket bolts and bracket.
4 Disconnect the shift cable from the transaxle (see Section 4), remove the cable bracket bolts and position the bracket and cable aside.
5 Detach the wiring harness retainers and ground cable from the transaxle.
6 On V6 models (6F50/6F55 transaxles), disconnect the transaxle main connector and any other transaxle connectors.
7 On all models, remove the upper transaxle-to-engine bolts.
8 Remove the starter (see Chapter 5, Section 8) and starter insulator.
9 On four-cylinder models, mark the relationship of the torque converter to the driveplate through the starter opening (see illustration). Remove the converter mounting nuts.
10 On 2016 and later four-cylinder models, remove the oil filter and oil cooler adapter. Inspect the O-ring and use a NEW O-ring if damaged.
11 Attach an engine support fixture to the transaxle end of the engine (see illustration).
Note: *On V6 models, it may be necessary to remove the upper intake manifold (see Chapter 2B, Section 5) to allow installation of the support fixture.*
12 Remove the front subframe from the vehicle (see Chapter 10, Section 19).
13 Remove the anti-roll mount(s) and bracket(s) from the transaxle.
14 If needed, drain the transaxle.
15 Remove the driveaxles (see Chapter 8, Section 2).
16 On AWD models, remove the transfer case (see Chapter 7B, Section 7).
17 On 2016 and later four-cylinder models, remove the catalytic converter bracket.
18 On V6 models, remove the converter cover and mark the relationship of the torque converter to the driveplate. Remove the converter mounting bolts.
19 On all models, disconnect the harness from the rear of the transaxle.
20 On all models, remove the transaxle mount (see Section 9).
21 On four-cylinder models, disconnect the transmission main connector and any other transmission connectors.
22 On 2016 and later four-cylinder models, remove the crankcase vent tube and intercooler piping that runs over the front of the engine.
23 On all models, disconnect the transaxle cooling lines.
24 Support the transaxle with a jack - preferably a transmission jack made for this purpose (available at most tool rental yards). Safety chains will help steady the transaxle on the jack.
25 Remove the lower and rear transaxle-to-engine bolts.
26 Ensure all harness connectors, clips and transaxle lines are disconnected from the transaxle.
27 Move the transaxle away from the engine to disengage it from the engine block dowel pins. Carefully lower the transmission jack to the floor and remove the transaxle.
Note: *If the transmission is being replaced, remove the transaxle cooling tubes from the transaxle.*

Installation

28 Installation is the reverse of removal, but note the following points:
a) As the torque converter is reinstalled, ensure that the drive tangs at the center of the torque converter hub engage with the recesses in the automatic transaxle fluid pump inner gear. This can be confirmed by turning the torque converter while pushing it towards the transaxle. If it isn't fully engaged, it will clunk into place.
b) When installing the transaxle, make sure the matchmarks you made on the torque converter and driveplate line up.
c) Install all of the driveplate-to-torque converter nuts before tightening any of them.
d) Tighten the driveplate-to-torque converter nuts to the specified torque.
e) Tighten the transaxle mounting bolts to the specified torque.
f) Tighten the suspension crossmember mounting bolts to the torque values listed in this Chapter's Specifications.
g) Tighten the driveaxle/hub nuts to the torque value listed in the Chapter 8 Specifications.
h) Tighten the wheel lug nuts to the torque listed in the Chapter 1 Specifications.
i) Fill the transaxle with the correct type and amount of automatic transmission fluid (see Chapter 1).
j) On completion, adjust the shift cable (see Section 4).

29 After installation, the Misfire Monitor Neutral Profile Correction procedure will need to be performed using a scan tool. If the valve body is replaced, the valve (solenoid) body strategy will need to be updated using a scan tool. These procedures will need to be performed by a qualified technician or repair facility using the proper diagnostic and repair equipment.

9.10 Transaxle mount details
1. Through-bolt
2. Mount-to-chassis bolts
3. Mount bracket-to-transaxle fasteners

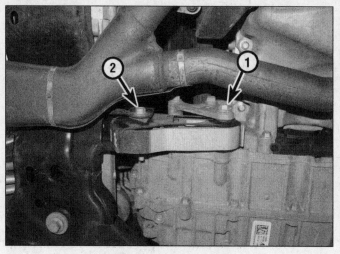

9.14 Anti-roll mount details
1. Mount-to-transaxle bolt
2. Mount-to-crossmember bolt

7 Automatic transaxle overhaul - general information

1 In the event of a problem occurring, it will be necessary to establish whether the fault is electrical, mechanical or hydraulic in nature, before repair work can be contemplated. Diagnosis requires detailed knowledge of the transaxle's operation and construction, as well as access to specialized test equipment, and so is deemed to be beyond the scope of this manual. It is therefore essential that problems with the automatic transaxle are referred to a dealer service department or other qualified repair facility for assessment.

2 Note that a faulty transaxle should not be removed before the vehicle has been diagnosed by a knowledgeable technician equipped with the proper tools, as troubleshooting must be performed with the transaxle installed in the vehicle.

8 Driveaxle oil seals - replacement

Note: *On AWD models, the right transaxle oil seal is not accessible without removing the transfer case assembly. See Chapter 7B for complete information.*

1 Oil leaks frequently occur due to wear of the driveaxle oil seals. Replacement of these seals is relatively easy, since the repair can be performed without removing the transaxle from the vehicle.

2 Driveaxle oil seals are located at the sides of the transaxle, where the driveaxles are attached. If leakage at the seal is suspected, raise the vehicle and support it securely on jackstands. If the seal is leaking, lubricant will be found on the sides of the transaxle, below the seals.

3 Remove the driveaxles (see Chapter 8).

4 Use a screwdriver or prybar to carefully pry the oil seal out of the transaxle bore.

5 If the oil seal cannot be removed with a screwdriver or prybar, a special oil seal removal tool (available at auto parts stores) will be required.

6 Using a large section of pipe or a large deep socket (slightly smaller than the outside diameter of the seal) as a drift, install the new oil seal. Drive it into the bore squarely and make sure it's completely seated. Coat the seal lip with transaxle lubricant.

7 Install the driveaxle(s). Be careful not to damage the lip of the new seal with the inboard splines of the driveaxle.

9 Transaxle mount - replacement

Transaxle mount

1 Insert a large screwdriver or prybar between the mount and the transaxle and pry up.

2 The transaxle should not move excessively away from the mount. If it does, replace the mount.

3 Remove the battery and battery tray (see Chapter 5).

4 Remove the air filter housing (see Chapter 4).

5 Remove the battery tray support bracket.

6 On four-cylinder models with the Transmission Control Module (TCM) mounted next to the battery, disconnect the TCM connectors and remove the TCM and bracket.

Note: *On V6 models, the upper intake manifold my require removal (see Chapter 2B, Section 5), however, attempt mount removal prior to removing the upper intake manifold.*

7 On four-cylinder models, remove the intercooler pipe that runs over the transaxle.

8 On 3.5L turbocharged models, remove the strut tower brace nuts and brace and the intercooler piping as necessary to access the mount.

9 On all models, support the transaxle with a floor jack and jack slightly to remove the load from the mount.

10 Remove the through-bolt and the mount-to-chassis nuts, then remove the mount (see illustration).

Note: *It may be necessary to raise the transaxle slightly to provide enough clearance to remove the mount.*

11 Installation is the reverse of removal.

Note: *Install all of the mount fasteners before tightening any of them to the specifications listed in this Chapter's Specifications. Use thread locker on the bolt/nut threads.*

Anti-roll mount

12 Raise and support the vehicle on jackstands.

13 Remove the vehicle undercover if necessary, to access the mount.

14 Remove the bolts and the bracket securing the mount to the transaxle (see illustration).

15 Remove the bolt attaching the mount to the bracket on the crossmember.

16 Installation is reverse of removal.

Note: *Align the keyed notches in the bracket with the mount and use thread locker on the bolts. Use thread locker on the bolt/nut threads.*

17 During installation, install all bolts and brackets finger tight before tightening to the torque listed in this Chapter's Specifications.

ND
Chapter 7 Part B
Transfer case

Contents

	Section		Section
All-Terrain Control Switch/Module (ACTM) - replacement	3	Transfer case cooler - replacement	7
All-Wheel Drive (AWD) module - replacement	2	Transfer case driveaxle oil seal - removal and installation	5
General information	1	Transfer case fluid pump - replacement	6
Transfer case - removal and installation	8	Transfer case rear output shaft oil seal - removal and installation	4

Specifications

Transfer case fluid type .... See Chapter 1

Torque specifications Ft-lbs (unless otherwise indicated)   Nm

Note: One foot-pound (ft-lb) of torque is equivalent to 12 inch-pounds (in-lbs) of torque. Torque values below approximately 15 foot-pounds are expressed in inch-pounds, because most foot-pound torque wrenches are not accurate at these smaller values.

Transfer case-to-transaxle bolts	66	90
Transfer case support bracket-to-engine block mounting bolts	35	48
Transfer case support bracket-to-transfer case mounting bolts		
2016 and earlier models	35	48
2017 models	59	80
Transfer case cooler bolts	106 in-lbs	12
Roll-restrictor-to transaxle bracket bolts	66	90
Roll-restrictor bolts	76	103
Output shaft pinion nut		
2.3L models	173	235
V6 models	Tighten to torque measured during the removal procedure	

1 General information

Note: *The transfer case may also be referred to as the Power Transfer Unit (PTU).*

1   Optional All Wheel Drive (AWD) is available. The transfer case sends power through a driveshaft that delivers the torque to a rear differential and two rear driveaxles. The AWD control module will shift into AWD when the PCM initiates the need due to differences in wheelslip speed between the front and rear wheels. The system requires no driver input to operate, thus there is no select button on the instrument panel.

2   Due to the complexity of the transfer case covered in this manual and the need for specialized equipment to perform most service operations, this Chapter contains only routine maintenance and removal and installation procedures.

3   If the transfer case requires major repair work, it must be replaced as a unit.

Transfer case oil life monitor reset

4   At a specific interval "Change AWD Power Transfer Unit Lube" message displays. After servicing the transfer case fluid, to reset the oil life monitor, perform the following steps:

 a) With the transaxle in Park (P), turn the ignition ON.
 b) Within 10 seconds of turning the ignition ON, firmly press and release the brake pedal 4 times within the 10 seconds.
 c) Within 10 seconds of pressing the brake pedal 4 times, press the gas pedal to the floor and release 4 times within the 10 seconds.
 d) Within 10 seconds of pressing the gas pedal 4 times, press the brake and gas pedals to the floor and hold until "AWD Power Transfer Unit Lube Set to New" message appears.

2 All-Wheel Drive (AWD) module - replacement

Note: *The AWD module is located in the left rear of the vehicle. behind the trim panel. The AWD module may also be referred to as the AWD relay module.*

1 Disconnect the negative battery cable.
2 Remove the driver's side rear interior trim panel.
3 Disconnect the AWD module electrical connector.
4 Remove the plastic fasteners and the AWD module from the vehicle.
5 Installation is reverse of removal.

3 All-Terrain Control Switch/Module (ATCM) - replacement

Note: *The All-Terrain Control Switch is part of the ATCM and is located on the center console.*

1 Remove the center console trim (see Chapter 11, Section 21).
2 Disconnect the ATCM connector.
3 Press in the retaining tabs and push the ATCM up and out of the center console trim panel.
4 Installation is reverse of removal.

4 Transfer case rear output shaft oil seal - removal and installation

Four-cylinder models

Removal

1 Raise the vehicle and support it securely on jackstands.
2 Remove the driveshaft (see Chapter 8).
3 On 2.0L models, remove the front subframe. On 2.3L models, remove the catalytic converter (see Chapter 6).

2.0L models

4 Use a beam- or dial-type inch-pound torque wrench to determine the torque required to rotate the pinion (vehicle in Neutral). Record it for use later.
5 Count the number of threads visible between the end of the nut and the end of the pinion shaft and record it for use later.

All models

6 Mark the relative positions of the pinion and flange.
7 Remove the flange mounting nut using a two-pin spanner to hold the pinion flange while loosening the locknut.
8 Remove the companion flange, using a puller whose bolt-holes can be attached to two, three or all four of the flange's driveshaft bolt holes. Tighten the puller center bolt to press the flange off the shaft.
9 Pry out the seal with a screwdriver or a seal removal tool. Don't damage the seal bore.
10 On 2.0L models, remove the oil slinger, outer pinion bearing and collapsible bearing spacer.

Installation

11 On 2.0L models, install a new bearing spacer, existing outer pinion bearing and oil slinger.
12 Lubricate the lips of the new seal with multi-purpose grease and tap it evenly into position with a seal installation tool or a large socket. Make sure it enters the housing squarely and is tapped in to its full depth.

2.0L models

13 Tighten the NEW nut carefully until the original number of threads are exposed and the marks are aligned.
14 Measure the torque required to rotate the pinion and tighten the nut in small increments until it matches the figure recorded earlier in this procedure.
Caution: *If you exceed the original torque reading by more than 3 in-lbs, the assembly will need to be taken apart again to install a new collapsible spacer.*

2.3L models

15 Align the mating marks made before disassembly and install the companion flange. Tighten the pinion nut to the torque listed in this Chapter's Specifications.

All models

16 Connect the driveshaft (see Chapter 8).
17 The remainder of installation is the reverse of removal.
18 Check the transfer case lubricant level and add some, if necessary, to bring it to the appropriate level (see Chapter 6).

V6 models

Removal

19 Raise the vehicle and support it securely on jackstands.
20 Drain the transfer case lubricant (see Chapter 1).
21 Remove the driveshaft (see Chapter 8).
22 Remove the front subframe (see Chapter 10, Section 19).
23 Mark the relative positions of the pinion, nut and flange.
24 Use a beam- or dial-type inch-pound torque wrench to determine the torque required to rotate the pinion (vehicle in Neutral). Record it for use later.
25 Count the number of threads visible between the end of the nut and the end of the pinion shaft and record it for use later.
26 Remove the flange mounting nut using a two-pin spanner to hold the pinion flange while loosening the locknut.
27 Remove the companion flange, using a puller whose bolt-holes can be attached to two, three or all four of the flange's driveshaft bolt holes. Tighten the puller center bolt to press the flange off the shaft.
28 Pry out the seal with a screwdriver or a seal removal tool. Don't damage the seal bore.
29 Remove the oil slinger, outer pinion bearing and collapsible bearing spacer.

Installation

30 Install a new bearing spacer, existing outer pinion bearing and oil slinger.
31 Lubricate the lips of the new seal with multi-purpose grease and tap it evenly into position with a seal installation tool or a large socket. Make sure it enters the housing squarely and is tapped in to its full depth.
32 Align the mating marks made before disassembly and install the companion flange. If necessary, tighten the pinion nut to draw the flange into place.
33 Tighten the NEW nut carefully until the original number of threads are exposed and the marks are aligned.
34 Measure the torque required to rotate the pinion and tighten the nut in small increments until it matches the figure recorded earlier in this procedure.
Caution: *If you exceed the original torque reading by more than 3 in-lbs, the assembly will need to be taken apart again to install a new collapsible spacer.*
35 Install the subframe (see Chapter 10).
36 Connect the driveshaft (see Chapter 8), add the specified lubricant to the transfer case (see Chapter 1).
Caution: *All driveshaft flange and support bearing bolts must be replaced with new fasteners.*

5 Transfer case driveaxle oil seal - removal and installation

Output shaft seal - passenger side

1 Loosen the wheel lug nuts, raise the vehicle and support it securely on jackstands. Remove the right wheel and the splash panel from the right fenderwell.
2 Remove the right driveaxle and intermediate shaft (see Chapter 8).
3 Remove the transfer case support bracket as necessary to access the seal.
4 Carefully remove the seal (there is a non-serviceable bearing behind it). There are several types of seal removal tools. The manufacturer recommends using a slide-hammer tool with an internal-expanding type of seal-puller. Using a long, threaded extension on the slide-hammer allows the seal to be removed without removing the exhaust system.
5 Using a seal installer or a large deep socket as a drift, install the new oil seal. Drive it into the bore squarely and make sure it's completely seated.
6 Lubricate the lip of the new seal with multi-purpose grease, then install a new dust shield.
7 Install the intermediate shaft and driveaxle (see Chapter 8).
8 The remainder of installation is the reverse of removal. Check the transfer case lubricant level and add some, if neces-

sary, to bring it to the appropriate level (see Chapter 1).

Input shaft seal

Note: *This procedure applies to V6 models. There are two input shaft seals. The input shaft seals are recessed in the transfer case behind a compression seal. 4-cylinder models require dissembly of the transfer case.*

9   Remove the transfer case (see Section 8).
10   Remove and discard the compression seal.
11   Carefully remove the outer and inner seals as they are recessed inside of the transaxle. There are several types of seal removal tools. The manufacturer recommends using a tool that presses the seal into the bearing race and screws into both seals, then use a slide-hammer tool to remove. Using a long, threaded extension on the slide-hammer may allow each seal to be removed individually.
12   Using a seal installer, install the new oil seals. Drive it into the bore squarely.

Note: *There is approximately a 0.197 in (5mm) distance between both seals and approximately 0.039 in (1mm) distance between the inner seal and the bearing race.*

13   Lubricate the lip of the new seals with multi-purpose grease before installation.
14   The remainder of installation is the reverse of removal. Check the transfer case lubricant level and add some, if necessary, to bring it to the appropriate level.

6   Transfer case fluid pump - replacement

Note: *2016 and later V6 models with oil-to-air coolers are equipped with a fluid pump. The fluid pump is mounted to the rear of the front crossmember.*

1   Raise and support the vehicle on jack stands.
2   Drain the transfer case fluid (see Chapter 1).
3   Disconnect the fluid pump electrical connector.
4   Remove the fluid pump line connector covers.
5   Remove the cooler line connection clips and disconnect the lines.
6   Remove the bolt attaching the fluid pump bracket to the crossmember and remove the fluid pump.
7   The remainder of installation is the reverse of removal. Check the transfer case lubricant level and add some, if necessary, to bring it to the appropriate level (see Chapter 1).

7   Transfer case cooler - replacement

Note: *Not all models are equipped with a transfer case oil cooler. The following procedure is for models with oil-to-coolant coolers.*

1   Raise and support the vehicle on jackstands.
2   Drain the transfer case fluid (see Chapter 1).
3   Remove the bolts and the transfer case temperature sensor. Disconnect the sensor harness connector and remove the sensor (unscrew) and harness from the transfer case.
4   Clamp off both transfer case cooler hoses, remove the spring clamps, and disconnect the hoses from the cooler lines near the left driveaxle.
5   Remove the bolt attaching the other end of the cooler lines to the transfer case cooler. Disconnect the lines and discard the O-rings.
6   Remove the power steering heat shield.
7   Remove the transfer case cooler bolts and remove the cooler assembly from the transfer case.
8   Remove and discard the cooler gasket.
9   Installation is reverse of removal. Use a new gasket and O-rings.
10   Tighten the cooler bolts to the torque listed in this Chapter's Specifications.
11   Check the transfer case lubricant level and add some, if necessary, to bring it to the appropriate level.

8   Transfer case - removal and installation

1   Raise the front of the vehicle and support it securely on jackstands. Remove the right wheel.
2   Unbolt the front portion of the driveshaft (see Chapter 8) and suspend it from a piece of wire (don't let it hang from the center support bearing).
3   Remove the right driveaxle and intermediate shaft (see Chapter 8). Remove the right-side catalytic converter (see Chapter 6).
4   On 3.5L turbocharged models, remove the flexible exhaust pipe on the left.
5   On models equipped with a transfer case oil-to-coolant cooler, remove the bolt attaching the other end of the cooler lines to the transfer case cooler. Disconnect the lines and discard the O-rings.
6   On 2016 and later V6 models equipped with a transfer case oil-to-air cooler, remove the bolt attaching the cooler line bracket to the transfer case. Remove the cooler connection clips and disconnect the lines.
7   On all models, disconnect the transfer case electrical connectors and detach the harness clips.
8   Remove the roll-restrictor mount fasteners and mount (see Chapter 7A).
9   Remove the bolts securing the transfer case to the support bracket.
10   Remove the bolts securing the transfer case support bracket to the engine.

Note: *The front crossmember may require removal before the bracket can be removed.*

11   Disconnect the transfer case vent hose (if equipped).
12   On 3.5L turbocharged models, if necessary, loosen the clamp for the rear turbo inlet hose and disconnect for access during removal.
13   If equipped, remove the transfer case heat shield, above the output shaft.
14   Remove the front crossmember (see Chapter 10, Section 19).
15   Remove the transaxle-to-transfer case mounting bolts.
16   To remove the transfer case from the vehicle, rock the transfer case until it comes free of the transaxle.
17   Installation is the reverse of removal, noting the following points:
a) *Install a new compression seal to the case.*
b) *Tighten the NEW driveshaft fasteners to the torque listed in the Chapter 8 Specifications.*
c) *Tighten the transfer case fasteners to the torque listed in this Chapter's Specifications.*
d) *Refill the transfer case with the proper type and amount of lubricant (see Chapter 1).*
e) *Tighten the wheel lug nuts to the torque listed in the Chapter 1 Specifications.*

Notes

Chapter 8
Driveline

Contents

	Section		Section
Differential (AWD models) - removal and installation	5	Driveshaft (AWD models) - removal and installation	3
Differential pinion oil seal - replacement	6	General information	1
Driveaxles - removal and installation	2	Rear driveaxle oil seals (AWD models) - replacement	4

Specifications

Torque specifications
Ft-lbs (unless otherwise indicated)   Nm

Note: *One foot-pound (ft-lb) of torque is equivalent to 12 inch-pounds (in-lbs) of torque. Torque values below approximately 15 foot-pounds are expressed in inch-pounds, because most foot-pound torque wrenches are not accurate at these smaller values.*

Driveaxles
	Ft-lbs	Nm
Driveaxle/hub nut (front and rear)	258	350
Driveaxle support bracket nuts (right driveaxle)	18	25
Control arm balljoint nut	See Chapter 10	
Tie-rod end nut	See Chapter 10	

Driveshaft (AWD models)
	Ft-lbs	Nm
Driveshaft flange bolts at transfer case		
4 bolt flange	52	70
6 bolt flange	26	35
Driveshaft bracket		
Nuts	18	25
Studs	133 in-lbs	15
Driveshaft center support bearing		
Inner bolts	177 in-lbs	20
Outer bracket bolts	22	30
Rear driveshaft-to-differential pinion flange bolts	18	25

Rear differential (AWD models)
	Ft-lbs	Nm
Differential-to-front insulator bracket bolts	66	90
Differential-to-side insulator bracket bolts	66	90
Differential insulator bracket-to-subframe	66	90
Filler plug	21	29
Differential pinion nut	180	244

1 General information

1 The information in this Chapter deals with the components from the rear of the engine to the drive wheels, except for the transaxle, which is dealt with in the previous Chapter.
2 Since nearly all the procedures covered in this Chapter involve working under the vehicle, make sure it's securely supported on sturdy jackstands or on a hoist where the vehicle can be easily raised and lowered.

Active Torque Coupling (ATC) solenoid

3 The PCM controls the amount of torque applied to the rear wheels by commanding the ATC to control the clutch in the rear differential. The ATC solenoid is part of the rear differential and non-serviceable.

2 Driveaxles - removal and installation

Warning: *Wait until the engine is completely cool before beginning this procedure.*

Front

Note: *The intermediate shaft is part of the driveaxle assembly and serviced as a unit.*

Removal

1 Loosen the wheel lug nuts, raise the vehicle and support it securely on jackstands. Remove the wheel. Remove the pushpins securing the splash shield in the fenderwell and the vehicle under cover.
2 Insert a punch into the disc cooling vanes and allow it to rest against the brake caliper mounting bracket to keep the driveaxle from turning, then remove the driveaxle/hub nut (see illustration).
3 Separate the tie-rod end from the steering knuckle (see Chapter 10).
4 Remove the nut from the control arm balljoint (see Chapter 10) and separate the control arm from the steering knuckle (see illustration).
5 Use a drive hub remover or other suitable puller to separate the driveaxle from the hub (see illustration).
6 Swing the knuckle/hub assembly out (away from the vehicle) until the end of the driveaxle is free of the hub. Use a length of wire to secure the knuckle out of the way to the rear. Support the outer end of the driveaxle with a piece of wire to avoid unnecessary strain on the inner CV joint.
7 For the left driveaxle, use a driveaxle remover and slide hammer to remove the driveaxle from the transaxle. If a removal tool is not available, pry the inner CV joint out of the transaxle using a large screwdriver or pry bar positioned between the transaxle and the CV joint housing (see illustration). Be careful not to damage the differential seal.

2.2 With a punch inserted into the disc cooling vanes, remove the driveaxle/hub nut

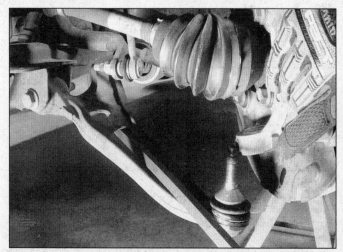

2.4 Separate the lower control arm from the steering knuckle

2.5 Using a drive hub remover to push the driveaxle out of the hub splines

2.7 Carefully pry the inner end of the left-side driveaxle from the transaxle

Chapter 8 Driveline

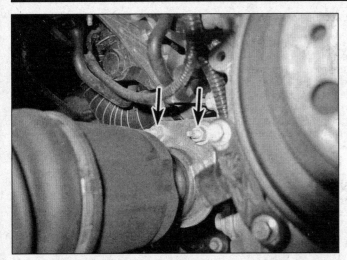

2.8 Remove the two nuts securing the driveaxle bracket

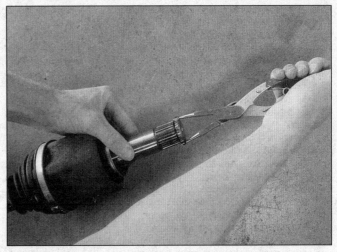

2.10 Pry the old spring clip from the inner end of the driveaxle

8   If you are removing the right side driveaxle, remove the two nuts securing the driveaxle bracket (see illustration). Carefully pull the driveaxle (with heat shield if equipped) from the transaxle. On AWD models, use care not to damage the seal in the seal in the transfer case when removing.
9   Support the CV joints and carefully remove the driveaxle from the vehicle.

Installation

10   Pry the old spring clip from the inner end of the driveaxle (left side) or outer end of the intermediate shaft (right side) and install a new one (see illustration).
11   Lubricate the differential or intermediate shaft seal with multi-purpose grease and raise the driveaxle into position while supporting the CV joints.
Note: *Position the spring clip with the opening facing down; this will ease insertion of the driveaxle and prevent damage to the clip.*
12   Push the splined end of the inner CV joint into the differential side gear and make sure the spring clip locks in its groove.
Caution: *Be careful to not let the sharp edges on the driveaxle splines cut or damage the oil seals when inserting the driveaxle.*
13   Check that the spring clip has engaged by tugging outward on the inboard CV joint housing.
Note: *To make insertion of the left-side driveaxle easier, position the gap in the spring clip in the 6 o-clock position.*
14   Apply a light coat of multi-purpose grease to the outer CV joint splines, pull out on the steering knuckle assembly and install the stub axle into the hub.
15   Reconnect the control arm to the steering knuckle and tighten all fasteners to the torque listed in the Chapter 10 Specifications.
16   Install new driveaxle/hub nut. With the inserted into the disc to keep the driveaxle from turning, tighten the nut to the torque listed in this Chapter's Specifications.
Caution: *The driveaxle/hub nut must be tight-*

2.24 Carefully pry the inner end of the driveaxle from the differential

ened before lowering the vehicle or damage to the wheel bearings may occur.
Note: *The driveaxle/hub nut is coated with a one-use locking chemical and cannot be reused for final assembly. The locking chemical is activated by heat created during tightening. The nut must be tightened to the specified torque within five minutes of starting it on the threads or the locking chemical will not activate properly.*
17   Install the wheel and lug nuts, then lower the vehicle.
18   Tighten the lug nuts to the torque listed in the Chapter 1 Specifications.

Rear (AWD models)
Warning: *The manufacturer recommends replacing all suspension fasteners with new ones during installation.*

Removal

19   Block the front wheels to prevent the vehicle from rolling. Loosen the wheel lug nuts, raise the rear of the vehicle and support it securely on jackstands. Remove the wheel.

Insert a punch into the disc cooling vanes and allow it to rest against the brake caliper mounting bracket, then unscrew the driveaxle/hub nut.
20   Remove the wheel speed sensor and brake line bracket from the wheel knuckle, then remove the brake caliper and brake disc (see Chapter 9). Hang the caliper with a length of wire and do not disconnect the brake hose.
21   Remove the driveaxle/hub nut but do not discard yet.
22   Use a drive hub remover to separate the driveaxle from the wheel hub (see illustration 2.5).
23   Remove the rear knuckle assembly (see Chapter 10, Section 12).
24   Carefully pry the inner end of the driveaxle out of the differential (see illustration) and remove from the vehicle.

Installation

25   Pry the old spring clip from the inner end of the driveaxle and install a new one.
26   Apply a light film of grease to the area on the inner CV joint stub shaft where the seal

rides, then insert the splined end of the inner CV joint into the differential. Make sure the spring clip locks in its groove.

Caution: Take care not to damage the differential seal with the driveaxle splines. It is recommended to use a plastic seal protector.

27 Install the rear knuckle assembly (see Chapter 10, Section 12).
28 Install the brake disc and caliper (see Chapter 9).
29 Install new driveaxle/hub nut. With the punch inserted into the disc cooling vanes to keep the driveaxle from turning, tighten the nut(s) to the specified torque.

Caution: The new driveaxle/hub nut(s) must be tightened before lowering the vehicle or damage to the wheel bearings may occur.

Note: The driveaxle/hub nuts are coated with a one-use locking chemical and cannot be reused for final assembly. The locking chemical is activated by heat created during tightening. The nut must be tightened to the specified torque within five minutes of starting it on the threads or the locking chemical will not activate properly.

30 The remainder of installation is the reverse of removal. Be sure to use new fasteners and tighten them to the specified torque.
31 Install the wheel and lug nuts, then lower the vehicle. Tighten the lug nuts to the torque listed in the Chapter 1 Specifications.

3 Driveshaft (AWD models) - removal and installation

Caution: The manufacturer recommends replacing driveshaft fasteners with new ones when installing the driveshaft.

1 Raise the vehicle and support it securely on jackstands. Place the shift lever in Neutral.
2 Remove the vehicle under cover.
3 Remove the muffler and tailpipe assembly.
4 Use chalk or a scribe to mark the relationship of the driveshaft flange to the differential pinion yoke. Do the same at the front U-joint flange where it bolts to the transfer case yoke. This ensures correct alignment when the driveshaft is reinstalled.
5 Working at the front of the driveshaft, remove the bolts securing the driveshaft to the transfer case.

Note: Some driveshafts have four bolts, some have six at the transfer case end

Caution: DO NOT detach the driveshaft from the transfer case flange by pulling on the driveshaft tube or the CV joint will be damaged.

6 Remove the bolts securing the universal joint flange to the differential pinion yoke.

Caution: DO NOT detach the driveshaft from the differential pinion yoke by pulling on the driveshaft tube or the CV joint will be damaged.

7 Remove the under body brace.
8 Remove the bolts securing the center support bearing.
9 With help from an assistant, carefully remove the driveshaft from the vehicle.
10 Installation is the reverse of removal. Use your index marks to keep the U-joints in alignment and make sure the mating surfaces are clean so the universal joint flanges seat properly in the yokes. Tighten the fasteners to the torque listed in this Chapter's Specifications.

4 Rear driveaxle oil seals (AWD models) - replacement

1 Raise the rear of the vehicle and support it securely on jackstands. Place the transaxle in Neutral with the parking brake off. Block the front wheels to prevent the vehicle from rolling.
2 Remove the driveaxle(s) (see Section 2).
3 Carefully pry out the driveaxle oil seal with a seal removal tool or a large screwdriver. Be careful not to damage or scratch the seal bore or the bearing inside the case.
4 Using a seal installer or a large deep socket as a drift, install the new oil seal. Drive it into the bore squarely and make sure it's completely seated.
5 Lubricate the lip of the new seal with multi-purpose grease, then install the driveaxle. Be careful not to damage the lip of the new seal.
6 Check the differential lubricant level and add some, if necessary, to bring it to the appropriate level (see Chapter 1).

5 Differential (AWD models) - removal and installation

Note: On AWD models, the manufacturer uses an Active Torque Control Coupling (ATCC) to manage power distribution to the front and rear wheels. Normally, the front wheels receive the majority of the torque produced by the engine. When the ATCC system senses wheel slip, it automatically increases the torque to the rear wheels. The system is always active, continuously monitoring vehicle conditions and adjusting the torque being delivered to the rear wheels by controlling the current sent to an electric clutch device inside the rear axle. The ATCC system has no mode selector nor does it require any driver input. The ATCC solenoid in the rear axle is not repairable and must be replaced with the rear axle assembly as a unit.

1 Raise the rear of the vehicle and support it securely on jackstands. Block the front wheels to prevent the vehicle from rolling.
2 Remove the driveshaft (see Section 3).
3 Remove the driveaxles (see Section 2).
4 Remove the rear stabilizer bar (see Chapter 10)
5 Support the rear axle housing with a floor jack.
6 Disconnect the electrical connector on the crossmember.
7 There are four rear differential mounting brackets securing the differential to the subframe, one at either side at the top-front of the assembly and two at the top-rear. Remove the four bolts (two each side) securing the differential to each front bracket.
8 Loosen the two front mounting brackets from the subframe and swing the brackets out to allow clearance to lower the rear differential.
9 Remove the five bolts securing the differential to the rear mounting brackets.
10 Lower the differential with the floor jack.
11 Installation is the reverse of removal. Tighten all fasteners to the torque listed in this Chapter's Specifications.

6 Differential pinion oil seal - replacement

1 Loosen the wheel lug nuts. Raise the rear of the vehicle and support it securely on jackstands. Block the opposite set of wheels to keep the vehicle from rolling off the stands. Remove the wheels.
2 Disconnect the driveshaft from the differential companion flange and fasten it out of the way (see Section 3).
3 Mark the relationship of the pinion flange to the shaft, then count and write down the number of exposed threads on the shaft.
4 A flange holding tool will be required to keep the companion flange from moving while the self-locking pinion nut is loosened. A chain wrench will also work.
5 Remove the pinion nut.
6 Withdraw the flange. It may be necessary to use a two-jaw puller engaged behind the flange to draw it off. Do not attempt to pry or hammer behind the flange or hammer on the end of the pinion shaft.
7 Pry out the old seal and discard it.
8 Inspect the seal bore to ensure it is clean and smooth.
9 Lubricate the lips of the new seal and fill the space between the seal lips with wheel bearing grease, then tap it evenly into position with a seal installation tool or a large socket. Make sure it enters the housing squarely and is tapped in to its full depth.
10 Install the pinion flange, lining up the marks made in Step 3. If necessary, tighten the pinion nut to draw the flange into place. Do not try to hammer the flange into position.
11 Install a new pinion nut, then tighten it to the torque listed in this Chapter's Specifications.
12 The remainder of installation is the reverse of removal.

Chapter 9
Brakes

Contents

	Section		Section
Anti-lock Brake System (ABS) - general information	3	Disc brake caliper - removal and installation	5
Brake booster vacuum pump (four-cylinder models) - removal and installation	12	Disc brake pads - replacement	4
		General information	1
Brake disc - inspection, removal and installation	6	Master cylinder - removal and installation	7
Brake hoses and lines - inspection and replacement	8	Power brake booster - check removal and installation	10
Brake hydraulic system - bleeding	9	Troubleshooting	2
Brake light switch - replacement	11		

Specifications

General

Brake fluid type	See Chapter 1

Disc brakes

Minimum pad lining thickness	See Chapter 1
Disc lateral runout limit	0.002 inch (0.050 mm)
Disc minimum thickness	Cast into disc

Torque specifications

Ft-lbs (unless otherwise indicated)   Nm

Note: One foot-pound (ft-lb) of torque is equivalent to 12 inch-pounds (in-lbs) of torque. Torque values below approximately 15 foot-pounds are expressed in inch-pounds, because most foot-pound torque wrenches are not accurate at these smaller values.

	Ft-lbs	Nm
Brake line banjo bolt (front)	35	47
Brake line-to-caliper fitting (rear)	18	25
2012 and earlier models	18	25
2013 and later models	20	30
Brake tubing to flexible brake hose	150 in-lbs	17
Caliper mounting bracket bolts*		
Front		
2012 and earlier models	111	150
2013 and later models	122	165
Rear	76	103
Caliper mounting (guide pin) bolts		
Front - 13 inch discs	53	72
Front - 14 inch discs	55	75
Rear	24	33
Master cylinder mounting nuts	177 in-lbs	20
Brake line nuts (to master cylinder)	16	22
Power brake booster mounting nuts	18	25
Brake booster vacuum pump mounting bolts	89 in-lbs	10
Wheel speed sensor bolt	133 in-lbs	15
Wheel lug nuts	See the Chapter 1 Specifications	

*Replace with new fasteners

Chapter 9 Brakes

1 General Information

1   The vehicles covered by this manual are equipped with hydraulically operated front and rear brake systems. The front brakes are disc type and the rear brakes are disc or drum type. Both the front and rear brakes are self adjusting. The disc brakes automatically compensate for pad wear, while the drum brakes incorporate an adjustment mechanism that is activated as the parking brake is applied.

Hydraulic system

2   The hydraulic system consists of two separate circuits. The master cylinder has separate reservoir chambers for the two circuits, and, in the event of a leak or failure in one hydraulic circuit, the other circuit will remain operative. A dynamic proportioning valve, integral with the ABS hydraulic unit, provides brake balance to each individual wheel.

Power brake booster and vacuum pump

3   The power brake booster, utilizing engine manifold vacuum, and atmospheric pressure to provide assistance to the hydraulically operated brakes, is mounted on the firewall in the engine compartment. An auxiliary vacuum pump, mounted below the power brake booster in the left side of the engine compartment, provides additional vacuum to the booster under certain operating conditions.

Parking brake

4   The parking brake operates the rear brakes only, through cable actuation. It's activated by a lever mounted in the center console.

Service

5   After completing any operation involving disassembly of any part of the brake system, always test drive the vehicle to check for proper braking performance before resuming normal driving. When testing the brakes, perform the tests on a clean, dry, flat surface. Conditions other than these can lead to inaccurate test results.
6   Test the brakes at various speeds with both light and heavy pedal pressure. The vehicle should stop evenly without pulling to one side or the other. Avoid locking the brakes, because this slides the tires and diminishes braking efficiency and control of the vehicle.
7   Tires, vehicle load and wheel alignment are factors which also affect braking performance.

Precautions

8   There are some general cautions and warnings involving the brake system on this vehicle:

a) *Use only brake fluid conforming to DOT 3 specifications.*
b) *The brake pads and linings contain fibers that are hazardous to your health if inhaled. Whenever you work on brake system components, clean all parts with brake system cleaner. Do not allow the fine dust to become airborne. Also, wear an approved filtering mask.*
c) *Safety should be paramount whenever any servicing of the brake components is performed. Do not use parts or fasteners that are not in perfect condition, and be sure that all clearances and torque specifications are adhered to. If you are at all unsure about a certain procedure, seek professional advice. Upon completion of any brake system work, test the brakes carefully in a controlled area before putting the vehicle into normal service. If a problem is suspected in the brake system, don't drive the vehicle until it's fixed.*
d) *Used brake fluid is considered a hazardous waste and it must be disposed of in accordance with federal, state and local laws. DO NOT pour it down the sink, into septic tanks or storm drains, or on the ground. Clean up any spilled brake fluid immediately and then wash the area with large amounts of water. This is especially true for any finished or painted surfaces.*

2 Troubleshooting

PROBABLE CAUSE	CORRECTIVE ACTION
No brakes - pedal travels to floor	
1 Low fluid level 2 Air in system	1 and 2 Low fluid level and air in the system are symptoms of another problem a leak somewhere in the hydraulic system. Locate and repair the leak
3 Defective seals in master cylinder	3 Replace master cylinder
4 Fluid overheated and vaporized due to heavy braking	4 Bleed hydraulic system (temporary fix). Replace brake fluid (proper fix)
Brake pedal slowly travels to floor under braking or at a stop	
1 Defective seals in master cylinder	1 Replace master cylinder
2 Leak in a hose, line, caliper or wheel cylinder	2 Locate and repair leak
3 Air in hydraulic system	3 Bleed the system, inspect system for a leak
Brake pedal feels spongy when depressed	
1 Air in hydraulic system	1 Bleed the system, inspect system for a leak
2 Master cylinder or power booster loose	2 Tighten fasteners
3 Brake fluid overheated (beginning to boil)	3 Bleed the system (temporary fix). Replace the brake fluid (proper fix)
4 Deteriorated brake hoses (ballooning under pressure)	4 Inspect hoses, replace as necessary (it's a good idea to replace all of them if one hose shows signs of deterioration)

Chapter 9 Brakes

PROBABLE CAUSE	CORRECTIVE ACTION
Brake pedal feels hard when depressed and/or excessive effort required to stop vehicle	
1 Power booster faulty	1 Replace booster
2 Engine not producing sufficient vacuum, or hose to booster clogged, collapsed or cracked	2 Check vacuum to booster with a vacuum gauge. Replace hose if cracked or clogged, repair engine if vacuum is extremely low
3 Brake linings contaminated by grease or brake fluid	3 Locate and repair source of contamination, replace brake pads or shoes
4 Brake linings glazed	4 Replace brake pads or shoes, check discs and drums for glazing, service as necessary
5 Caliper piston(s) or wheel cylinder(s) binding or frozen	5 Replace calipers or wheel cylinders
6 Brakes wet	6 Apply pedal to boil-off water (this should only be a momentary problem)
7 Kinked, clogged or internally split brake hose or line	7 Inspect lines and hoses, replace as necessary
Excessive brake pedal travel (but will pump up)	
1 Drum brakes out of adjustment	1 Adjust brakes
2 Air in hydraulic system	2 Bleed system, inspect system for a leak
Excessive brake pedal travel (but will not pump up)	
1 Master cylinder pushrod misadjusted	1 Adjust pushrod
2 Master cylinder seals defective	2 Replace master cylinder
3 Brake linings worn out	3 Inspect brakes, replace pads and/or shoes
4 Hydraulic system leak	4 Locate and repair leak
Brake pedal doesn't return	
1 Brake pedal binding	1 Inspect pivot bushing and pushrod, repair or lubricate
2 Defective master cylinder	2 Replace master cylinder
Brake pedal pulsates during brake application	
1 Brake drums out-of-round	1 Have drums machined by an automotive machine shop
2 Excessive brake disc runout or disc surfaces out-of-parallel	2 Have discs machined by an automotive machine shop
3 Loose or worn wheel bearings	3 Adjust or replace wheel bearings
4 Loose lug nuts	4 Tighten lug nuts
Brakes slow to release	
1 Malfunctioning power booster	1 Replace booster
2 Pedal linkage binding	2 Inspect pedal pivot bushing and pushrod, repair/lubricate
3 Malfunctioning proportioning valve	3 Replace proportioning valve
4 Sticking caliper or wheel cylinder	4 Repair or replace calipers or wheel cylinders
5 Kinked or internally split brake hose	5 Locate and replace faulty brake hose
Brakes grab (one or more wheels)	
1 Grease or brake fluid on brake lining	1 Locate and repair cause of contamination, replace lining
2 Brake lining glazed	2 Replace lining, deglaze disc or drum

Chapter 9 Brakes

Troubleshooting (continued)

PROBABLE CAUSE	CORRECTIVE ACTION
Vehicle pulls to one side during braking	
1 Grease or brake fluid on brake lining	1 Locate and repair cause of contamination, replace lining
2 Brake lining glazed	2 Deglaze or replace lining, deglaze disc or drum
3 Restricted brake line or hose	3 Repair line or replace hose
4 Tire pressures incorrect	4 Adjust tire pressures
5 Caliper or wheel cylinder sticking	5 Repair or replace calipers or wheel cylinders
6 Wheels out of alignment	6 Have wheels aligned
7 Weak suspension spring	7 Replace springs
8 Weak or broken shock absorber	8 Replace shock absorbers
Brakes drag (indicated by sluggish engine performance or wheels being very hot after driving)	
1 Brake pedal pushrod incorrectly adjusted	1 Adjust pushrod
2 Master cylinder pushrod (between booster and master cylinder)	2 Adjust pushrod incorrectly adjusted
3 Obstructed compensating port in master cylinder	3 Replace master cylinder
4 Master cylinder piston seized in bore	4 Replace master cylinder
5 Contaminated fluid causing swollen seals throughout system	5 Flush system, replace all hydraulic components
6 Clogged brake lines or internally split brake hose(s)	6 Flush hydraulic system, replace defective hose(s)
7 Sticking caliper(s) or wheel cylinder(s)	7 Replace calipers or wheel cylinders
8 Parking brake not releasing	8 Inspect parking brake linkage and parking brake mechanism, repair as required
9 Improper shoe-to-drum clearance	9 Adjust brake shoes
10 Faulty proportioning valve	10 Replace proportioning valve
Brakes fade (due to excessive heat)	
1 Brake linings excessively worn or glazed	1 Deglaze or replace brake pads and/or shoes
2 Excessive use of brakes	2 Downshift into a lower gear, maintain a constant slower speed (going down hills)
3 Vehicle overloaded	3 Reduce load
4 Brake drums or discs worn too thin	4 Measure drum diameter and disc thickness, replace drums or discs as required
5 Contaminated brake fluid	5 Flush system, replace fluid
6 Brakes drag	6 Repair cause of dragging brakes
7 Driver resting left foot on brake pedal	7 Don't ride the brakes
Brakes noisy (high-pitched squeal)	
1 Glazed lining	1 Deglaze or replace lining
2 Contaminated lining (brake fluid, grease, etc.)	2 Repair source of contamination, replace linings
3 Weak or broken brake shoe hold-down or return spring	3 Replace springs
4 Rivets securing lining to shoe or backing plate loose	4 Replace shoes or pads
5 Excessive dust buildup on brake linings	5 Wash brakes off with brake system cleaner
6 Brake drums worn too thin	6 Measure diameter of drums, replace if necessary
7 Wear indicator on disc brake pads contacting disc	7 Replace brake pads
8 Anti-squeal shims missing or installed improperly	8 Install shims correctly

PROBABLE CAUSE	CORRECTIVE ACTION
Brakes noisy (scraping sound)	
1 Brake pads or shoes worn out; rivets, backing plate or brake	1 Replace linings, have discs and/or drums machined (or replace) shoe metal contacting disc or drum
Brakes chatter	
1 Worn brake lining	1 Inspect brakes, replace shoes or pads as necessary
2 Glazed or scored discs or drums	2 Deglaze discs or drums with sandpaper (if glazing is severe, machining will be required)
3 Drums or discs heat checked	3 Check discs and/or drums for hard spots, heat checking, etc. Have discs/drums machined or replace them
4 Disc runout or drum out-of-round excessive	4 Measure disc runout and/or drum out-of-round, have discs or drums machined or replace them
5 Loose or worn wheel bearings	5 Adjust or replace wheel bearings
6 Loose or bent brake backing plate (drum brakes)	6 Tighten or replace backing plate
7 Grooves worn in discs or drums	7 Have discs or drums machined, if within limits (if not, replace them)
8 Brake linings contaminated (brake fluid, grease, etc.)	8 Locate and repair source of contamination, replace pads or shoes
9 Excessive dust buildup on linings	9 Wash brakes with brake system cleaner
10 Surface finish on discs or drums too rough after machining	10 Have discs or drums properly machined (especially on vehicles with sliding calipers)
11 Brake pads or shoes glazed	11 Deglaze or replace brake pads or shoes
Brake pads or shoes click	
1 Shoe support pads on brake backing plate grooved or	1 Replace brake backing plate excessively worn
2 Brake pads loose in caliper	2 Loose pad retainers or anti-rattle clips
3 Also see items listed under Brakes chatter	
Brakes make groaning noise at end of stop	
1 Brake pads and/or shoes worn out	1 Replace pads and/or shoes
2 Brake linings contaminated (brake fluid, grease, etc.)	2 Locate and repair cause of contamination, replace brake pads or shoes
3 Brake linings glazed	3 Deglaze or replace brake pads or shoes
4 Excessive dust buildup on linings	4 Wash brakes with brake system cleaner
5 Scored or heat-checked discs or drums	5 Inspect discs/drums, have machined if within limits (if not, replace discs or drums)
6 Broken or missing brake shoe attaching hardware	6 Inspect drum brakes, replace missing hardware
Rear brakes lock up under light brake application	
1 Tire pressures too high	1 Adjust tire pressures
2 Tires excessively worn	2 Replace tires
3 Defective proportioning valve	3 Replace proportioning valve
Brake warning light on instrument panel comes on (or stays on)	
1 Low fluid level in master cylinder reservoir (reservoirs with fluid level sensor)	1 Add fluid, inspect system for leak, check the thickness of the brake pads and shoes
2 Failure in one half of the hydraulic system	2 Inspect hydraulic system for a leak
3 Piston in pressure differential warning valve not centered	3 Center piston by bleeding one circuit or the other (close bleeder valve as soon as the light goes out)

Chapter 9 Brakes

Troubleshooting (continued)

Brake warning light on instrument panel comes on (or stays on) (continued)

PROBABLE CAUSE	CORRECTIVE ACTION
4 Defective pressure differential valve or warning switch	4 Replace valve or switch
5 Air in the hydraulic system	5 Bleed the system, check for leaks
6 Brake pads worn out (vehicles with electric wear sensors - small	6 Replace brake pads (and sensors) probes that fit into the brake pads and ground out on the disc when the pads get thin)

Brakes do not self adjust

Disc brakes

1 Defective caliper piston seals	1 Replace calipers. Also, possible contaminated fluid causing soft or swollen seals (flush system and fill with new fluid if in doubt)
2 Corroded caliper piston(s)	2 Same as above

Drum brakes

1 Adjuster screw frozen	1 Remove adjuster, disassemble, clean and lubricate with high-temperature grease
2 Adjuster lever does not contact star wheel or is binding	2 Inspect drum brakes, assemble correctly or clean or replace parts as required
3 Adjusters mixed up (installed on wrong wheels after brake job)	3 Reassemble correctly
4 Adjuster cable broken or installed incorrectly (cable-type adjusters)	4 Install new cable or assemble correctly

Rapid brake lining wear

1 Driver resting left foot on brake pedal	1 Don't ride the brakes
2 Surface finish on discs or drums too rough	2 Have discs or drums properly machined
3 Also see Brakes drag	

3 Anti-lock Brake System (ABS) - general information

General information

1 The anti-lock brake system is designed to maintain vehicle steerability, directional stability and optimum deceleration under severe braking conditions on most road surfaces. It does so by monitoring the rotational speed of each wheel and controlling the brake line pressure to each wheel during braking. This prevents the wheels from locking up.

2 The ABS system has three main components - the wheel speed sensors, the electronic control unit (ECU) and the hydraulic unit (see illustration). Four wheel-speed sensors - one at each wheel - send a variable voltage signal to the control unit, which monitors these signals, compares them to its program and determines whether a wheel is about to lock up. When a wheel is about to lock up, the control unit signals the hydraulic unit to reduce hydraulic pressure (or not increase it further) at that wheel's brake caliper. Pressure modulation is handled by electrically-operated solenoid valves.

3 If a problem develops within the system, an "ABS" warning light will glow on the dashboard. Sometimes, a visual inspection of the ABS system can help you locate the problem. Carefully inspect the ABS wiring harness. Pay particularly close attention to the harness and connections near each wheel. Look for signs of chafing and other damage caused by incor-

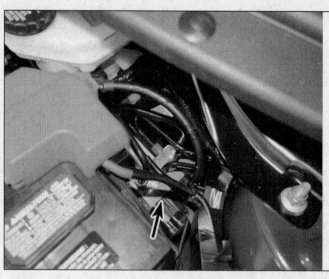

3.2 The ABS hydraulic unit and ECU are located under the master cylinder

rectly routed wires. If a wheel sensor harness is damaged, the sensor must be replaced.

Warning: *Do NOT try to repair an ABS wiring harness. The ABS system is sensitive to even the smallest changes in resistance. Repairing the harness could alter resistance values and cause the system to malfunction. If the ABS wiring harness is damaged in any way, it must be replaced.*

Caution: *Make sure the ignition is turned off before unplugging or reattaching any electrical connections.*

4   Some vehicles come equipped with traction control, which helps prevent wheel spin under acceleration. The traction control system (TCS) utilizes some of the same components as the ABS system. The wheel sensors monitor wheel rotation and transmit data through the same wiring to the ABS control module. The control module compares rotation speeds, then controls the amount of power supplied to the wheels by adjusting fuel settings to the engine.

Diagnosis and repair

5   If a dashboard warning light comes on and stays on while the vehicle is in operation, the ABS system requires attention. Although special electronic ABS diagnostic testing tools are necessary to properly diagnose the system, you can perform a few preliminary checks before taking the vehicle to a dealer service department.

a) Check the brake fluid level in the reservoir.
b) Verify that the computer electrical connectors are securely connected.
c) Check the electrical connectors at the hydraulic control unit.
d) Check the fuses.
e) Follow the wiring harness to each wheel and verify that all connections are secure and that the wiring is undamaged.

6   If the above preliminary checks do not rectify the problem, the vehicle should be diagnosed by a dealer service department or other qualified repair shop. Due to the complex nature of this system, all actual repair work must be done by a qualified automotive technician.

Wheel speed sensor - removal and installation

7   Loosen the wheel lug nuts, raise the vehicle and support it securely on jackstands. Remove the wheel.
8   Make sure the ignition key is turned to the Off position.
9   If needed, remove the inner fender splash shield (see Chapter 11). Trace the wiring back from the sensor, detaching all brackets and clips while noting its correct routing, then disconnect the electrical connector.
10   Remove the mounting bolt and carefully pull the sensor out from the steering knuckle.
11   Installation is the reverse of the removal procedure. Tighten the mounting bolt to the torque listed in this Chapter's Specifications.
12   Install the wheel and lug nuts, tightening them securely. Lower the vehicle and tighten the lug nuts to the torque listed in the Chapter 1 Specifications.

4   Disc brake pads - replacement

Warning: *Disc brake pads must be replaced on both front or both rear wheels at the same time - never replace the pads on only one wheel. Also, the dust created by the brake system is harmful to your health. Never blow it out with compressed air and don't inhale any of it. An approved filtering mask should be worn when working on the brakes. Do not, under any circumstances, use petroleum-based solvents to clean brake parts. Use brake system cleaner only!*

1   Remove the cap from the brake fluid reservoir. To prevent brake fluid from possibly overflowing the reservoir when the calipers are fully retracted, use a syringe or suction gun to remove brake fluid until the reservoir is approximately half full.
2   Loosen the wheel lug nuts, raise the vehicle and support it securely on jackstands. Block the wheels at the opposite end.
3   Remove the wheels. Work on one brake assembly at a time, using the assembled brake for reference if necessary.
4   Inspect the brake disc carefully as outlined in Section 6. If machining is necessary, follow the information in that Section to remove the disc, at which time the pads can be removed as well.
5   Before disassembling the brake, wash it thoroughly with brake system cleaner and allow it to dry (see illustration). Position a drain pan under the brake to catch the residue - DO NOT use compressed air to blow off the brake dust.

Front brake pads

6   Follow the accompanying photos (illustrations 4.6a through 4.6k) for the pad replacement procedure. Be sure to stay in order and read the caption under each illustration.
7   Repeat the same procedure for the opposite side.

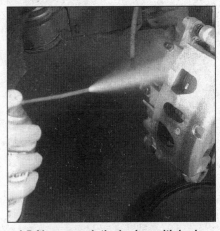

4.5 Always wash the brakes with brake cleaner before disassembling anything

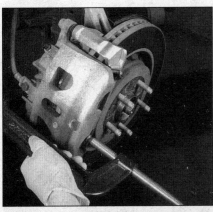

4.6a Depress the piston into the bottom of its bore in the caliper with a large C-clamp to make room for the new pads (alternate from one end to the other to depress the pistons evenly) - make sure the fluid in the master cylinder reservoir doesn't overflow

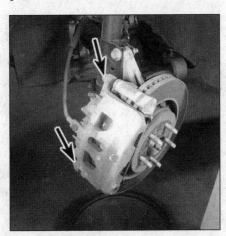

4.6b Remove the caliper mounting (guide pin) bolts

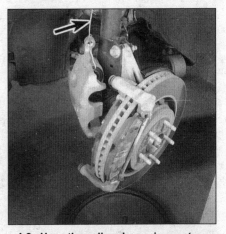

4.6c Hang the caliper by a wire or strap in such a way that it is not pulling or dangling from the brake hose

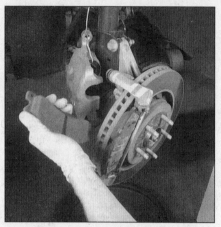
4.6d Remove the inner brake pad and shim

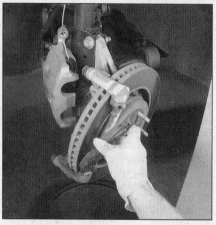
4.6e Remove the outer brake pad and shim

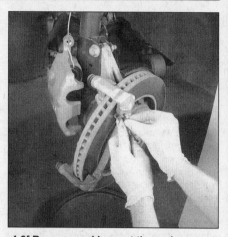
4.6f Remove and inspect the pad support plates, then reinstall (or replace if necessary) and lubricate them with a light film of high-temperature brake grease

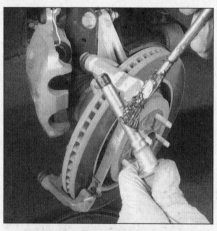
4.6g Remove and clean the caliper guide pins, then apply a coat of high-temperature brake grease to the sliding surface

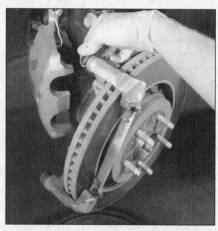
4.6h Reinstall the rubber boots to the caliper guide pins. Replace if needed

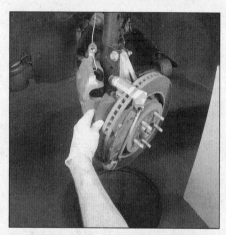
4.6i Install the pads onto the caliper bracket. Be sure to use a light amount of high-temperature brake grease on the areas where the ends of the pads touch the support plates

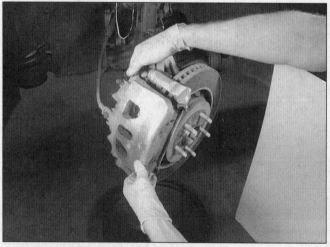
4.6j Work the caliper over the brake pads and install the caliper bolts

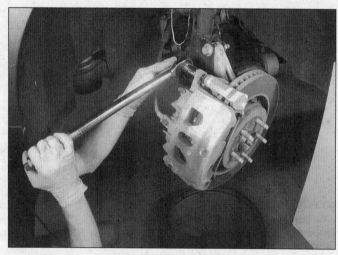
4.6k Tighten the caliper mounting bolts to the torque listed in this Chapter's Specifications and proceed to Step 9

Chapter 9 Brakes

Rear brake pads

8 For the brake pad replacement sequence, follow Steps 1 through 5, then see illustrations 4.8a through 4.8i. Note the following points:
a) When removing and installing the brake noise dampener, use only hand tools or the dampener will be damaged.
b) The parking brake cable is attached to the caliper. The cable does not have to be separated from the caliper in order to change the pads, but be careful not to twist or damage the cable ends when positioning and supporting the caliper to the side.
c) The manufacturer states that the brake pads are one-time use only. Anytime the brake pads are separated from the caliper, they must be replaced with new pads even if they are not worn out. Brake shudder and noise will occur if pads are reused.
d) The caliper guide pins are pressed into the mounting bracket (anchor plate). Do not try to remove them or they will be damaged and the mounting bracket will have to be replaced.
e) The inboard brake pad has a pin on the back. Turn the caliper piston so the notch will line-up with the pin on the backside of the pad.

Caution: *Make sure the pin on the inner pad engages with the notch in the caliper piston*

Front or rear pads

9 When reinstalling the caliper, be sure to tighten the mounting bolts to the torque listed in this Chapter's Specifications. After the job has been completed, firmly depress the brake pedal a few times to bring the pads into contact with the disc. Check the level of the brake fluid, adding some if necessary. Check the operation of the brakes carefully before placing the vehicle into normal service.

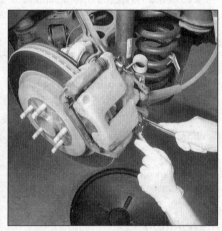

4.8a Hold the caliper guide pins with one wrench while loosening the caliper mounting bolts

4.8b Remove the caliper and hang it with a length of wire - don't let it hang by the hose

4.8c Use needle-nose pliers or a caliper piston rotating tool to rotate the piston clockwise while pushing it back in the caliper - make sure the fluid in the master cylinder reservoir doesn't overflow

4.8d One of the notches must be at the bottom position to align with the pin on the inner pad

4.8e Remove the inner brake pad

4.8f Remove the outer brake pad

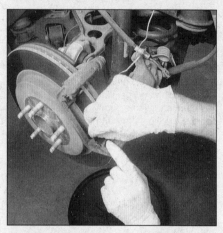

4.8g Remove and inspect the pad support plates, then reinstall (or replace if necessary) and lubricate them with a light film of high-temperature brake grease

4.8h Pull out the caliper guide pins and clean them, then apply a coat of high-temperature grease to the pins and reinstall them in the caliper bracket. Then install the new brake pads into the mounting bracket

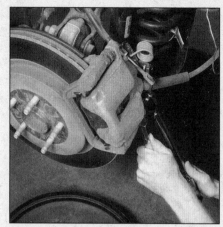

4.8i Install the pads and caliper, tightening the caliper mounting bolts to the torque listed in this Chapter's Specifications. Be sure to coat the areas where the pad touches the support plates and calipers with a thin film of high-temperature brake grease

5 Disc brake caliper - removal and installation

Warning: *Dust created by the brake system is harmful to your health. Never blow it out with compressed air and don't inhale any of it. An approved filtering mask should be worn when working on the brakes. Do not, under any circumstances, use petroleum-based solvents to clean brake parts. Use brake system cleaner only.*

Note: *If replacement is indicated (usually because of fluid leakage or damage to a piston boot), it is recommended that the calipers be replaced, not overhauled. New and factory rebuilt units are available on an exchange basis, which makes this job quite easy. Always replace the calipers in pairs - never replace just one of them.*

Removal

1 Loosen the wheel lug nuts, raise the vehicle (front or rear) and place it securely on jackstands. Remove the wheels.

2 If you're working on a front caliper, remove the banjo fitting bolt and disconnect the brake hose from the caliper (see illustration). Discard the sealing washers from each side of the hose fitting. Plug the brake hose to keep contaminants out of the brake system and to prevent losing any more brake fluid than is necessary.

Note: *If the caliper is being removed for access to another component, don't disconnect the hose.*

Note: *For the rear calipers, have an assistant pull down on the front parking brake cable until the sector at the parking brake pedal comes to its stop, then insert a 5/32-inch (4 mm) x 6-inch (152 mm) rod can be inserted into the service hole in the parking brake pedal mechanism. This will relieve tension on the cables.*

3 If you're removing a rear caliper, loosen the brake hose from the caliper (don't attempt to unscrew it yet).

4 Start at illustration 4.6a (front) or illustration 4.8a (rear) for the caliper removal procedure. If the caliper is being removed just to access another component, use a piece of wire to securely hang it out of the way (see illustration 4.6c).

Caution: *Do not let the caliper hang by the brake hose.*

5 If you're removing a rear caliper, disconnect the parking brake cable from the caliper, then unscrew the caliper from the brake hose.

Installation

6 Install the caliper by reversing the removal procedure, tightening the mounting bolts to the torque listed in this Chapter's Specifications. On front calipers, be sure to use new sealing washers and tighten the banjo fitting bolt to the torque listed in this Chapter's Specifications.

7 Bleed the brake circuit according to the procedure in Section 9. Make sure there are no leaks from the hose connections. Test the brakes carefully before returning the vehicle to normal service.

6 Brake disc - inspection, removal and installation

Inspection

1 Loosen the wheel lug nuts, raise the vehicle and support it securely on jackstands.

2 Remove the brake caliper as outlined in Section 5. It isn't necessary to disconnect the brake hose. After removing the caliper bolts, suspend the caliper out of the way with a piece of wire (see illustration 4.7b).

3 Visually inspect the disc surface for score marks and other damage. Light scratches and shallow grooves are normal after use and may not always be detrimental to brake operation, but deep scoring requires disc removal and

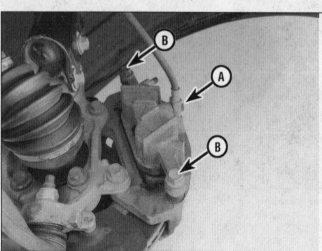

5.2 Remove the brake line banjo bolt (A), then remove the caliper mounting bolts (B)

Chapter 9 Brakes

6.3 The brake pads on this vehicle were obviously neglected, as they wore down completely and cut deep grooves into the disc - wear this severe means the disc must be replaced

6.4a To check disc runout, mount a dial indicator as shown and rotate the disc

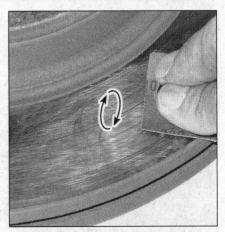

6.4b Using a swirling motion, remove the glaze from the disc surface with sandpaper or emery cloth

6.5a The minimum thickness dimension is cast into the front or back side of the disc

6.5b Use a micrometer to measure disc thickness

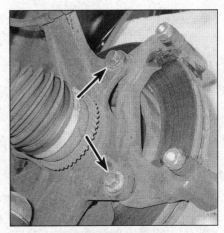

6.6a Caliper mounting bracket-to-knuckle bolts

refinishing by an automotive machine shop. Be sure to check both sides of the disc (see illustration). If pulsating has been noticed during application of the brakes, suspect disc runout.

4   To check disc runout, reinstall the lug nuts (inverted) and place a dial indicator at a point about 1/2-inch from the outer edge of the disc (see illustration). Set the indicator to zero and turn the disc. The indicator reading should not exceed the specified allowable runout limit. If it does, the disc should be refinished by an automotive machine shop.
Note: *The discs should be resurfaced regardless of the dial indicator reading, as this will impart a smooth finish and ensure a perfectly flat surface, eliminating any brake pedal pulsation or other undesirable symptoms related to questionable discs. At the very least, if you elect not to have the discs resurfaced, remove the glaze from the surface with emery cloth or sandpaper, using a swirling motion (see illustration).*

5   It's absolutely critical that the disc not be machined to a thickness under the specified minimum thickness. The minimum (or discard) thickness is cast or stamped into the inside of the disc (see illustration). The disc thickness can be checked with a micrometer (see illustration).

Removal

6   Remove the caliper mounting bracket (see illustration). Also remove the lug nuts if they were reinstalled for the runout, then remove the disc (see illustration).
Note: *If equipped, remove the brake disc retaining screw. If necessary, insert an M10 bolt into the threaded hole, then tighten the bolt to separate the disc from the hub.*

Installation

7   While the disc is off, wire-brush the backside of the center portion that contacts the wheel hub. Also clean off any rust or dirt

6.6b On models so equipped, remove the screw that secures the disc to the hub

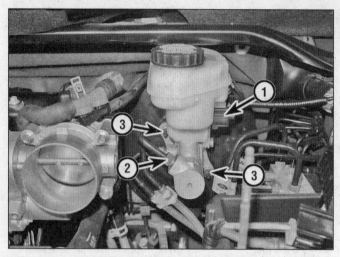
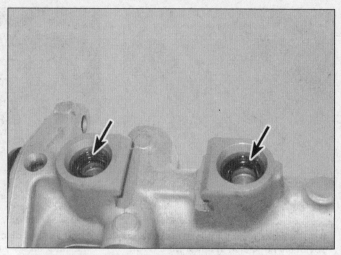

7.5 Master cylinder mounting details
1. Fluid level warning switch connector
2. Brake line fittings
3. Mounting nuts

7.10 After the reservoir has been removed, replace the O-rings with new ones

on the hub face.
8 Apply small dots of high-temperature anti-seize around the circumference of the hub, and around the raised center portion.
9 Place the disc in position over the threaded studs. If equipped, install the disc retaining screw and tighten it securely.
10 Install the caliper mounting bracket and caliper, tightening the bolts to the torque values listed in this Chapter's Specifications.
11 Install the wheel, then lower the vehicle to the ground. Tighten the lug nuts to the torque listed in the Chapter 1 Specifications . Depress the brake pedal a few times to bring the brake pads into contact with the disc. Bleeding won't be necessary unless the brake hose was disconnected from the caliper. Check the operation of the brakes carefully before driving the vehicle.
12 Check the operation of the brakes carefully before driving the vehicle.

7 Master cylinder - removal and installation

Removal

1 The master cylinder is located in the engine compartment, mounted to the power brake booster.
2 Remove the battery and the battery tray (see Chapter 5).
3 Remove the intake duct between the air filter housing and throttle body (see Chapter 4). On turbocharged V6 models, remove the charge air cooler inlet pipe
4 Remove as much fluid as you can from the reservoir with a syringe, such as an old turkey baster.
Warning: *If a baster is used, never again use it for the preparation of food.*
Caution: *Brake fluid will damage paint. Cover all painted surfaces and avoid spilling fluid during this procedure. If you do spill fluid, wash it off with water immediately.*

5 Disconnect the brake booster check valve from the brake booster to remove any vacuum from the booster (see illustration) .
6 Disconnect the electrical connector at the brake fluid level switch on the master cylinder reservoir.
7 Place rags under the fluid fittings and prepare caps or plastic bags to cover the ends of the lines once they are disconnected.
Caution: *Brake fluid will damage paint. Cover all body parts and be careful not to spill fluid during this procedure.*
8 Loosen the fittings at the ends of the brake lines where they enter the master cylinder. To prevent rounding off the corners on these nuts, the use of a flare-nut wrench, which wraps around the nut, is preferred. Pull the brake lines slightly away from the master cylinder and plug the ends to prevent contamination.
9 Remove the nuts attaching the master cylinder to the power booster. Pull the master cylinder off the studs and out of the engine compartment. Again, be careful not to spill the fluid as this is done.
10 If a new master cylinder is being installed, remove the reservoir from the old master cylinder and transfer it to the new master cylinder.
Note: *Be sure to install new seals when transferring the reservoir(see illustration).*

Installation

11 Bench bleed the new master cylinder before installing it. Mount the master cylinder in a vise, with the jaws of the vise clamping on the mounting flange.
12 Attach a pair of master cylinder bleeder tubes to the outlet ports of the master cylinder (see illustration).
13 Fill the reservoir with brake fluid of the recommended type (see Chapter 1).
14 Slowly push the pistons into the master

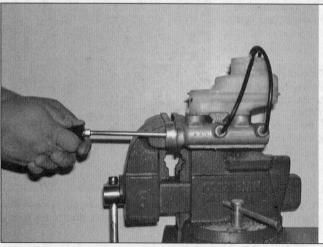

7.12 The best way to bleed air from the master cylinder before installing it on the vehicle is with a pair of bleeder tubes that direct brake fluid into the reservoir during bleeding

Chapter 9 Brakes

9-13

7.17 Install a new O-ring onto the master cylinder sleeve

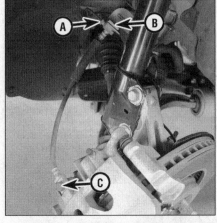

8.3a To remove a front brake hose, unscrew the brake line fitting (A) while holding the bracket with a pair of pliers, remove the retaining clip (B), then unscrew the banjo bolt (C)

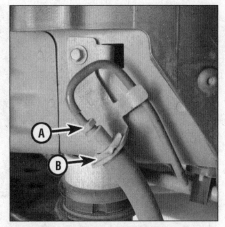

8.3b To remove a rear brake hose, unscrew the brake line fitting (A) while holding the bracket with a pair of pliers, remove the clip (B)…

cylinder (a large Phillips screwdriver can be used for this) - air will be expelled from the pressure chambers and into the reservoir. Because the tubes are submerged in fluid, air can't be drawn back into the master cylinder when you release the pistons.

15   Repeat the procedure until no more air bubbles are present.

16   Remove the bleed tubes, one at a time, and install plugs in the open ports to prevent fluid leakage and air from entering. Install the reservoir cap.

17   Install the master cylinder over the studs on the power brake booster and tighten the attaching nuts only finger tight at this time.

Note: *Be sure to install a new O-ring into the sleeve of the master cylinder (see illustration).*

18   Thread the brake line fittings into the master cylinder. Since the master cylinder is still a bit loose, it can be moved slightly in order for the fittings to thread in easily. Do not strip the threads as the fittings are tightened.

19   Tighten the mounting nuts to the torque listed in this Chapter's Specifications, then tighten the brake line fittings securely.

20   Fill the master cylinder reservoir with fluid, then bleed the master cylinder and the brake system as described in Section 9. To bleed the cylinder on the vehicle, have an assistant depress the brake pedal and hold the pedal to the floor. Loosen the fitting to allow air and fluid to escape. Repeat this procedure on both fittings until the fluid is clear of air bubbles.

Caution: *Have plenty of rags on hand to catch the fluid - brake fluid will ruin painted surfaces. After the bleeding procedure is completed, rinse the area under the master cylinder with clean water.*

21   Test the operation of the brake system carefully before placing the vehicle into normal service.

Warning: *Do not operate the vehicle if you are in doubt about the effectiveness of the brake system. It is possible for air to become trapped in the anti-lock brake system hydraulic control unit, so, if the pedal continues to feel spongy after repeated bleedings or the BRAKE or ANTI-LOCK light stays on, have the vehicle towed to a dealer service department or other qualified shop to be bled with the aid of a scan tool.*

8   Brake hoses and lines - inspection and replacement

Caution: *Brake fluid will damage paint. Cover all painted surfaces and avoid spilling fluid during this procedure. If you do spill fluid, wash it off with water immediately.*

1   About every six months, with the vehicle raised and placed securely on jackstands, the flexible hoses which connect the steel brake lines with the front and rear brake assemblies should be inspected for cracks, chafing of the outer cover, leaks, blisters and other damage. These are important and vulnerable parts of the brake system and inspection should be complete. A light and mirror will be needed for a thorough check. If a hose exhibits any of the above defects, replace it with a new one.

Flexible hoses

2   Clean all dirt away from the ends of the hose.

3   To remove a brake hose, unscrew the tube nut with a flare-nut wrench, if available, to prevent rounding-off the corners of the nut, then remove the bolt(s) or clip(s) securing the hose to the body (and any suspension components) (see illustrations).

4   Disconnect the hose from the caliper, discarding the sealing washers on either side of the fitting.

5   Using new sealing washers, attach the new brake hose to the caliper. Tighten the banjo fitting bolt to the torque listed in this Chapter's Specifications.

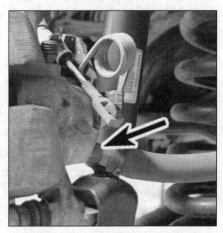

8.3c … then unscrew the hose from the caliper

6   Reverse the removal procedure to install the hose, making sure it isn't twisted.

7   Carefully check to make sure the suspension or steering components don't make contact with the hose. Have an assistant push down on the vehicle and also turn the steering wheel lock-to-lock during inspection.

8   Bleed the brake system (see Section 9).

Metal brake lines

9   When replacing brake lines, be sure to use the correct parts. Don't use copper tubing for any brake system components. Purchase steel brake lines from a dealer parts department or auto parts store.

10   Prefabricated brake line, with the tube ends already flared and fittings installed, is available at auto parts stores and dealer parts departments. These lines can be bent to the proper shapes using a tubing bender.

11   When installing the new line make sure it's well supported in the brackets and has plenty of clearance between moving or hot components.

12   After installation, check the master cyl-

9.8 When bleeding the brakes, a hose is connected to the bleed screw at the caliper and submerged in brake fluid - air will be seen as bubbles in the tube and container (all air must be expelled before moving to the next wheel)

inder fluid level and add fluid as necessary. Bleed the brake system as outlined in Section 9 and test the brakes carefully before placing the vehicle into normal operation.

9 Brake hydraulic system - bleeding

Warning: *If air has found its way into the hydraulic control unit, the system must be bled with the use of a scan tool. If the brake pedal feels spongy even after bleeding the brakes, or the ABS light on the instrument panel does not go off, or if you have any doubts whatsoever about the effectiveness of the brake system, have the vehicle towed to a dealer service department or other repair shop equipped with the necessary tools for bleeding the system.*
Warning: *Wear eye protection when bleeding the brake system. If the fluid comes in contact with your eyes, immediately rinse them with water and seek medical attention.*
Note: *Bleeding the brake system is necessary to remove any air that's trapped in the system when it's opened during removal and installation of a hose, line, caliper, wheel cylinder or master cylinder.*
Caution: *Brake fluid will damage paint. Cover all painted surfaces and avoid spilling fluid during this procedure. If you do spill fluid, wash it off with water immediately.*

1   It will probably be necessary to bleed the system at all four brakes if air has entered the system due to low fluid level, or if the brake lines have been disconnected at the master cylinder.
2   If a brake line was disconnected only at a wheel, then only that caliper or wheel cylinder must be bled.
3   If a brake line is disconnected at a fitting located between the master cylinder and any of the brakes, that part of the system served by the disconnected line must be bled.
4   Remove any residual vacuum (or hydraulic pressure) from the brake power booster by applying the brake several times with the engine off.

5   Remove the master cylinder reservoir cap and fill the reservoir with brake fluid. Reinstall the cap.
Note: *Check the fluid level often during the bleeding operation and add fluid as necessary to prevent the fluid level from falling low enough to allow air bubbles into the master cylinder.*
6   Have an assistant on hand, as well as a supply of new brake fluid, an empty clear plastic container, a length of plastic, rubber or vinyl tubing to fit over the bleeder valve and a wrench to open and close the bleeder valve.
7   Beginning at the right rear wheel, loosen the bleeder screw slightly, then tighten it to a point where it's snug but can still be loosened quickly and easily.
8   Place one end of the tubing over the bleeder screw fitting and submerge the other end in brake fluid in the container (see illustration).
9   Have the assistant slowly depress the brake pedal and hold it in the depressed position.
10   While the pedal is held depressed, open the bleeder screw just enough to allow a flow of fluid to leave the valve. Watch for air bubbles to exit the submerged end of the tube. When the fluid flow slows after a couple of seconds, tighten the screw and have your assistant release the pedal.
11   Repeat Steps 9 and 10 until no more air is seen leaving the tube, then tighten the bleeder screw and proceed to the left rear wheel, the right front wheel and the left front wheel, in that order, and perform the same procedure. Be sure to check the fluid in the master cylinder reservoir frequently.
12   Never use old brake fluid. It contains moisture which can boil, rendering the brake system inoperative.
13   Refill the master cylinder with fluid at the end of the operation.
14   Check the operation of the brakes. The pedal should feel solid when depressed, with no sponginess. If necessary, repeat the entire process.
Warning: *Do not operate the vehicle if you are in doubt about the effectiveness of the brake system. It is possible for air to become trapped in the anti-lock brake system hydraulic control unit, so, if the pedal continues to feel spongy after repeated bleedings or the BRAKE or ANTI-LOCK light stays on, have the vehicle towed to a dealer service department or other qualified shop to be bled with the aid of a scan tool.*

10 Power brake booster - check, removal and installation

Operating check

1   Depress the brake pedal several times with the engine off and make sure that there is no change in the pedal reserve distance.
2   Depress the pedal and start the engine. If the pedal goes down slightly, operation is normal.

Airtightness check

3   Start the engine and turn it off after one or two minutes. Depress the brake pedal several times slowly. If the pedal goes down farther the first time but gradually rises after the second or third depression, the booster is airtight.
4   Depress the brake pedal while the engine is running, then stop the engine with the pedal depressed. If there is no change in the pedal reserve travel after holding the pedal for 30 seconds, the booster is airtight.

Removal and installation

5   Disassembly of the power unit requires special tools and is not ordinarily performed by the home mechanic. If a problem develops, it's recommended that a new or factory rebuilt unit be installed.
6   Before disconnecting the brake pedal, brake booster or booster pushrod, the brake light switch must be removed. With the brake pedal in the at-rest position, pull very lightly rearward on the brake pedal, rotate the switch clockwise 45-degrees and remove the switch.
7   Remove the strut tower brace.
8   Remove the master cylinder (see Section 7).
9   Unbolt the ABS hydraulic control unit mounting bracket so it can be repositioned aside when the booster is removed.
10   Disconnect the vacuum hose from the power brake booster.
11   Under the dash, remove the lower instrument panel cover.
12   Remove the pushrod clevis lock pin, then disconnect the pushrod from the brake pedal (see illustration).
Note: *These models utilize a one-piece clevis locking pin to connect the push-rod to the brake pedal. Use a 12-point, 11 mm wrench or socket to release the tabs on the pin. The pin is a one-time-use part and must be discarded and replaced if removed.*
13   Remove the booster mounting nuts from inside, then carefully lift the booster unit away

Chapter 9 Brakes

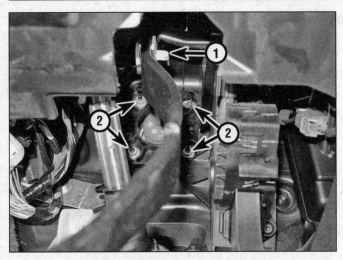

10.12 Power brake booster mounting details

1 Pushrod clevis lock pin 2 Booster mounting nuts

11.1 Disconnect the electrical connector (A) then rotate the switch (B) 45-degrees clockwise to remove it

12.2 Squeeze the sides of the connector and detach the vacuum hose from the brake booster pump

12.3 Brake booster vacuum pump bolts

from the firewall and out of the engine compartment.

14 To install the booster, reverse the removal steps and test the operation of the brakes before placing the vehicle in normal service.

15 When installing the brake light switch, insert the switch into the bracket and rotate the switch 45-degrees counterclockwise. Be careful not to move the brake pedal when installing the switch or damage to the switch can occur.

11 Brake light switch - replacement

1 Disconnect the electrical connector at the switch (see illustration).

2 With the brake pedal in the at-rest position, pull very lightly rearward on the brake pedal, rotate the switch clockwise 45-degrees and remove the switch.

3 Insert the switch into the bracket and rotate the switch 45-degrees counterclockwise. Be careful not to move the brake pedal when installing the switch or damage to the switch can occur. The brake light switch is self-adjusting.

Note: *Do not press on the brake pedal during installation of the switch.*

4 Connect the switch electrical connector.

12 Brake booster vacuum pump (four-cylinder models) - removal and installation

1 Remove the air intake duct (see Chapter 4).

2 Detach the vacuum hose from the fitting on the pump (see illustration).

3 Remove the mounting bolts and detach the pump from the cylinder head (see illustration).

4 Check the condition of the pump O-ring; if it's in good condition, it can be re-used.

5 Installation is the reverse of the removal procedure. Be sure to align the pump with the slot on the camshaft, and tighten the bolts to the torque listed in this Chapter's Specifications.

Notes

Chapter 10
Suspension and steering

Contents

	Section		Section
Coil spring (rear) - removal and installation	10	Steering gear - removal and installation	18
Control arm (front) - removal, inspection and installation	5	Steering gear boots - replacement	17
General information and precautions	1	Steering knuckle and hub - removal and installation	6
Hub and bearing assembly (front) - removal and installation	7	Steering wheel - removal and installation	14
Hub and bearing assembly (rear) - removal and installation	11	Strut/coil spring assembly (front) - removal, inspection and installation	2
Rear knuckle - removal and installation	12		
Shock absorber (rear) - removal and installation	8	Strut/coil spring assembly (front) - replacement	3
Stabilizer bar, links and bushings (front) - removal and installation	4	Subframe - removal and installation	19
		Suspension arms (rear) - removal and installation	9
Stabilizer bar, links and bushings (rear) - removal and installation	13	Tie-rod ends - removal and installation	16
		Wheel alignment - general information	21
Steering column - removal and installation	15	Wheels and tires - general information	20

Specifications

Warning: *The manufacturer recommends replacing all suspension and steering fasteners with new ones whenever they are removed.*

Torque specifications   Ft-lbs (unless otherwise indicated)   Nm
Note: *One foot-pound (ft-lb) of torque is equivalent to 12 inch-pounds (in-lbs) of torque. Torque values below approximately 15 foot-pounds are expressed in inch-pounds, because most foot-pound torque wrenches are not accurate at these smaller values.*

Front suspension
Lower control arms
  Balljoint-to-steering knuckle nut
    2015 and earlier models 148   200
    2016 and later models 155   200
  Lower arm-to-subframe
    Front bolts 195   265
    Rear bolts 73   99
Stabilizer bar
  Bracket bolts 41   55
  Upper and lower link nuts 111   150
Front wheel speed sensor .... See Chapter 9
Front brake caliper bolts .... See Chapter 9
Driveaxle/hub nut .... See Chapter 8
Strut/coil spring assembly
  Upper mounting nuts
    2015 and earlier models 41   55
    2016 and later models 46   63
  Lower mounting bolts/nuts
    2015 and earlier models 184   250
    2016 and later models 166   225
  Strut rod nut 98   133
Subframe
  Subframe bracket bolts 41   55
  Subframe forward bolts 148   200
  Subframe rear bolts 148   200

Rear suspension
Shock absorber
　　Lower bolt/nut... 129 　　175
　　Upper nuts... 41 　　55
Rear subframe
　　Bracket bolts... 41 　　55
　　Forward and rear bolts.. 148 　　200
Wheel speed sensor bolt.. See Chapter 9
Rear brake caliper bolts... See Chapter 9
Driveaxle/hub nuts (AWD models).. See Chapter 8
Upper control arm
　　Bushing-to-control arm bolt.. 148 　　200
　　Bushing-to-subframe bolts... 111 　　150
　　Upper arm to subframe bolts... 111 　　150
Lower control arm
　　Lower arm to knuckle... 195 　　265
　　Lower arm to subframe bolts... 159 　　215
Trailing arm
　　Pivot bolt/nut... 111 　　150
　　Trailing arm to subframe.. 122 　　165
Toe link
　　Inner bolt... 111 　　150
　　Outer bolt.. 111 　　150
Stabilizer bar
　　Bracket nuts.. 41 　　55
　　Upper and lower link nuts.. 41 　　55
Rear hub and bearing assembly bolts... 122 　　165
Rear wheel hub bearing nut... 258 　　350
Subframe
　　Main bolts.. 148 　　200
　　Bracket bolts... 41 　　55

Steering
Tie-rod end-to-steering knuckle nut... 111 　　150
Steering shaft U-joint pinch bolt... 19 　　26
Steering gear mounting bolts... 122 　　165
Subframe steering gear support brackets.. 41 　　55
Steering column pinch bolt... 19 　　26
Steering column mounting bolts.. 18 　　25

Ride height
Front FWD.. 1.34 inches 　　34mm
Front AWD.. 1.73 inches 　　44mm
Rear FWD... 1.2 inches 　　30mm
Rear AWD... 1.6 inches 　　40mm
Allowable tolerance variation.. +/- 0.5 inches 　　+/- 12mm

Chapter 10 Suspension and steering systems

10-3

1.1 Front suspension and steering components

1 Balljoint
2 Steering knuckle
3 Tie-rod end
4 Control arm
5 Strut/coil spring assembly
6 Subframe
7 Front subframe bolts
8 Steering gear assembly mounting bolts
9 Rear subframe bolts and brackets
10 Steering gear

1.2 Rear suspension components

1 Lower control arm	4 Toe link	7 Stabilizer bushing bracket
2 Rear knuckle	5 Coil spring	8 Upper control arm
3 Trailing arm	6 Stabilizer bar	

Chapter 10   Suspension and steering systems

1   General information and precautions

Front suspension
1   The front suspension is made up of a strut assembly, control arm, steering knuckle/hub assembly and a stabilizer bar (see illustration). The strut assembly is made up of a shock absorber and a coil spring. The strut combines several functions into one by supporting the weight of the vehicle, functioning as the spring dampener and providing the pivot point for the steering knuckle.

Rear suspension
2   The rear suspension employs upper and lower control arms, a trailing arm, a toe link, and a coil spring and shock absorber on each side (see illustration).

Steering
3   The rack-and-pinion steering gear is located on the front suspension subframe and actuates the tie-rods, which are attached to the steering knuckles. The inner ends of the tie-rods are protected by rubber boots which should be inspected periodically for secure attachment, tears and leaking lubricant (which would indicate a failed rack seal).
4   All models are equipped with electric power-assisted rack-and-pinion steering systems. The steering gear is bolted to the crossmember and is connected to the steering knuckles by a pair of tie-rods.
5   The EPAS (Electronic Power Assist Steering) system takes the place of the conventional hydraulic power steering system. The EPAS communicates by way of a PSCM (Power Steering Control Module) on a high-speed data line with various other modules. The PSCM sends these signals to the EPAS and operates the reversible electric motor at various torque and reaction levels for the different driving conditions and road hazards.
6   The PSCM compensates for the speed as well as pull or drift experienced in normal driving and maintains steering stability. Along with drift control, the electronic steering system allows for greater parking lot maneuverability by increasing the torque-to-turn ratio at slow speeds as well as decreasing the level of power assist for straight ahead highway driving.
7   The steering wheel operates the steering shaft, which actuates the steering gear through universal joints. Looseness in the steering can be caused by wear in the steering shaft universal joints, the steering gear, the tie-rod ends and loose retaining bolts.

Precautions
8   Frequently, when working on the suspension or steering system components, you may come across fasteners which seem impossible to loosen. These fasteners on the underside of the vehicle are continually subjected to water, road grime, mud, etc., and can become rusted or frozen, making them extremely difficult to remove. In order to unscrew these stubborn fasteners without damaging them (or other components), be sure to use lots of penetrating oil and allow it to soak in for a while. Using a wire brush to clean exposed threads will also ease removal of the nut or bolt and prevent damage to the threads. Sometimes a sharp blow with a hammer and punch will break the bond between a nut and bolt threads, but care must be taken to prevent the punch from slipping off the fastener and ruining the threads. Heating the stuck fastener and surrounding area with a torch sometimes helps too, but isn't recommended because of the obvious dangers associated with fire. Long breaker bars and extension, or cheater, pipes will increase leverage, but never use an extension pipe on a ratchet - the ratcheting mechanism could be damaged. Sometimes tightening the nut or bolt first will help to break it loose. Fasteners that require drastic measures to remove should always be replaced with new ones.
9   Since most of the procedures dealt with in this Chapter involve jacking up the vehicle and working underneath it, a good pair of jackstands will be needed. A hydraulic floor jack is the preferred type of jack to lift the vehicle, and it can also be used to support certain components during various operations.
Warning: *Never, under any circumstances, rely on a jack to support the vehicle while working on it. Whenever any of the suspension or steering fasteners are loosened or removed they must be inspected and, if necessary, replaced with new ones of the same part number or of original equipment quality and design. Torque specifications must be followed for proper reassembly and component retention. Never attempt to heat or straighten any suspension or steering components. Instead, replace any bent or damaged part with a new one.*

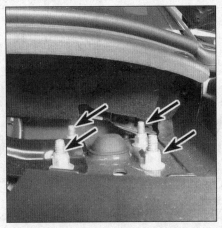

2.1 Loosen the strut upper mounting nuts

2.6 Unscrew the nuts, then drive the bolts out with a hammer (don't turn the bolts - they have serrated shoulders)

2   Strut/coil spring assembly (front) - removal, inspection and installation

Warning: *The manufacturer recommends replacing all fasteners with new ones during installation.*

Removal
1   Remove the strut tower cover, if equipped, then loosen (but do not remove) the four strut upper mounting nuts (see illustration).
2   Loosen the wheel lug nuts, raise the vehicle and support it securely on jackstands. Remove the wheel.
3   Remove the nut and detach the stabilizer bar link from the strut (see Section 4).
4   Detach the wheel speed sensor harness from the strut.
5   Support the steering knuckle with a floor jack.
6   Unscrew the nuts securing the strut to the steering knuckle, then tap the bolts out with a hammer (see illustration).
Caution: *Only turn the nuts; the bolts have serrated shoulders, and to turn them could damage the steering knuckle.*
7   Remove the strut upper mounting nuts.
8   Use the floor jack to lower the knuckle to allow room to remove the strut/coil spring assembly, but be careful not to overextend the inner CV joint. Guide the strut assembly from the fenderwell.

Inspection
9   Check the strut body for leaking fluid, dents, cracks and other obvious damage which would warrant replacement.
10   Check the coil spring for chips or cracks in the spring coating (this will cause premature spring failure due to corrosion). Inspect the spring seat for cuts, hardness and general deterioration.

3.3 Make sure the spring compressor tool is on securely

3.5a Remove the upper mount and spring seat . . .

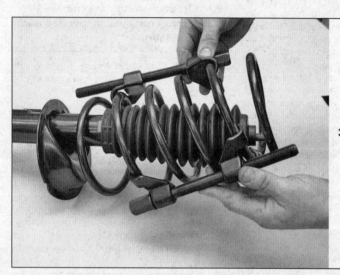

3.5b . . . then remove the spring . . .

11 If any undesirable conditions exist, proceed to the strut/coil spring disassembly procedure (see Section 3).

Installation

12 Guide the assembly up into the fenderwell and insert the upper mounting studs through the holes in the body. Once the studs protrude, install the new nuts so the shock won't fall back through but don't tighten them yet. This is most easily accomplished with the help of an assistant, as the strut is quite heavy and awkward.
13 Use the jack to position the steering knuckle and install the two new lower mounting bolts and nuts, then tighten the nuts to the torque listed in this Chapter's Specifications.
14 Connect the stabilizer bar link to the strut and install a new nut, tightening it to the torque listed in this Chapter's Specifications.
15 Reconnect the wheel speed sensor harness to the strut.
16 Install the wheel and lug nuts, then lower the vehicle and tighten the lug nuts to the torque listed in the Chapter 1 Specifications.
17 Tighten the upper mounting nuts to the torque listed in this Chapter's Specifications.
18 Have the front end alignment checked, and if necessary, adjusted.

3   Strut/coil spring assembly (front) - replacement

Warning: *Before attempting to disassemble the shock absorber/coil spring assembly, a tool to hold the coil spring in compression must be obtained. Do not attempt to use makeshift methods. Uncontrolled release of the spring could cause damage and personal injury or even death. Use a high-quality spring compressor, and carefully follow the tool manufacturer's instructions provided with it. After removing the coil spring with the compressor still installed, place it in a safe, isolated area. The strut is pressurized; do not apply heat or flame to the assembly or it could explode.*

Warning: *The manufacturer recommends replacing all fasteners with new ones during installation.*

1 If the front suspension shock absorber/coil springs exhibit signs of wear (leaking fluid, loss of damping capability, sagging or cracked coil springs) then they should be disassembled and overhauled as necessary. The shock absorbers themselves cannot be serviced, and should be replaced if faulty; the springs and related components can be replaced individually. To maintain balanced characteristics on both sides of the vehicle, the components on both sides should be replaced at the same time.
2 With the assembly removed from the vehicle (see Section 2), clean away all external dirt but be careful not to damage or remove the protective coating from the spring as this helps prevent corrosion damage.
3 Install the coil spring compressor tools (ensuring that they are fully engaged), and compress the spring until all tension is relieved from the upper mount (see illustration).
4 Hold the shock absorber piston rod with a wrench and unscrew the nut. Do not use an impact wrench as this could damage the shock absorber.
5 Remove the upper mount and spring, followed by the boot, bump stop and lower spring seat (see illustrations).
6 If a new spring is to be installed, the original spring must now be carefully released from the compressor. If it is to be re-used, the spring can be left in compression.
7 With the strut assembly now completely disassembled, examine all the components for wear and damage. Replace components as necessary.
8 Examine the shock for signs of fluid leakage. Check the piston rod for signs of pitting along its entire length, and check the shock body for signs of damage. Test the operation of the shock, while holding it in an upright position, by moving the piston through a full stroke, then through short strokes

Chapter 10 Suspension and steering systems

3.5c ... followed by the boot ...

3.5d ... and the bump stop

3.10 When installing the spring, make sure the ends fit into the recessed portion of the seats, and keep the gap at 0.39-inch or less

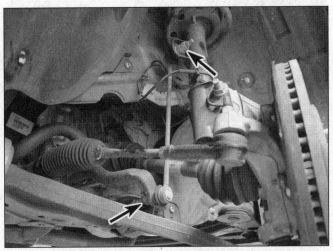

4.2 Stabilizer bar link ballstud nuts

of 2 to 4 inches. In both cases, the resistance felt should be smooth and continuous. If the resistance is jerky or uneven, or if there is any visible sign of wear or damage, replacement is necessary.

9   To reassemble, install the lower spring seat first, then the bump stop and dust boot onto the shock absorber. Next install the spring and upper mount/upper seat. The arrow or notch on the upper mount must be aligned with the lower mount (pointing out).

10   With the spring compressed, install the new rod nut. Position the ends of the spring to within 0.39-inch (10 mm) of the step on both the top and bottom mounts (see illustration). Tighten the nut to the specified torque while maintaining the position at both ends.

11   Release the tension on the spring and re-install the spring assembly in the vehicle.

4   Stabilizer bar, links and bushings (front) - removal and installation

Warning: *The manufacturer recommends replacing all fasteners with new ones during installation.*

Removal

1   Loosen the front wheel lug nuts. Raise the front of the vehicle and support it securely on jackstands. Apply the parking brake and block the rear wheels to keep the vehicle from rolling off the stands. Remove the front wheels. Also remove the under-vehicle air duct and the under-vehicle splash shield. Disconnect the cable from the negative terminal of the battery (see Chapter 5).

2   Remove the lower stabilizer bar link fasteners (see illustration). If the ballstud turns with the nut, use a wrench to hold the stud.

3   Disconnect the oxygen sensor connectors and remove hold down clips on the subframe for the oxygen sensors.

4   Disconnect the tie-rod ends from the steering knuckles (see Section 16).

5   Detach the control arm-to-subframe fasteners on each side and position the control arms out of the way.

6   Remove the fasteners securing the inner

Chapter 10 Suspension and steering systems

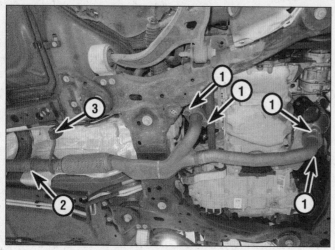

4.7 Front exhaust pipe details (non-turbocharged V6 model shown)

1. Flange nuts
2. Clamp sleeve
3. Exhaust hanger

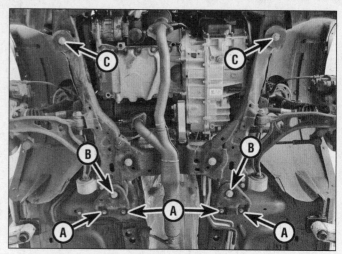

4.8 Subframe bracket bolts (A), subframe rear bolts (B) and front bolts (C)

4.9 Engine roll restrictor-to-subframe bolt

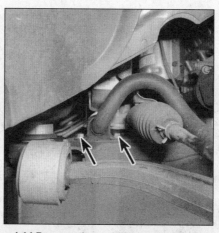

4.14 Remove the stabilizer bar bushing clamp nuts and bolts

5.2a Remove the nut...

fender splash shields to the subframe on each side.

7 On V6 models, remove the exhaust Y pipe (see illustration). On four-cylinder models, remove the front (flexible) portion of the exhaust pipe.

8 Mark the location of the subframe to the underbody at the mounting locations. Support the subframe with a pair of floor jacks (see illustration).

9 Remove the engine roll restrictor heat shield, if equipped (the lower nut need only be loosened), then remove the bolt securing the engine roll restrictor to the subframe (see illustration).

10 Remove the two steering gear assembly bolts (see Section 18).

11 Remove the subframe rear main bolts and rear bracket bolts.

12 Remove the subframe front bolts.

13 Very carefully lower the subframe by about two inches.

14 Unbolt the stabilizer bar bushing clamps (see illustration).

15 Guide the stabilizer bar out towards the right side of the vehicle.

16 While the stabilizer bar is off the vehicle, slide off the retainer bushings off and inspect them. If they're cracked, worn or deteriorated, replace them.

17 Clean the bushing area of the stabilizer bar with a stiff wire brush to remove any rust or dirt. Lubricate the inside and outside of the new bushing with vegetable oil (used in cooking) to facilitate reassembly.

Caution: *Don't use petroleum or mineral-based lubricants or brake fluid - they will lead to deterioration of the bushings.*

Installation

18 When installing the subframe, loosely attach the subframe brackets with the bolts. Use your marks to align the subframe to the body. Tighten the subframe bolts to the torque listed in this Chapter's Specifications, then tighten the remaining support bracket fasteners.

19 It's a good idea to have the front wheel alignment checked and, if necessary, adjusted after this job has been performed.

5 Control arm (front) - removal, inspection and installation

Warning: *The manufacturer recommends replacing all fasteners with new ones during installation.*

Removal

1 Loosen the wheel lug nuts on the side to be disassembled, raise the front of the vehicle, support it securely on jackstands and remove the wheel.

2 Remove the nut and pry the balljoint from the steering knuckle (see illustrations).

Chapter 10 Suspension and steering systems

5.2b ... then pry the control arm down to detach the balljoint from the steering knuckle

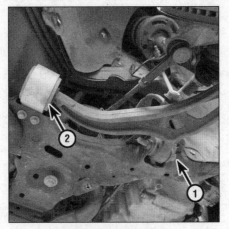

5.3 Control arm details
1. Front pivot bolt
2. Rear bushing bracket bolts

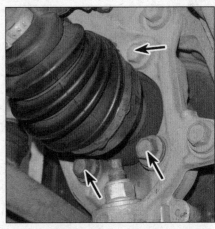

7.5 Hub and bearing-to-steering knuckle bolts (one bolt not visible here)

3 Remove the control front pivot bolt and spacer (see illustration).
4 Remove the two rear bushing bracket-to-frame bolts and remove the lower arm.

Inspection
5 Check the control arm for distortion and the bushings for wear, replacing parts as necessary. Do not attempt to straighten a bent control arm.

Installation
6 Installation is the reverse of removal. Be sure to install new fasteners and tighten them to the torque listed in this Chapter's Specifications.
Caution: *Don't tighten the forward control arm pivot bolt until the weight of the vehicle is on its wheels. As an alternative, raise the outer end of the control arm to simulate normal ride height, then tighten the pivot bolt.*
7 It's a good idea to have the front wheel alignment checked and, if necessary, adjusted after this job has been performed.

6 Steering knuckle and hub - removal and installation

Warning: *Dust created by the brake system is harmful to your health. Never blow it out with compressed air and don't inhale any of it. Do not, under any circumstances, use petroleum-based solvents to clean brake parts. Use brake system cleaner only.*
Warning: *The manufacturer recommends replacing all fasteners with new ones during installation.*

Removal
1 Loosen the wheel lug nuts. Raise the vehicle and support it securely on jackstands, then remove the wheel.
2 With the brake applied to prevent the driveaxle from rotating, remove the driveaxle/hub nut, washer and hub seal. Discard the old seal but do not discard the nut and washer yet.
3 Remove the ABS wheel speed sensor and the wiring harness bracket bolt (see Chapter 9).
4 Remove the brake caliper (don't disconnect the hose), the caliper mounting bracket and the brake disc (see Chapter 9). Hang the caliper from the coil spring with a piece of wire - don't let it hang by the brake hose.
5 Separate the control arm from the steering knuckle (see Section 5).
6 Separate the tie-rod end from the steering knuckle arm (see Section 16).
7 Separate the driveaxle from the hub using a drive hub remover or an equivalent puller (see Chapter 8).
8 Detach the lower end of the strut from the steering knuckle (see Section 2), then remove the knuckle.

Installation
9 Guide the knuckle and hub assembly into position, inserting the driveaxle into the hub.
10 Install and tighten the lower strut bolts and nuts.
11 Connect the balljoint to the knuckle and tighten the pinch bolt nut to the torque listed in this Chapter's Specifications.
12 Attach the tie-rod end to the steering knuckle arm (see Section 16). Tighten the nut to the torque listed in this Chapter's Specifications.
13 Install the wheel speed sensor and harness bracket.
14 Install the brake disc and caliper/mounting bracket assembly (see Chapter 9).
15 With the brake applied to prevent the driveaxle from rotating, seat the driveaxle by using the old driveaxle/hub nut and washer and tighten the nut to the torque listed in the Chapter 8 Specifications. Remove and discard the old nut and washer, replace them with new parts and retighten to the specified torque.
Note: *The driveaxle/hub nut must be tightened before lowering the vehicle or damage to the wheel bearings may occur.*
Note: *The driveaxle/hub nut is coated with a one-use locking chemical and cannot be re-used for final assembly. The locking chemical is activated by heat created during tightening. The nut must be tightened to the specified torque within five minutes of starting it on the threads or the locking chemical will not activate properly.*
16 Install the wheel and lug nuts. Lower the vehicle and tighten the lug nuts to the torque listed in the Chapter 1 Specifications.
17 Have the front-end alignment checked and, if necessary, adjusted.

7 Hub and bearing assembly (front) - removal and installation

Warning: *The manufacturer recommends replacing all fasteners with new ones during installation.*

1 Loosen the wheel lug nuts. Raise the vehicle and support it securely on jackstands, then remove the wheel.
2 With an assistant applying the brake to prevent the driveaxle from rotating, remove the driveaxle/hub nut. Obtain a new nut but do not discard the old nut yet.
3 Remove the brake caliper, the caliper mounting bracket and the brake disc from the hub (see Chapter 9).
Caution: *Suspend the caliper to the strut coil spring with a piece of wire. DO NOT let the caliper hang by the brake hose.*
4 Push the driveaxle into the hub using a drive hub remover or an equivalent puller (see Chapter 8).
Caution: *Just push it into the hub as far as necessary to allow you to put a socket on the hub and bearing assembly bolts.*
5 Remove the hub and bearing-to-steering knuckle bolts (see illustration). Separate the

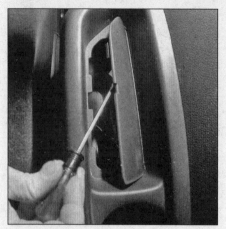

8.1 Use a small flat tipped screwdriver to pry the access panel open

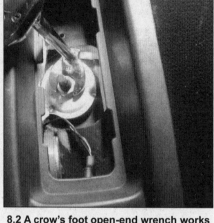

8.2 A crow's foot open-end wrench works great for this job

8.4a Removing the stabilizer bar link from the upper bracket will give you more room to access the lower shock absorber fastener

8.4b With the stabilizer bar link out of the way you can reach the lower bolt to the shock absorber

hub and bearing assembly from the knuckle as you draw the assembly off of the driveaxle.
Caution: *Be careful not to overextend the inner CV joint.*
6 Make sure that the mounting surface inside the steering knuckle and on the driveaxle splines is smooth and free of burrs and nicks prior to installing the hub/bearing assembly.
7 Install the hub/bearing assembly to the steering knuckle using NEW mounting bolts. Tighten the bolts to the torque listed in this Chapter's Specifications.
8 Install the brake disc, the caliper mounting bracket and the caliper; tighten the fasteners to the torque listed in the Chapter 9 Specifications.
9 With the brake applied to prevent the driveaxle from rotating, seat the driveaxle by using the old driveaxle/hub nut and washer and tighten the nut to the torque listed in the Chapter 8 Specifications. Remove and discard the old nut, replace it with the new nut and tighten to the specified torque.
10 Install the wheel and lug nuts, lower the vehicle and tighten the lug nuts to the torque listed in the Chapter 1 Specifications.

8   Shock absorber (rear) - removal and installation

Warning: *The manufacturer recommends replacing all fasteners with new ones during installation.*
1 From inside the rear cargo area, remove the interior quarter trim access panel (see illustration).
2 Remove the shock absorber upper mounting nut (see illustration).
3 Loosen the wheel lug nuts, then raise the vehicle and support it securely on jackstands. Block the front wheels to prevent the vehicle from rolling and remove the rear wheel.
4 Support the rear lower control arm with a floor jack, then remove the shock absorber lower nut and bolt (see illustrations).
5 Maneuver the shock out of the vehicle.
6 Installation is the reverse of the removal procedure.
Caution: *New shock absorbers are gas-filled and come compressed and retained with a fiberglass strap. Do NOT remove the strap until the shock is installed.*
7 When installing the shock absorber, install the upper and lower shock mounting fasteners but leave them loose until after the vehicle has been lowered. Tighten the lower bolt and nut to the specified torque with the full weight of the vehicle on the ground, then tighten the upper mounting nuts to the torque listed in this Chapter's Specifications.
Note: *As an alternative, raise the lower control arm to simulate normal ride height, then tighten the fasteners.*

9   Suspension arms (rear) - removal and installation

Warning: *The manufacturer recommends replacing all fasteners with new ones during installation.*
Note: *To prevent possible damage to the bushings, always tighten bushing fasteners*

while the full weight of the vehicle is resting on the wheels.
1 Loosen the wheel lug nuts, raise the vehicle and support it securely on jackstands. Block the front wheels to prevent the vehicle from rolling. Remove the wheel.
Note: *If you're working on a 2015 or earlier model, remove both rear wheels.*

Upper control arm

2 Detach the wheel speed sensor harness from the bracket on the control arm.

2015 and earlier models

Note: *On these models, the rear subframe must be lowered in order to remove the control arm forward pivot bolt.*
3 Remove the rear brake calipers and suspend them with lengths of wire (see Chapter 9).
4 Remove the parking brake cable bracket bolts (both sides).
5 Detach the stabilizer bar links from the stabilizer bar (both sides) (see Section 13).
6 Place a floor jack under each rear knuckle, where the lower control arm connects. Raise the jacks slightly.
7 Remove the shock absorber lower mounting bolts (see Section 8).
8 Remove the upper control arm-to-knuckle nut and bolt (see illustration).
9 Remove the upper control arm rear bushing-to-subframe bolts.
10 Loosen the control arm-to-subframe forward pivot bolt.
11 Remove the subframe front bracket bolts and main bolts (see illustration). Loosen the subframe rear bolts a few turns.
12 Lower the jacks far enough to unscrew the forward pivot bolt, then remove the control arm.
13 If necessary, unscrew the rear pivot bolt and remove the bushing.
14 Installation is the reverse of removal. Tighten the subframe mounting bolts to the torque listed in this Chapter's Specifications. Raise the control arms with the floor jacks to

Chapter 10 Suspension and steering systems

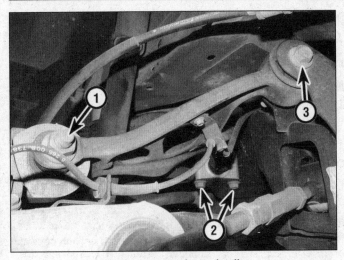

9.8 Upper control arm details

1. Control arm-to-knuckle nut and bolt
2. Control arm rear bushing-to-subframe bolts
3. Control arm forward pivot bolt

9.11 Rear subframe front mounts (rear mounts not visible in photo)

1. Bracket bolts
2. Subframe front mounting bolts

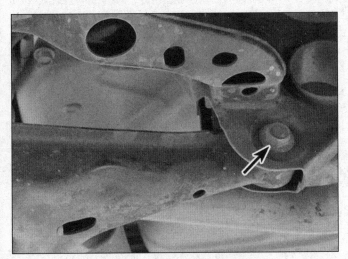

9.22 Lower control arm pivot bolt

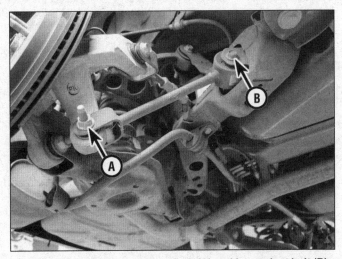

9.25 Trailing arm-to-knuckle nut/bolt (A) and inner pivot bolt (B)

simulate normal ride height, then tighten the fasteners to the torque listed in this Chapter's Specifications.

2016 and later models

15 Position a floor jack under the rear knuckle, where the control arm connects.
16 Remove the upper control arm-to-knuckle nut and bolt (see illustration 9.8).
17 Remove the control arm-to-subframe forward pivot bolt.
18 Remove the upper control arm rear bushing-to-subframe bolts, then remove the control arm.
19 If necessary, unscrew the rear pivot bolt and remove the bushing.
20 Installation is the reverse of removal. Tighten the subframe mounting bolts to the torque listed in this Chapter's Specifications. Raise the control arms with the floor jacks to simulate normal ride height, then tighten the fasteners to the torque listed in this Chapter's Specifications.

Lower control arm

21 Remove the rear coil spring (see Section 10).
22 Remove the lower control arm-to-subframe pivot bolt (see illustration). Remove the arm from the vehicle.
23 Installation is the reverse of removal. Raise the lower control arm with a floor jack to simulate normal ride height, then tighten the fasteners to the torque listed in this Chapter's Specifications.

Trailing arm

24 Position a floor jack under the rear knuckle, where the lower control arm connects. Raise the jack to simulate normal ride height.

Warning: *The lower arm is under considerable pressure from the rear spring. Use caution not to remove or dislodge the jack, and at any time you are working around a loaded spring.*

25 Remove the trailing arm-to-knuckle nut and bolt and the trailing arm-to-subframe bolt, then remove the trailing arm (see illustration).
26 Remove the nut and bolt from the trailing arm pivot to disconnect the trailing arm from the pivot bracket.
27 Installation is the reverse of the removal procedure. Be sure the lower control arm is raised to normal ride height before tightening the fasteners to the torque listed in this Chapter's Specifications.

Toe link

28 Position a floor jack under the rear knuckle, where the lower control arm connects. Raise the jack to simulate normal ride height.

10-12 Chapter 10 Suspension and steering systems

9.29a Outer toe link fastener

9.29b Inner toe link fastener

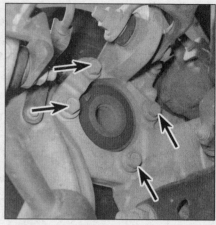

11.5 Remove the four wheel hub bolts and remove the hub (FWD model shown)

29 Remove the fasteners from the inboard and outboard ends of the toe link (see illustrations).
30 Installation is the reverse of the removal procedure. Be sure the lower control arm is raised to normal ride height before tightening the fasteners to the torque listed in this Chapter's Specifications.
31 Have the rear wheel alignment checked and, if necessary, adjusted.

10 Coil spring (rear) - removal and installation

Warning: *Always replace the springs as a set - never replace just one of them.*
Warning: *The manufacturer recommends replacing all fasteners with new ones during installation.*

1 Loosen the wheel lug nuts, raise the vehicle and support it securely on jackstands. Block the front wheels to prevent the vehicle from rolling. Remove the wheel.

Removal

2 Support the lower control arm with a floor jack.
3 Remove the nuts and detach the stabilizer bar links from the stabilizer bar (both sides) (see Section 13). Rotate the ends of the stabilizer bar downward.
4 Remove the lower control arm-to-rear knuckle nut and bolt (see Section 9).
5 Loosen the lower control arm pivot bolt.
6 Slowly lower the jack and allow the coil spring to extend fully. Remove the spring.
7 Check the spring for cracks and chips, replacing the springs as a set if any defects are found. Also check the upper and lower seats for damage and deterioration, replacing if necessary.

Installation

8 If a new lower seat is installed, position it in the lower arm by aligning the recess in the seat with the ridge in the lower arm.
9 Install the spring on the lower arm and position the lower end of the spring in the step in the lower spring seat.
10 Raise the jack and connect the lower control arm to the knuckle using new fasteners (see Section 12). With the control arm at normal ride height, tighten the control arm fasteners to the torque listed in this Chapter's Specifications.
11 The remainder of installation is the reverse of removal.

11 Hub and bearing assembly (rear) - removal and installation

Warning: *The manufacturer recommends replacing all fasteners with new ones during installation.*

Removal

1 Loosen the rear wheel lug nuts. Raise the rear of the vehicle and support it securely on jackstands, then remove the wheel.
2 Remove the brake disc (see Chapter 9).
3 If you're working on an all-wheel drive model, remove the driveaxle/hub nut (see Chapter 8).
4 Remove the wheel speed sensor from the knuckle (see Chapter 9).
5 Remove the bolts and detach the wheel bearing and hub assembly from the wheel knuckle (see illustration). If you're working on an all-wheel drive model, simultaneously push the driveaxle from the hub with a suitable puller.

Installation

6 Install the wheel hub/bearing assembly onto the rear knuckle. If you're working on an all-wheel drive model, guide the end of the driveaxle into the splines as the hub is being installed. Tighten the hub assembly-to-rear knuckle bolts to the torque listed in this Chapter's Specifications.

Caution: *All-wheel drive models: With the brake applied to prevent the driveaxle from rotating, seat the driveaxle by using the driveaxle/hub nut and tightening it to the torque listed in the Chapter 8 Specifications. Remove and discard the old nut, install the new nut and tighten to the specified torque.*
Caution: *The driveaxle/hub nut must be tightened before lowering the vehicle or damage to the wheel bearings may occur.*
Caution: *The driveaxle/hub nut is coated with a one-use locking chemical and cannot be re-used for final assembly. The locking chemical is activated by heat created during tightening. The nuts must be tightened to the specified torque within five minutes of starting it on the threads or the locking chemical will not activate properly.*

7 The remainder of installation is the reverse of removal. Tighten the brake fasteners to the torque values listed in the Chapter 9 Specifications. If you're working on an all-wheel drive model, tighten the driveaxle/hub nut to the torque listed in the Chapter 8 Specifications.
8 Install the wheel and lug nuts, then lower the vehicle. Tighten the lug nuts to the torque listed in the Chapter 1 Specifications.

12 Rear knuckle - removal and installation

Warning: *The manufacturer recommends replacing all fasteners with new ones during installation.*

1 Before beginning the disassembly, with the full weight of the vehicle on the ground, take a measurement from the center of the rear hub to the lower edge of the bodywork at the top of the wheel opening. This measurement is the normal ride height of the vehicle and will be used when re-assembling the rear suspension components (see illustration).
2 Loosen the rear wheel lug nuts. Raise the rear of the vehicle and support it securely on jackstands, then remove the wheel.

Chapter 10 Suspension and steering systems

12.1 Measure and record the ride height

3   Support the lower control arm with a floor jack.
4   Remove the hub and bearing assembly (see Section 11).
5   Disconnect the lower end of the shock absorber from the knuckle (see Section 8).
6   Detach the outer end of the upper control arm from the knuckle (see Section 9).
7   Detach the toe link from the knuckle (see Section 9).
8   Detach the trailing arm from the knuckle (see Section 9).
9   Remove the lower arm-to-knuckle bolt, then detach the knuckle from the lower arm.
10 Installation is the reverse of the removal procedure, tightening all suspension fasteners to the torque values listed in this Chapter's Specifications with the suspension at normal ride height (this can be simulated by raising the suspension with the floor jack). Tighten the brake fasteners to the torque values listed in the Chapter 9 Specifications. If you're working on an all-wheel drive model, tighten the driveaxle/hub nut to the torque listed in the Chapter 8 Specifications.
11 Install the wheel and lug nuts, then lower the vehicle. Tighten the lug nuts to the torque listed in the Chapter 1 Specifications.

13 Stabilizer bar, links and bushings (rear) - removal and installation

Warning: *The manufacturer recommends replacing all fasteners with new ones during installation.*

Removal

1   Lift the vehicle on a hoist (if available). Otherwise, back onto a set of ramps and safely block and secure the vehicle, or raise the rear of the vehicle and support it securely on jackstands. Block the front wheels to keep the vehicle from rolling off the stands.
2   Remove the left and right side stabilizer bar links by removing the nuts from the upper and lower ends of the links (see illustration).
3   Remove the nuts holding the stabilizer bar bushing brackets and remove the brackets and the stabilizer bar (see illustration).
4   While the stabilizer bar is off the vehicle, inspect the bushings and replace if cracked, worn or deteriorated.
5   Clean the bushing area of the stabilizer bar with a stiff wire brush to remove any rust or dirt. Lubricate the inside and outside of the new bushing with vegetable oil (used in cooking) to simplify reassembly.
Caution: *Don't use petroleum or mineral-based lubricants or brake fluid - they will lead to deterioration of the bushings.*
6   Installation is the reverse of removal. Tighten the fasteners to the torque values listed in this Chapter's Specifications.

14 Steering wheel - removal and installation

Warning: *These models are equipped with a Supplemental Restraint System (SRS), more commonly known as airbags. Always disable the airbag system before working in the vicinity of any airbag system component to avoid the possibility of accidental deployment of the airbag(s), which could cause personal injury (see Chapter 12).*
Warning: *Do not use a memory saving device to preserve the PCM or radio memory when working on or near airbag system components.*
Warning: *The manufacturer recommends replacing all fasteners with new ones during installation.*

Removal

1   Turn the ignition key to Off, then disconnect the cable from the negative terminal of the battery (see Chapter 5). Wait at least two minutes before proceeding.
2   Turn the steering wheel so the wheels are pointing straight ahead and remove the key.
3   Use a small screwdriver or a 3 mm Allen wrench through the holes in the backside of the steering wheel to release the airbag mod-

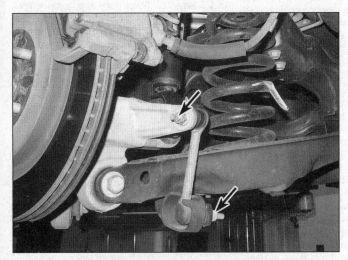

13.2 Rear stabilizer bar link nuts

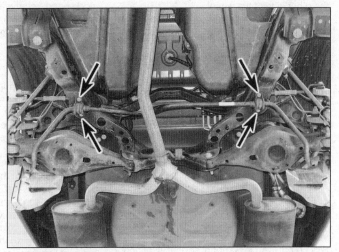

13.3 Remove the stabilizer bar bushing bracket nuts

10-14   Chapter 10 Suspension and steering systems

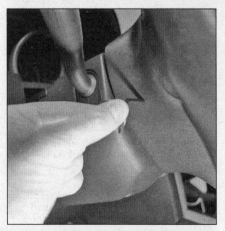

14.3 Depress the clips through the holes on each side of the steering wheel to release the airbag

14.4 Pry up the yellow locking tabs, then disconnect the airbag connectors

14.6a Remove the steering wheel bolt...

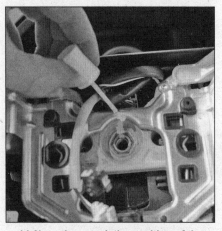

14.6b ... then mark the position of the steering wheel to the shaft

14.8 Apply a piece of tape to the clockspring to prevent it from rotating

ule clips (see illustration). Gently pull the airbag module away from the wheel on the side you have released, then repeat this Step on the other side of the wheel.

4   Pry UP the yellow connector locks, then disconnect the airbag module electrical connectors (see illustration).

5   Remove the airbag and set the module in a safe, isolated area.

Warning: *Carry the airbag module with the trim side facing away from you, and set the steering wheel/airbag module down with the trim side facing up. Don't place anything on top of the steering wheel/airbag module.*

6   Remove the steering wheel bolt, then make match marks on the wheel and the steering shaft (see illustrations).

7   Try rocking the wheel lightly to remove it from the shaft. If this does not work, use a suitable puller to remove the steering wheel. Do not hit the steering shaft with a hammer or you will damage the shaft.

8   Tape the clockspring so that it cannot rotate (see illustration).

Caution: *Do not rotate the steering shaft at any time while the wheel and airbag are removed or the clockspring could be damaged.*

9   If it is necessary to remove the airbag clockspring, remove the steering column covers (see Chapter 11), disconnect the electrical connectors, then remove the screws and take the clockspring off the steering column.

Installation

10   If the clockspring was removed, install it onto the steering column, making sure the pins on the back of the clockspring engage with the holes in the steering wheel rotation sensor. Install the screws, tightening them securely.

11   New clocksprings come equipped with an anti-rotation pin that holds them in the pre-centered position. If you are installing a new clockspring, don't remove the pin until you are ready to install the steering wheel after the steering column covers have been installed. If you are installing the old clockspring, or are not sure whether or not the clockspring has become uncentered, center the clockspring:

 a) Turn the inner hub of the clockspring counterclockwise until you feel resistance.

Caution: *Don't apply too much force to the hub or you could damage the clockspring ribbon.*

 b) Turn the hub so the electrical connector is in the 12 o'clock position, then rotate it clockwise five turns, returning the electrical connector to the 12 o'clock position.

 c) Turn the hub counterclockwise two turns, returning the electrical connector to the 12 o'clock position

12   Install the steering column covers, if removed. If you installed a new clockspring, remove the anti-rotation pin.

13   Install the steering wheel by reversing the removal procedure. When placing the steering wheel onto the shaft, be sure to align the match marks you made in Step 6.

14   Tighten the steering wheel bolt to the torque listed in this Chapter's Specifications.

Caution: *Make sure the connectors fit correctly on the airbag module (with the safety tabs still UP). When connecting the airbag connectors, match the key on the connector to the corresponding keyway on the module. Do not push the connectors on if the clips are down. After the clips are secured, push the airbag module down toward the wheel until the retaining clips lock. When seated properly, the gap between the airbag module trim cover and the steering wheel should be even and consistent.*

15 Steering column - removal and installation

Warning: *These models are equipped with a Supplemental Restraint System (SRS), more commonly known as airbags. Always disable the airbag system before working in the vicinity of any airbag system component to avoid the possibility of accidental deployment of the airbag(s), which could cause personal injury (see Chapter 12).*

Chapter 10 Suspension and steering systems

15.7 Mark the U-joint to the steering column shaft, then remove the pinch bolt

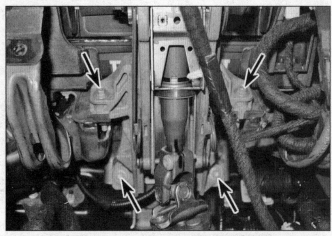

15.8 Steering column mounting bolts

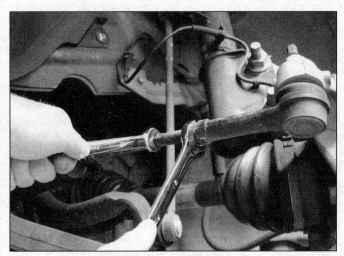

16.2a Loosen the jam nut . . .

16.2b . . . then mark the position of the tie-rod end

Warning: *Do not use a memory saving device to preserve the PCM or radio memory when working on or near airbag system components.*
Warning: *The manufacturer recommends replacing all fasteners with new ones during installation.*

Removal

1 Park the vehicle with the wheels pointing straight ahead. Disconnect the cable from the negative terminal of the battery (see Chapter 5).
2 Remove the steering wheel (see Section 14) and tape the airbag clockspring to prevent it from turning.
Caution: *If this is not done, the airbag clockspring could be damaged.*
3 Remove the steering column covers (see Chapter 11).
4 Remove the knee bolster and reinforcement (see Chapter 11). If the vehicle is equipped with an in-vehicle temperature sensor, disconnect the electrical connector and

hose from the sensor.
5 Disconnect the electrical connectors from both sides of the column, then detach the retainer holding the wiring harness and position the harness out of the way.
6 Remove the cover from the lower steering shaft joint.
7 Make a match-mark on the steering column shaft as it relates to the U-joint, then remove the pinch bolt (see illustration).
8 Remove the four steering column mounting fasteners (see illustration). Lower the column and pull it off of the intermediate shaft, then remove the column.

Installation

9 Guide the steering column into position, engaging the U-joint with the bottom of the steering shaft, then install the steering column mounting fasteners and tighten them to the torque listed in this Chapter's Specifications.
Note: *An assistant will be very helpful in this step.*
10 Install a new pinch bolt, tightening it to the

torque listed in this Chapter's Specifications.
11 The remainder of installation is the reverse of removal.

16 Tie-rod ends - removal and installation

Warning: *The manufacturer recommends replacing all fasteners with new ones during installation.*

Removal

1 Loosen the front wheel lug nuts. Apply the parking brake, raise the front of the vehicle and support it securely on jackstands. Remove the front wheel.
2 Hold the tie-rod with a pair of locking pliers or wrench and loosen the jam nut enough to mark the position of the tie-rod end in relation to the threads (see illustrations).
3 Loosen (but don't remove) the nut on the tie-rod end stud.

Chapter 10 Suspension and steering systems

16.4a Loosen the tie-rod end nut from the steering knuckle...

16.4b ... then, using a dead-blow mallet, tap the tie rod loose from the steering knuckle. Now remove the nut

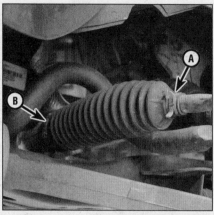

17.3 The outer ends of the steering gear boots are secured by band-type clamps (A); they're easily removed with a pair of pliers. The inner ends are retained by boot clamps (B) which must be cut off and discarded

4   Disconnect the tie-rod end from the steering knuckle arm with a tie-rod end remover or a hard dead blow mallet (a dead blow mallet works extremely well on this model) (see illustrations). Remove the nut and detach the tie-rod end.
5   Unscrew the tie-rod end from the tie-rod.

Installation

6   Thread the tie-rod end on to the marked position and insert the stud into the steering knuckle arm. Tighten the jam nut securely.
7   Install the nut on the stud and tighten it to the torque listed in this Chapter's Specifications. Install a new cotter pin.
8   Install the wheel and lug nuts. Lower the vehicle and tighten the lug nuts to the torque listed in the Chapter 1 Specifications.
9   Have the wheel alignment checked and, if necessary, adjusted.

17 Steering gear boots - replacement

Warning: *The manufacturer recommends replacing all fasteners with new ones during installation.*

1   Loosen the lug nuts, raise the vehicle and support it securely on jackstands. Remove the wheel.

2   Remove the tie-rod end and jam nut (see Section 16).
3   Remove the outer steering gear boot clamp with a pair of pliers (see illustration). Cut off the inner boot clamp with a pair of diagonal cutters. Slide off the boot.
4   Before installing the new boot, wrap the threads and serrations on the end of the steering rod with a layer of tape so the small end of the new boot isn't damaged.
5   Slide the new boot into position on the steering gear until it seats in the groove in the steering rod and install new clamps.
6   Remove the tape and install the tie-rod end (see Section 16).
7   Install the wheel and lug nuts. Lower the vehicle and tighten the lug nuts to the torque listed in the Chapter 1 Specifications.

18 Steering gear - removal and installation

Warning: *These models are equipped with airbags. Always disable the airbag system before working in the vicinity of airbag system components (see Chapter 12). Make sure the steering column shaft is not turned while the steering gear is removed or you could damage the airbag system clockspring. To prevent the shaft from turning, turn the ignition key to*

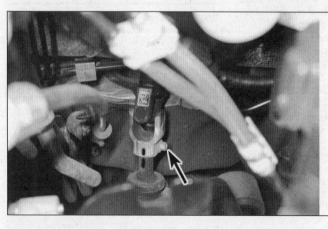

18.5 Index the shaft to the joint, then remove the pinch bolt

the lock position with the steering wheel in the straight ahead position before beginning work.
Warning: *The manufacturer recommends replacing all fasteners with new ones during installation.*
Note: *If a new steering gear is to be installed, the information stored in the Power Steering Control Module (PSCM) must be uploaded into a scan tool before proceeding.*

Removal

1   Disconnect the cable from the negative terminal of the battery (see Chapter 5).
2   Loosen the front wheel lug nuts. Raise the vehicle and place it securely on jackstands. Remove both front wheels.
3   Detach the tie-rod ends from the steering knuckles (see Section 16)
4   Remove the front stabilizer bar bracket bolts (see Section 4).
Note: *On V6 models, remove the exhaust Y pipe (see Section 4).*
5   Index mark the steering shaft to the steering gear input shaft, then disconnect the steering shaft from the steering gear input shaft (see illustration).
Note: *Be sure to secure the steering wheel from rotating once the steering shaft has been separated from the steering gear input shaft. This can be done by passing the seat belt through the steering wheel and clipping it into its latch.*
6   Place a floor jack under the subframe in such a way that it is not going to hinder the removal of the steering gear assembly.
7   Remove the two rear subframe bolts and support brackets (see Section 4, illustration 4.8).
8   Loosen but do not remove the front two subframe bolts.
9   Position the stabilizer bar up far enough to gain clearance for the steering gear assembly removal.
10 Lower the subframe on the jack to gain

Chapter 10 Suspension and steering systems

18.11 Remove the heat shield to gain access to the electrical connections

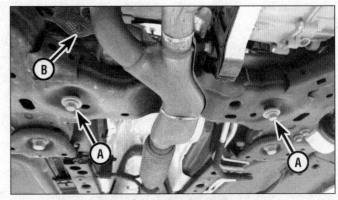

18.12 Steering gear is mounted on top of the subframe and secured by two large bolts

A   Steering gear mounting bolts
B   Steering gear (protective exhaust shield surrounds the steering gear. Be sure to reinstall shielding)

more room to remove the steering gear assembly.
11 Disconnect the steering gear electrical connections (see illustration).
12 Remove the two steering gear assembly bolts (discard bolts and replace them with new bolts) (see illustration).
13 Guide the steering gear assembly out of the right side of the vehicle.

Installation

14 Installation is the reverse of removal, noting the following points:
  a) New fasteners should be used in the reassembly.
  b) If you're installing a new steering gear, be sure to install a new steering gear turn tube heat wrap.
  c) Tighten the steering gear mounting bolts to the specified torque in sequence: left rear, right rear, left front, right front.
  d) When connecting the steering column shaft to the steering gear input shaft, be sure to align the matchmarks, and install a new pinch bolt.
  e) Have the front end alignment checked and, if necessary, adjusted.

19 Subframe - removal and installation

1 The fasteners used in conjunction with the subframe, suspension, and steering are a critical part to the safe operation of the vehicle. Use new replacement fasteners of the same quality as the original components. Do not use substandard replacement parts or parts of a different design. Always tighten all fasteners to the specified torque settings.

Front

Front ride height measurement

Note: *The tires should be fully inflated and the fuel tank should be full to properly measure the vehicle ride height. No extra weight should be in the cab or in the cargo area.*

2 Measure and record the distance between the center of the control arm forward pivot bolt and the ground.
3 Measure and record the distance from the bottom of the balljoint to the ground.
4 Subtract the second measurement from the first; this is the ride height.
5 See this Chapter's Specifications for the standard ride height values.

Removal and installation

6 Position the front wheels straight ahead. Now, immobilize the steering wheel from moving. Use a suitable holding device to maintain the steering wheel position, or pass the seat belt through the steering wheel and clip it into its latch.
7 Loosen the front wheel lug nuts, then raise the front of the vehicle and support it on jackstands. Be sure the jackstands are not impeding the removal of the subframe. Remove the front wheels.
8 If equipped, remove the lower air duct system. (Three retainers hold it in place.)
9 If equipped, remove the lower under body splash shield. (Four retainers hold the splash shield in place.)
10 Remove the right and left lower engine splash shield retainer pins.
11 On the non-turbocharged V6 engines, remove the exhaust Y pipe (see Section 4, illustration 4.7).
12 On turbocharged V6 engines, remove the right and left hand flexible exhaust pipes.
13 Remove the stabilizer bracket bolts (see Section 4).
14 Detach the tie-rod ends from the steering knuckles (see Section 16).
15 Detach the balljoints from the steering knuckles (see Section 6).
16 Remove the heat shield flap bolt and position it out of the way to gain access to the steering gear electrical connectors (see Section 18).
17 Remove the wire harness bracket bolt and disconnect the electrical connections from the steering gear.
18 Remove the steering shaft to steering gear bolt. Do not allow the steering wheel, shaft, or gear box to turn once it is disconnected (see Section 18).
19 If equipped, remove the rear roll restrictor to subframe bolt (see illustration 4.9).
20 If equipped, remove the cooler hose retainer latch. Pry up on the oil cooler retainer to release it.
21 If equipped, remove the three PTU cooler line hold down brackets from the left (upper) side of the subframe.
22 Use a marker, or stencil to mark the relationship between the subframe and the mounting area. Mark it so that you can use these location to reposition the subframe for reinstalling.
23 Support the subframe with two floor jacks (one positioned on each side).
24 Remove the front and rear subframe mounting bolts and subframe brackets (see Section 4, illustration 4.8).
25 Lower the subframe from the vehicle.
26 To install, reverse the process of removal. Use all new fasteners and tighten them to the torque listed in this Chapter's Specifications.

Rear

27 With the vehicle on a hard flat surface measure the vehicle ride height and record it for later retrieval.

Ride height measurement

Note: *The tires should be fully inflated and the fuel tank should be full to properly measure the vehicle ride height. No extra weight should be in the cab or in the cargo area.*

28 Measure and record the distance between the center of the control arm forward pivot bolt and the ground.
29 Measure and record the distance between the center of the control arm-to-knuckle bolt and the ground.
30 Subtract the second measurement from the first; this is the ride height. See this Chapter's Specifications for the standard ride height values.

Removal and installation

31 Mark the orientation of the lower suspension to the subframe and under body mount-

10-18 Chapter 10 Suspension and steering systems

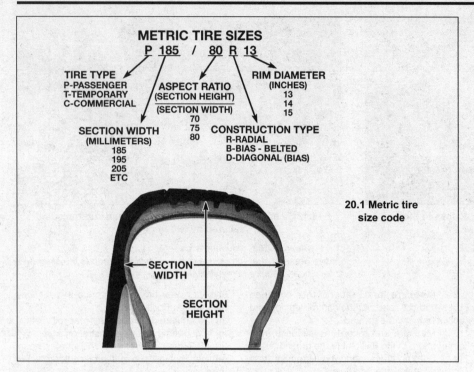

20.1 Metric tire size code

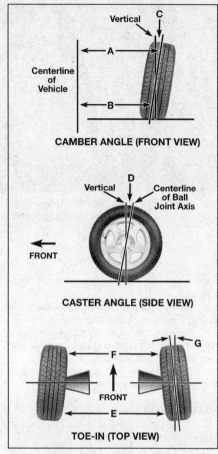

21.1 Camber, caster and toe-in angles

A minus B = C (degrees camber)
D = degrees caster
E minus F = toe-in (measured in inches)
G = toe-in (expressed in degrees)

ing locations. Use a wax pencil, paint, or a scribe to make a suitable mark.

32 Loosen the rear wheel lug nuts, then raise the rear of the vehicle and support it on jackstands. Be sure the jackstands do not interfere with any portion of the rear subframe. Remove both wheels.

33 Remove the muffler and tail pipe sections (see Chapter 4).

34 On AWD models, index mark the driveshaft, then remove bolts and disconnect the driveshaft from the rear differential.

35 Remove the nuts and detach the stabilizer bar links from the rear knuckle (see Section 13).

36 Use a floor jack to support the lower control arm. Remove the shock absorber lower mounting bolt (see Section 8), then slowly lower the floor jack. Repeat on the other shock absorber.

37 Disconnect the electrical connector attached to the rear subframe.

38 Have an assistant pull down on the forward parking brake cable until the sector gear on in the parking brake pedal assembly is fully rotated, then insert a six-inch long pin into the hole to prevent the sector from rotating back. This relieves tension on the parking brake cable.

39 Disconnect the right hand parking brake cable from the front parking brake cable.

40 Remove the rear parking brake cable bracket bolt and position the cable bracket off to the side.

41 Remove the right and left hand parking brake cable brackets and bolts. Position them off to the side.

42 Disconnect the parking brake cables from the the brake calipers.

43 Remove the brake calipers (see Chapter 9). Don't disconnect the brake hoses.

Caution: *Support the calipers with lengths of wire. Do not allow the calipers to hang by the hoses.*

44 Support the subframe with a pair of floor jacks.

45 Remove the rear subframe brackets and bolts.

46 Carefully lower the subframe, keeping an eye of any obstructions or attached components. Disconnect any components that have been missed and guide the subframe from the vehicle.

Note: *There are two rear drive unit (RDU) rear bushings in the rear subframe. Replacing the bushings will require the use of a press.*

47 To install, reverse the process of removal. Use all new fasteners and follow the torque settings listed in this Chapter's Specifications.

20 Wheels and tires - general information

1 All vehicles covered by this manual are equipped with metric-sized fiberglass or steel belted radial tires (see illustration). Use of other size or type of tires may affect the ride and handling of the vehicle. Don't mix different types of tires, such as radials and bias belted, on the same vehicle as handling may be seriously affected. It's recommended that tires be replaced in pairs on the same axle, but if only one tire is being replaced, be sure it's the same size, structure and tread design as the other.

2 Because tire pressure has a substantial effect on handling and wear, the pressure on all tires should be checked at least once a month or before any extended trips (see Chapter 1).

3 Wheels must be replaced if they are bent, dented, leak air, have elongated bolt holes, are heavily rusted, out of vertical symmetry or if the lug nuts won't stay tight. Wheel repairs that use welding or peening are not recommended.

4 Tire and wheel balance is important in the overall handling, braking and performance of the vehicle. Unbalanced wheels can adversely affect handling and ride characteristics as well as tire life. Whenever a tire is installed on a wheel, the tire and wheel should be balanced by a shop with the proper equipment.

21 Wheel alignment - general information

1 A wheel alignment refers to the adjustments made to the wheels so they are in proper angular relationship to the suspension and the ground. Wheels that are out of proper alignment not only affect vehicle control, but also increase tire wear. The front end angles normally measured are camber, caster and toe-in (see illustration). Toe-in and camber are adjustable; if the caster is not correct, check

Chapter 10 Suspension and steering systems

for bent components. Rear toe-in is also adjustable.

2 Getting the proper wheel alignment is a very exacting process, one in which complicated and expensive machines are necessary to perform the job properly. Because of this, you should have a technician with the proper equipment perform these tasks. We will, however, use this space to give you a basic idea of what is involved with a wheel alignment so you can better understand the process and deal intelligently with the shop that does the work.

3 Toe-in is the turning in of the wheels. The purpose of a toe specification is to ensure parallel rolling of the wheels. In a vehicle with zero toe-in, the distance between the front edges of the wheels will be the same as the distance between the rear edges of the wheels. The actual amount of toe-in is normally only a fraction of an inch. On the front end, toe-in is controlled by the tie-rod end position on the tie-rod. On the rear end, it's controlled by the threaded sleeve of the toe-link. Incorrect toe-in will cause the tires to wear improperly by making them scrub against the road surface.

4 Camber is the tilting of the wheels from vertical when viewed from one end of the vehicle. When the wheels tilt out at the top, the camber is said to be positive (+). When the wheels tilt in at the top the camber is negative (-). The amount of tilt is measured in degrees from vertical and this measurement is called the camber angle. This angle affects the amount of tire tread which contacts the road and compensates for changes in the suspension geometry when the vehicle is cornering or traveling over an undulating surface. On the front end it is not adjustable.

5 Caster is the tilting of the front steering axis from the vertical. A tilt toward the rear is positive caster and a tilt toward the front is negative caster. Caster isn't adjustable on these vehicles.

Notes

Chapter 11
Body

Contents

Section		Section	
Body repair - major damage	4	General information	1
Body repair - minor damage	3	Hood - removal, installation and adjustment	7
Bumper covers - removal and installation	9	Hood latch and release cable - removal and installation	8
Center console - removal and installation	21	Instrument panel - removal and installation	23
Cowl panels - removal and installation	12	Liftgate - removal and installation	19
Dashboard trim panels - removal and installation	22	Liftgate/Trunk Module (LTM) and power liftgate motor - removal and installation	20
Door - removal, installation and adjustment	14	Mirrors - removal and installation	18
Door latch, lock cylinder and handles - removal and installation	15	Quarter trim panel - removal and installation	26
Door trim panel - removal and installation	13	Radiator grille and shutter - removal and installation	11
Door window glass - removal and installation	16	Repair of minor paint scratches	2
Door window regulator and motor - removal and installation	17	Steering column covers - removal and installation	24
Fastener and trim removal	5	Upholstery, carpets and vinyl trim - maintenance	6
Front fender - removal and installation	10		
Front seats - removal and installation	25		

Specifications

Torque specifications

Ft-lbs (unless otherwise indicated) **Nm**

Note: One foot-pound (ft-lb) of torque is equivalent to 12 inch-pounds (in-lbs) of torque. Torque values below approximately 15 ft-lbs are expressed in inch-pounds, because most foot-pound torque wrenches are not accurate at these smaller values.

	Ft-lbs	Nm
Window regulator bolt and nuts	93 in-lbs	10.5
Glass to window regulator clamps	71 in-lbs	8
Strut tower cross brace	41	55
Dash to instrument panel (located in cowling area)	18	25
Dash board left and right end bolts	30	40
Dash board center mounting bolts	18	25
Dash board ground terminal bolts	89 in-lbs	10
Seat frame to floor bolts	37	50
Seat belt buckle to floor	27	37

1 General Information

Warning: *The models covered by this manual are equipped with Supplemental Restraint Systems (SRS), more commonly known as airbags. Always disable the airbag system before working in the vicinity of any airbag system components to avoid the possibility of accidental deployment of the airbags, which could cause personal injury (see Chapter 12).*

1 Certain body components are particularly vulnerable to accident damage and can be unbolted and repaired or replaced. Among these parts are the hood, doors, tailgate, liftgate, bumpers and front fenders.

2 Only general body maintenance practices and body panel repair procedures within the scope of the do-it-yourselfer are included in this Chapter.

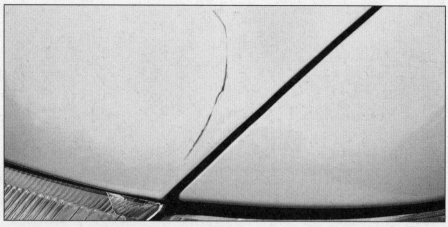

Make sure the damaged area is perfectly clean and rust free. If the touch-up kit has a wire brush, use it to clean the scratch or chip. Or use fine steel wool wrapped around the end of a pencil. Clean the scratched or chipped surface only, not the good paint surrounding it. Rinse the area with water and allow it to dry thoroughly

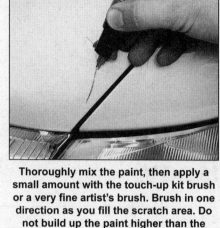

Thoroughly mix the paint, then apply a small amount with the touch-up kit brush or a very fine artist's brush. Brush in one direction as you fill the scratch area. Do not build up the paint higher than the surrounding paint

2 Repair of minor paint scratches

1 No matter how hard you try to keep your vehicle looking like new, it will inevitably be scratched, chipped or dented at some point. If the metal is actually dented, seek the advice of a professional. But you can fix minor scratches and chips yourself. Buy a touch-up paint kit from a dealer service department or an auto parts store. To ensure that you get the right color, you'll need to have the specific make, model and year of your vehicle and, ideally, the paint code, which is located on a special metal plate under the hood or in the door jamb.

3 Body repair - minor damage

Plastic body panels

1 The following repair procedures are for minor scratches and gouges. Repair of more serious damage should be left to a dealer service department or qualified auto body shop. Below is a list of the equipment and materials necessary to perform the following repair procedures on plastic body panels.

Wax, grease and silicone removing solvent
Cloth-backed body tape
Sanding discs
Drill motor with three-inch disc holder
Hand sanding block
Rubber squeegees
Sandpaper
Non-porous mixing palette
Wood paddle or putty knife
Wood paddle or putty knife
Curved-tooth body file
Flexible parts repair material

Flexible panels (bumper trim)

2 Remove the damaged panel, if necessary or desirable. In most cases, repairs can be carried out with the panel installed.

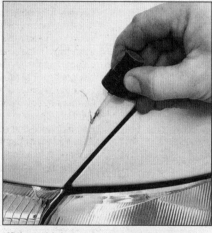

If the vehicle has a two-coat finish, apply the clear coat after the color coat has dried

3 Clean the area(s) to be repaired with a wax, grease and silicone removing solvent applied with a water-dampened cloth.
4 If the damage is structural, that is, if it extends through the panel, clean the backside of the panel area to be repaired as well. Wipe dry.
5 Sand the rear surface about 1-1/2 inches beyond the break.
6 Cut two pieces of fiberglass cloth large enough to overlap the break by about 1-1/2 inches. Cut only to the required length.
7 Mix the adhesive from the repair kit according to the instructions included with the kit, and apply a layer of the mixture approximately 1/8-inch thick on the backside of the panel. Overlap the break by at least 1-1/2 inches.
8 Apply one piece of fiberglass cloth to the adhesive and cover the cloth with additional adhesive. Apply a second piece of fiberglass cloth to the adhesive and immediately cover the cloth with additional adhesive in sufficient

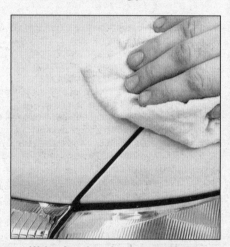

Wait a few days for the paint to dry thoroughly, then rub out the repainted area with a polishing compound to blend the new paint with the surrounding area. When you're happy with your work, wash and polish the area

quantity to fill the weave.
9 Allow the repair to cure for 20 to 30 minutes at 60-degrees to 80-degrees F.
10 If necessary, trim the excess repair material at the edge.
11 Remove all of the paint film over and around the area(s) to be repaired. The repair material should not overlap the painted surface.
12 With a drill motor and a sanding disc (or a rotary file), cut a "V" along the break line approximately 1/2-inch wide. Remove all dust and loose particles from the repair area.
13 Mix and apply the repair material. Apply a light coat first over the damaged area; then continue applying material until it reaches a level slightly higher than the surrounding finish.
14 Cure the mixture for 20 to 30 minutes at 60-degrees to 80-degrees F.

15 Roughly establish the contour of the area being repaired with a body file. If low areas or pits remain, mix and apply additional adhesive.
16 Block sand the damaged area with sandpaper to establish the actual contour of the surrounding surface.
17 If desired, the repaired area can be temporarily protected with several light coats of primer. Because of the special paints and techniques required for flexible body panels, it is recommended that the vehicle be taken to a paint shop for completion of the body repair.

Steel body panels

Repairing simple dents

18 When repairing dents, the first job is to pull the dent out until the affected area is as close as possible to its original shape. There is no point in trying to restore the original shape completely as the metal in the damaged area will have stretched on impact and cannot be restored to its original contours. It is better to bring the level of the dent up to a point that is about 1/8-inch below the level of the surrounding metal. In cases where the dent is very shallow, it is not worth trying to pull it out at all.
19 If the backside of the dent is accessible, it can be hammered out gently from behind using a soft-face hammer. While doing this, hold a block of wood firmly against the opposite side of the metal to absorb the hammer blows and prevent the metal from being stretched.
20 If the dent is in a section of the body which has double layers, or some other factor makes it inaccessible from behind, a different technique is required. Drill several small holes through the metal inside the damaged area, particularly in the deeper sections. Screw long, self-tapping screws into the holes just enough for them to get a good grip in the metal. Now pulling on the protruding heads of the screws with locking pliers can pull out the dent.
21 The next stage of repair is the removal of paint from the damaged area and from an inch or so of the surrounding metal. This is easily done with a wire brush or sanding disk in a drill motor, although it can be done just as effectively by hand with sandpaper. To complete the preparation for filling, score the surface of the bare metal with a screwdriver or the tang of a file or drill small holes in the affected area. This will provide a good grip for the filler material. To complete the repair, see the Section on filling and painting.

Repair of rust holes or gashes

22 Remove all paint from the affected area and from an inch or so of the surrounding metal using a sanding disk or wire brush mounted in a drill motor. If these are not available, a few sheets of sandpaper will do the job just as effectively.
23 With the paint removed, you will be able to determine the severity of the corrosion and decide whether to replace the whole panel, if possible, or repair the affected area. New body panels are not as expensive as most people think and it is often quicker to install a new panel than to repair large areas of rust.
24 Remove all trim pieces from the affected area except those which will act as a guide to the original shape of the damaged body, such as headlight shells, etc. Using metal snips or a hacksaw blade, remove all loose metal and any other metal that is badly affected by rust. Hammer the edges of the hole in to create a slight depression for the filler material.
25 Wire-brush the affected area to remove the powdery rust from the surface of the metal. If the back of the rusted area is accessible, treat it with rust inhibiting paint.
26 Before filling is done, block the hole in some way. This can be done with sheet metal riveted or screwed into place, or by stuffing the hole with wire mesh.
27 Once the hole is blocked off, the affected area can be filled and painted. See the following subsection on filling and painting.

Filling and painting

28 Many types of body fillers are available, but generally speaking, body repair kits which contain filler paste and a tube of resin hardener are best for this type of repair work. A wide, flexible plastic or nylon applicator will be necessary for imparting a smooth and contoured finish to the surface of the filler material. Mix up a small amount of filler on a clean piece of wood or cardboard (use the hardener sparingly). Follow the manufacturer's instructions on the package, otherwise the filler will set incorrectly.
29 Using the applicator, apply the filler paste to the prepared area. Draw the applicator across the surface of the filler to achieve the desired contour and to level the filler surface. As soon as a contour that approximates the original one is achieved, stop working the paste. If you continue, the paste will begin to stick to the applicator. Continue to add thin layers of paste at 20-minute intervals until the level of the filler is just above the surrounding metal.
30 Once the filler has hardened, the excess can be removed with a body file. From then on, progressively finer grades of sandpaper should be used, starting with a 180-grit paper and finishing with 600-grit wet-or-dry paper. Always wrap the sandpaper around a flat rubber or wooden block, otherwise the surface of the filler will not be completely flat. During the sanding of the filler surface, the wet-or-dry paper should be periodically rinsed in water. This will ensure that a very smooth finish is produced in the final stage.
31 At this point, the repair area should be surrounded by a ring of bare metal, which in turn should be encircled by the finely feathered edge of good paint. Rinse the repair area with clean water until all of the dust produced by the sanding operation is gone.
32 Spray the entire area with a light coat of primer. This will reveal any imperfections in the surface of the filler. Repair the imperfections with fresh filler paste or glaze filler and once more smooth the surface with sandpaper. Repeat this spray-and-repair procedure until you are satisfied that the surface of the filler and the feathered edge of the paint are perfect. Rinse the area with clean water and allow it to dry completely.
33 The repair area is now ready for painting. Spray painting must be carried out in a warm, dry, windless and dust free atmosphere. These conditions can be created if you have access to a large indoor work area, but if you are forced to work in the open, you will have to pick the day very carefully. If you are working indoors, dousing the floor in the work area with water will help settle the dust that would otherwise be in the air. If the repair area is confined to one body panel, mask off the surrounding panels. This will help minimize the effects of a slight mismatch in paint color. Trim pieces such as chrome strips, door handles, etc., will also need to be masked off or removed. Use masking tape and several thickness of newspaper for the masking operations.
34 Before spraying, shake the paint can thoroughly, then spray a test area until the spray painting technique is mastered. Cover the repair area with a thick coat of primer. The thickness should be built up using several thin layers of primer rather than one thick one. Using 600-grit wet-or-dry sandpaper, rub down the surface of the primer until it is very smooth. While doing this, the work area should be thoroughly rinsed with water and the wet-or-dry sandpaper periodically rinsed as well. Allow the primer to dry before spraying additional coats.
35 Spray on the top coat, again building up the thickness by using several thin layers of paint. Begin spraying in the center of the repair area and then, using a circular motion, work out until the whole repair area and about two inches of the surrounding original paint is covered. Remove all masking material 10 to 15 minutes after spraying on the final coat of paint. Allow the new paint at least two weeks to harden, then use a very fine rubbing compound to blend the edges of the new paint into the existing paint. Finally, apply a coat of wax

4 Body repair - major damage

1 Major damage must be repaired by an auto body shop specifically equipped to perform body and frame repairs. These shops have the specialized equipment required to do the job properly.
2 If the damage is extensive, the frame must be checked for proper alignment or the vehicle's handling characteristics may be adversely affected and other components may wear at an accelerated rate.
3 Due to the fact that all of the major body components (hood, fenders, etc.) are separate and replaceable units, any seriously damaged components should be replaced rather than repaired. Sometimes the components can be found in a wrecking yard that specializes in used vehicle components, often at considerable savings over the cost of new parts.

These photos illustrate a method of repairing simple dents. They are intended to supplement *Body repair - minor damage* in this Chapter and should not be used as the sole instructions for body repair on these vehicles.

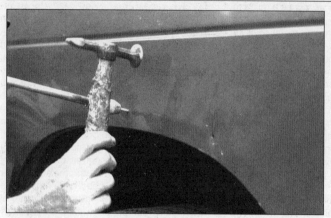

1 If you can't access the backside of the body panel to hammer out the dent, pull it out with a slide-hammer-type dent puller. Tap with a hammer near the edge of the dent to help 'pop' the metal back to its original shape, about 1/8-inch below the surface of the surrounding metal

2 Using coarse-grit sandpaper, remove the paint down to the bare metal. Clean the repair area with wax/silicone remover.

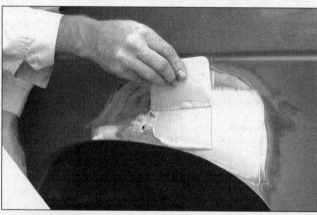

3 Following label instructions, mix up a batch of plastic filler and hardener, then quickly press it into the metal with a plastic applicator. Work the filler until it matches the original contour and is slightly above the surrounding metal

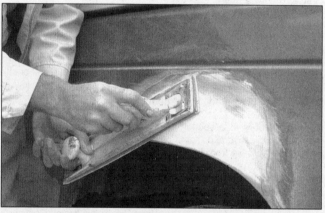

4 Let the filler harden until you can just dent it with your fingernail. File, then sand the filler down until it's smooth and even. Work down to finer grits of sandpaper - always using a board or block - ending up with 360 or 400 grit

5 When the area is smooth to the touch, clean the area and mask around it. Apply several layers of primer to the area. A professional-type spray gun is being used here, but aerosol spray primer works fine

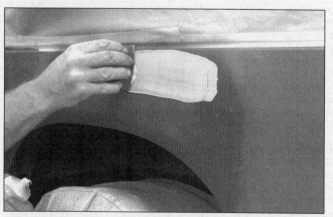

6 Fill imperfections or scratches with glazing compound. Sand with 360 or 400-grit and re-spray. Finish sand the primer with 600 grit, clean thoroughly, then apply the finish coat. Don't attempt to rub out or wax the repair area until the paint has dried completely (at least two weeks)

5 Fastener and trim removal

1 There is a variety of plastic fasteners used to hold trim panels, splash shields and other parts in place in addition to typical screws, nuts and bolts. Once you are familiar with them, they can usually be removed without too much difficulty.

2 The proper tools and approach can prevent added time and expense to a project by minimizing the number of broken fasteners and/or parts.

3 The accompanying illustration shows various types of fasteners that are typically used on most vehicles and how to remove and install them (see illustration). Replacement fasteners are commonly found at most auto parts stores, if necessary.

4 Trim panels are typically made of plastic and their flexibility can help during removal. The key to their removal is to use a tool to pry the panel near its retainers to release it without damaging surrounding areas or breaking-off any retainers. The retainers will usually snap out of their designated slot or hole after force is applied to them. Stiff plastic tools designed for prying on trim panels are available at most auto parts stores (see illustration). Tools that are tapered and wrapped in protective tape, such as a screwdriver or small pry tool, are also very effective when used with care.

6 Upholstery, carpets and vinyl trim - maintenance

Upholstery and carpets

1 Every three months remove the floormats and clean the interior of the vehicle (more frequently if necessary). Use a stiff whiskbroom to brush the carpeting and loosen dirt and dust, then vacuum the upholstery and carpets thoroughly, especially along seams and crevices.

2 Dirt and stains can be removed from carpeting with basic household or automotive carpet shampoos available in spray cans.

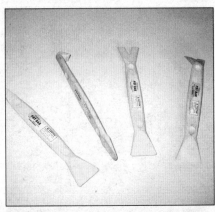

5.4 These small plastic pry tools are ideal for prying off trim panels

Follow the directions and vacuum again, then use a stiff brush to bring back the nap of the carpet.

3 Most interiors have cloth or vinyl upholstery, either of which can be cleaned and maintained with a number of material-specific

Fasteners

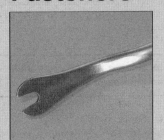

This tool is designed to remove special fasteners. A small pry tool used for removing nails will also work well in place of this tool

A Phillips head screwdriver can be used to release the center portion, but light pressure must be used because the plastic is easily damaged. Once the center is up, the fastener can easily be pried from its hole

Here is a view with the center portion fully released. Install the fastener as shown, then press the center in to set it

This fastener is used for exterior panels and shields. The center portion must be pried up to release the fastener. Install the fastener with the center up, then press the center in to set it

This type of fastener is used commonly for interior panels. Use a small blunt tool to press the small pin at the center in to release it . . .

. . . the pin will stay with the fastener in the released position

Reset the fastener for installation by moving the pin out. Install the fastener, then press the pin flush with the fastener to set it

This fastener is used for exterior and interior panels. It has no moving parts. Simply pry the fastener from its hole like the claw of a hammer removes a nail. Without a tool that can get under the top of the fastener, it can be very difficult to remove

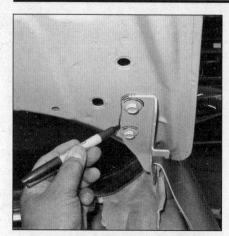

7.2 Draw alignment marks around the hood hinges to ensure proper alignment of the hood when it's reinstalled

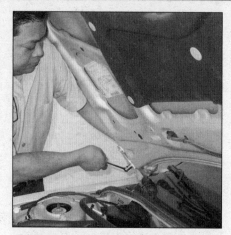

7.3 Support the hood with your shoulder while removing the hood bolts

7.8 There are two vertical height adjustment cushions on each side of the core support that adjust the hood

8.1 Hood latch bolts

cleaners or shampoos available in auto supply stores. Follow the directions on the product for usage, and always spot-test any upholstery cleaner on an inconspicuous area (bottom edge of a backseat cushion) to ensure that it doesn't cause a color shift in the material.

4 After cleaning, vinyl upholstery should be treated with a protectant.
Note: Make sure the protectant container indicates the product can be used on seats - some products may make a seat too slippery.
Caution: *Do not use protectant on vinyl-covered steering wheels.*

5 Leather upholstery requires special care. It should be cleaned regularly with saddle-soap or leather cleaner. Never use alcohol, gasoline, nail polish remover or thinner to clean leather upholstery.

6 After cleaning, regularly treat leather upholstery with a leather conditioner, rubbed in with a soft cotton cloth. Never use car wax on leather upholstery.

7 In areas where the interior of the vehicle is subject to bright sunlight, cover leather seating areas of the seats with a sheet if the vehicle is to be left out for any length of time.

Vinyl trim

8 Don't clean vinyl trim with detergents, caustic soap or petroleum-based cleaners. Plain soap and water works just fine, with a soft brush to clean dirt that may be ingrained. Wash the vinyl as frequently as the rest of the vehicle.

9 After cleaning, application of a high-quality rubber and vinyl protectant will help prevent oxidation and cracks. The protectant can also be applied to weather-stripping, vacuum lines and rubber hoses, which often fail as a result of chemical degradation, and to the tires.

7 Hood - removal, installation and adjustment

Note: *The hood is awkward to remove and install; at least two people should perform this procedure.*

Removal and installation

1 Open the hood, then place blankets or pads over the fenders and cowl area of the body. This will protect the body and paint as the hood is lifted off.

2 Make marks around the hood hinge to ensure proper alignment during installation (see illustration).

3 Have an assistant support one side of the hood. Grasp the lower corner of the hood and use your shoulder to brace the hood (see illustration). Take turns removing the hinge-to-hood bolts and lift off the hood.

4 Installation is the reverse of removal. Align the hinge bolts with the marks made in Step 2.

Adjustment

5 Fore-and-aft and side-to-side adjustment of the hood is done by moving the hinges after loosening the hinge-to-body bolts.

6 Loosen the bolts and move the hood into correct alignment. Move it only a little at a time. Tighten the hinge bolts and carefully lower the hood to check the position.

7 The hood can also be adjusted vertically so that it's flush with the fenders.

8 Turn each cushion clockwise to lower the hood or counterclockwise to raise the hood (see illustration).

9 The hood latch assembly, as well as the hinges, should be periodically lubricated with white, lithium-base grease to prevent binding and wear.

8 Hood latch and release cable - removal and installation

Hood latch

1 Open the hood and scribe a line around the latch to aid alignment when installing, then remove the retaining bolts securing the hood latch to the radiator support (see illustration). Remove the latch.

2 Squeeze the cable retainer to release the cable from the latch assembly, then disengage the cable end plug from the latch.

3 Installation is the reverse of removal.

Release cable

4 Working in the engine compartment, remove the hood latch and disconnect the hood release cable from the latch (see Steps 1 and 2).

5 Detach the two release cable clips from the radiator support.

6 Loosen the left front wheel lug nuts. Raise the vehicle and place it securely on jackstands. Remove the wheel. Remove the inner fender splash shield (see Section 10).

7 Open the two release cable clips in the upper part of the fender opening and detach the release cable from the clips.

8 Disengage the release cable clip from the wiring harness.
Note: *It may be necessary to remove the Smart Junction Box on some models.*

9 Remove the lower panel trim, then remove the hood release handle mounting

Chapter 11 Body

8.9a Remove the release handle mounting bolt, lift up slightly to dislodge the lower hook tab

8.9b Rotate the ball end (A) of the cable to the open slot and slide it out of the handle. Pinch the cable retaining splines (B) inwards to free the cable from the release handle

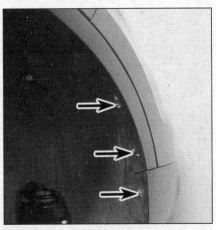

9.2a Remove the screws securing the trim to the fender and bumper

9.2b Pull the trim free from the bumper and front edge of the fender

9.2c Keep pulling the trim until it is completely out of the way of the bumper to gain access to the hidden screws, remove the screws

9.2d If equipped, disconnect the fog lights electrical connections

bolt (see illustrations). Pull out the handle, then disengage the release cable from the release handle and pull the release cable through the firewall.

10 Installation is the reverse of removal.

9 Bumper covers - removal and installation

Note: Refer to Section 6 for fastener and trim removal.

Front bumper cover

1   Loosen the front wheel lug nuts, raise the vehicle and support it securely on jackstands.
2   Remove the front wheels. Follow along with the illustrations for front bumper removal.
3   Before pulling the bumper completely free, verify that the bumper cover electrical

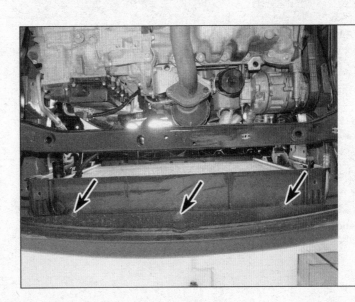

9.2e Remove the bumper cover lower fasteners

11-8 Chapter 11 Body

9.2f Remove the bumper cover upper fasteners

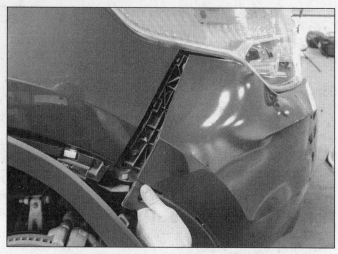

9.2g Pull out on the edge of the bumper to release the compression fittings holding the bumper edge to the fender

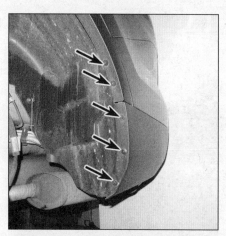

9.7a Remove the fasteners securing the fender trim molding and the inner fender to the rear bumper

9.7b Pull the fender trim molding out far enough to gain access to the hidden fasteners

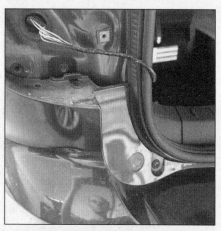

9.8 With the tail lamp assemblies removed you can access the bumper fasteners

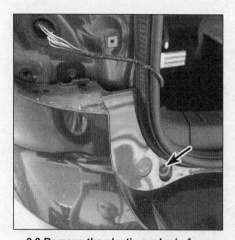

9.9 Remove the plastic push pin from each side of the liftgate opening

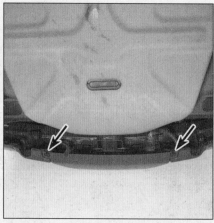

9.10 Remove the lower push pins securing the bumper to the inner bumper brace

connectors are disconnected and all wiring harnesses are safely out of the way.

Note: *As you remove the bumper cover, it will have a tendency to get caught around the headlamp assemblies. Use extra caution to maneuver the bumper cover free from those areas.*

4   With an assistant's help, remove the front bumper cover.
5   Installation is the reverse of removal.

Rear bumper cover

6   Loosen the rear wheel lug nuts, raise the vehicle and support it securely on jackstands. Remove the rear wheels.
7   Remove the rear fender molding (see illustration).
8   Remove the taillight assembly to expose the hidden fasteners (see illustration).
9   Remove the push pins in the liftgate opening (see illustration).
10   From under the vehicle, remove the plastic push pins (see illustration).

9.11 Disconnect the parking aid harness connector before removing the rear bumper

10.4 Remove the rocker panel molding end cap and all of the fasteners along the edge of the inner fender

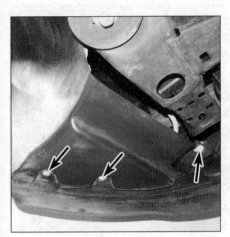

10.5 Inner fender lower fastener locations

10.6a Pry the plastic pressure clips off to remove the splash shield

10.6b Remove the lower fender bolts from the rocker panel

11 Disconnect the trailer tow connector and the parking aid harness connectors, if equipped (see illustration).
12 Disengage the corners of the bumper cover from the body panel.
13 Before removing, verify that the bumper cover is completely detached, all electrical connectors are disconnected and all wiring harnesses are safely out of the way.
14 With an assistant's help, remove the rear bumper cover.
15 Installation is the reverse of removal. Again, get help when putting the bumper cover back into position.

10 Front fender - removal and installation

Note: *Refer to Section 6 for fastener and trim removal details.*

1 Remove the headlight housing (see Chapter 12).

2 Carefully work the cowl panel filler trim off with a flat trim tool (see Section 12).
3 Loosen the front wheel lug nuts. Raise the vehicle, support it securely on jackstands and remove the front wheel.

Inner fender splash shield removal

4 Remove the rocker panel molding end cap mounting screws, and remove the end cap (see illustration).
5 Remove the fasteners that secure the inner fender to the underside of the vehicle (see illustration).

Fender removal

6 Remove the lower fender splash shield and fender bolts securing the fender to the rocker panel (see illustrations).
7 Remove the upper fender edge trim from under the hood (see illustration).
8 Open the door and remove the upper

10.7 Pry the trim panel off to gain access to the upper fender bolts

11-10   Chapter 11   Body

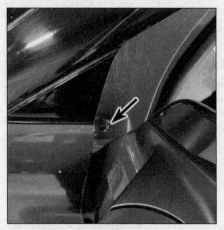

10.8 Access the upper fender bolt with the door open

10.9 Windshield applique uses hook and loop fasteners to hold it in place. The cowling bolt is also shown with the arrow in this view

10.11 Remove the two bolts from behind the horns

10.12 Fender upper mounting bolts

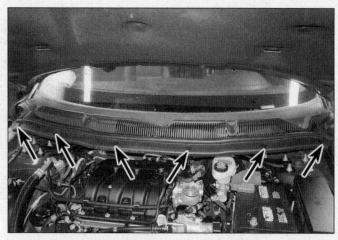

12.2 Upper cowling retainer locations

fender bolt securing the fender to the A pillar (see illustration).
9   Remove the windshield applique, working from the door jamb area (see illustration).
10   Remove the fender-to-cowl bolt.
11   Remove the two bolts from the fender to the fender apron located near the horns (see illustration).
12   Remove the two top fender-to-fender apron bolts (see illustration).
13   Carefully cut the foam seam along the body and foam covering the inner fender nut.
14   Lift off the fender. It's a good idea to have an assistant support the fender while it's being moved away from the vehicle to prevent damage to the surrounding body panels.
15   Using Flexible Foam repair or equivalent (available from your local dealer, auto parts store or automotive paint store), reseal the area along the seam at the body and around the front fender inner nut.
16   Installation is the reverse of removal. Check the alignment of the fender to the hood and front edge of the door before tightening the bolts.

11   Radiator grille and shutter - removal and installation

1   Remove the front bumper cover (see Section 9).

Radiator grille

Note: *If you're just removing the grille, release the 8 clips and remove the grille from the front bumper assembly.*
2   Remove the four radiator grille support nuts.
3   Release the six radiator grille retaining clips and separate the grille support from the front bumper assembly.
4   Remove the four radiator grille support bolts.
5   Release the grille to grille guard clips and separate it.
6   Installation is the reverse of removal.

Grille shutter assembly

7   Remove the grille shutter assembly mounting bolts from each corner of the shutter assembly.
8   Lower the assembly until the grille shutter actuator connector is exposed, then disconnect the electrical connector and remove the shutter assembly.
9   Installation is the reverse of removal.

12   Cowl panels - removal and installation

Note: *Refer to Section 6 for fastener and trim removal techniques.*

Upper cowl panel

1   Remove the wiper arms (see Chapter 12).
2   Remove the six pin-type retainers (see illustration).
3   With your fingers under the front edge and near the rear retaining clips gently pry up to release each of the seven rear edge retaining clips (see illustration).

Chapter 11 Body

12.3 Gently pry the edges up as you work around the cowling with a trim tool

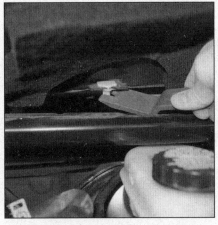

12.7a Lower cowl panel retainer clips

12.7b Some clips are easier to remove with a pair of needle-nose pliers

12.8 Pry the clip off to remove the washer hose, then separate the hose at the hose joint

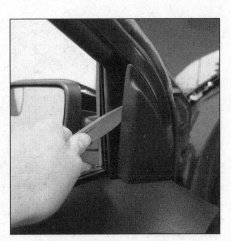

13.2a Using a flat trim tool, pry the sail trim off

13.2b A flat pocket screwdriver works best to remove the trim from behind the inside release handle

4 Remove the upper cowl panel.
5 Installation is the reverse of removal.

Lower cowl panel

6 Remove the upper cowl panel.
7 Remove the lower cowl mounting fasteners from inside the cowl panel (see illustrations).
8 Remove the retaining clips for the washer hose, and position it off to the side (see illustration).
9 Remove the cowl panel.
10 Installation is the reverse of removal.

13 Door trim panel - removal and installation

Caution: *Wear gloves when working inside the door openings to protect against sharp metal edges.*
Note: *Refer to Section 6 for fastener and trim removal.*

Removal

1 Disconnect the cable from the negative battery terminal (see Chapter 5).

Door trim panels

Note: *This procedure applies to both front and rear door trim panels. The only difference is that rear door trim panels do not have a screw at the center of the lower edge of the panel.*
2 Follow along with the illustrations 13.2a through 13.2j to remove the door panel.
3 Installation is the reverse of removal.

Liftgate trim panel

4 Support the liftgate in the open position, then remove the power liftgate rod nut cover and nut, if equipped.

13.2c Remove the screw hidden by the trim

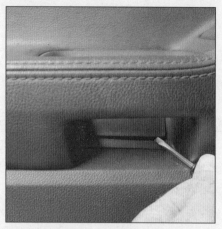

13.2d Remove the trim plate from below the arm rest

13.2e Remove the screws hidden behind the trim plate

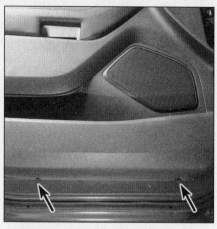

13.2f Remove the fasteners on the bottom of the door panel

13.2g Carefully pry the door panel away from the door

Caution: *Be sure to only pry at the clip locations.*

13.2h As the door panel comes loose reach over the door panel and disconnect the electrical connections

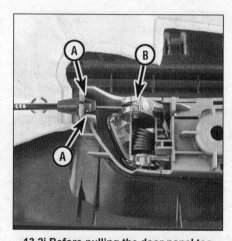

13.2i Before pulling the door panel too far away release the cable from the inner handle. Press in on the plastic tabs (A) to release the cable, then slide the cable ball end through the slot (B) and off of the inner handle

13.2j You now have access to the speaker, watershield and interior door components

13.5 Start by prying the upper panel off first

5 Remove the liftgate upper trim panel by releasing the retaining clips with a flat trim tool. Gently pry the panel up to release the seven retaining clips (see illustration).
6 Remove the liftgate assist handle cover, then remove the retaining bolts (see illustrations).
7 Disengage the liftgate trim panel clips using a door panel removal tool, and remove the panel (see illustration).
8 Disconnect the wire connector from the courtesy lamps and remove the trim panel.

Installation

9 Prior to installation of the door/liftgate trim panel(s), be sure to reinstall any clips which may have come out when you removed the panel.
10 Position the wire harness connectors for the power door lock switch and the power window switch (if equipped) on the back of the

Chapter 11 Body

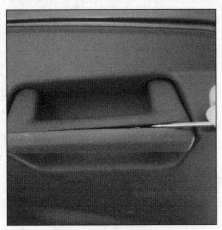

13.6a Pry the handle apart...

13.6b ... then remove the two retaining bolts

13.7 Work around the outside edges to free all the retaining clips

14.5 Door stop strut bolt

14.7a Upper hinge fastener

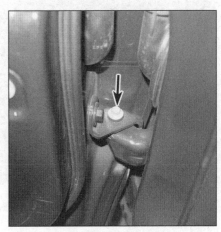

14.7b Lower hinge fastener

panel, then place the panel in position in the door. Press the door panel into place until the clips are seated.
11 The remainder of installation is the reverse of removal.

14 Door - removal, installation and adjustment

Warning: *The models covered by this manual are equipped with a Supplemental Restraint System (SRS), more commonly known as airbags. Always disable the airbag system before working in the vicinity of any airbag system component to avoid the possibility of accidental deployment of the airbag, which could cause personal injury (see Chapter 12).*

Removal and installation

Note: *The door is heavy and somewhat awkward to remove - at least two people should perform this procedure.*
Note: *This procedure applies to front and rear doors.*
1 Raise the window completely in the door.
2 Disconnect the cable from the negative battery terminal (see Chapter 5).
3 Open the door all the way and support it with a jack or blocks covered with rags to prevent damaging the outer surface.
4 Pull off the rubber conduit that protects the door's wiring harness, then disconnect the connector.
5 Remove the door stop strut mounting bolt (see illustration).
6 Mark around the door hinges with a pen or a scribe to facilitate realignment during reassembly.
7 With an assistant holding the door to steady it, remove the hinge-to-door fasteners (see illustrations) and lift off the door.
8 Installation is the reverse of removal.

Adjustment

9 Correct door-to-body alignment is a critical part of a well-functioning door assembly. First check the door hinge pins for excessive play. Fully open the door and lift up and down on the door without lifting the body. If a door has 1/16-inch or more excessive play, replace the hinges.
10 If you need to adjust or replace the hinges for a front door, remove the front fender (see Section 10).
11 Make door-to-body alignment adjustments by loosening the hinge-to-body bolts or hinge-to-door bolts and moving the door. When body alignment is correct, the top of each door is parallel with the roof section, the front door is flush with the fender, the rear door is flush with the rear quarter panel and the bottom of each door is aligned with the lower rocker panel. If you're unable to adjust the door correctly, you might be able to obtain body alignment shims that are inserted behind

14.12 Adjust the door latch striker by loosening the mounting screws and gently tapping the striker in the desired direction

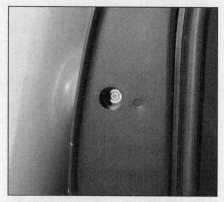

15.1 Outside door handle retaining screw (plastic dust cap has been removed already)

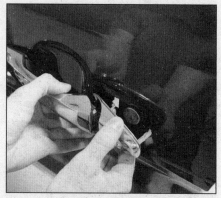

15.2 Remove the outside door handle cover

15.3 To disengage the outside door handle lever from the handle reinforcement inside the door, pull it out of the door and to the rear

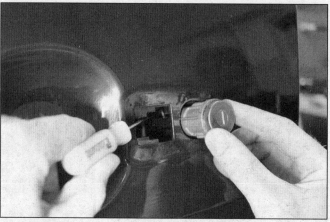

15.17 Release the tab by reaching into the door and pulling out on the tab

the hinges to correctly align the door.

12 To adjust the door-closed position, scribe a line or mark around the striker plate (see illustration) to provide a reference point, then verify that the door latch is contacting the center of the striker. If not, adjust the up and down position first. To move the striker, tap it gently with a small hammer.

13 Once the door latch is contacting the center of the striker, adjust the latch striker sideways position, so that the door panel is flush with the center pillar or rear quarter panel and provides positive engagement with the latch mechanism.

15 Door latch, lock cylinder and handles - removal and installation

Outside door handle lever (side doors)

Note: *This procedure only applies to the outside door handle lever. If you need to replace the outside door handle reinforcement (the actual mechanism for the outside door handle), see Steps 10 through 14.*

1 Remove the trim cap for the outside handle retaining screw (see illustration). Loosen, but don't remove, the screw.
2 Remove the cover from the outside handle (see illustration).
3 Pull the outside door handle out of the door and to the rear to disengage it from the handle reinforcement inside the door (see illustration).
4 Remove and inspect the outside door handle seals. If the seals are cracked, torn or otherwise deteriorated, replace them.
5 Installation is the reverse of removal.

Inside door handle

6 Remove the door trim panel (see Section 13).
7 Remove the five nuts securing the inner door handle to the door panel.
8 Remove the inner door handle from the door panel.
9 Installation is the reverse of removal.

Outside door handle reinforcement

10 Remove the outside door handle lever (see Steps 1 through 3), then remove the door trim panel (see Section 13) and the power window regulator (see Section 17).
11 Disengage the upper lock cylinder actuating rod from the handle reinforcement, if equipped.
12 Disengage the lower exterior door handle reinforcement actuating rod.
13 Disconnect the keyless entry pad harness retaining clip and move the harness out of the way.
14 Disengage the handle reinforcement-to-window regulator module locking clips and remove the handle reinforcement.
15 Installation is the reverse of removal.

Door lock cylinder

Note: *Before removing the door lock cylinder on these models, be aware that you cannot simply install a new door lock cylinder. Instead, you must discard the old cylinder and build a new unit with the appropriate lock repair package, which you can obtain from your dealer. The lock repair package includes instructions for how to build the new lock cylinder to the key code of the vehicle.*

16 Remove the outside door handle (see Steps 1 through 3).
17 Disengage the release tab and remove the lock cylinder (see illustration).

Chapter 11 Body

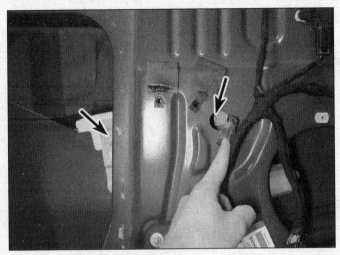

16.4 Front window glass clamp bolts

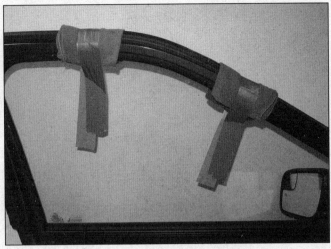

17.3 Some strong tape can suffice if you do not have any window suction cups to hold the glass up

18 When installing the rebuilt lock cylinder, align the D-slot in the lock cylinder with the D-slot in the lock cylinder lever.
19 Installation is otherwise the reverse of removal.

Liftgate latch

20 Remove the trim panel (see Section 13).
21 Disconnect the electrical connector from the liftgate latch.
22 Remove the liftgate latch bolts and remove the latch from the vehicle.
23 Installation is the reverse of removal. Perform the de-initialization/initialization procedures (see Section 16).

Liftgate release switch

Power release switch

24 Remove the liftgate trim panel (see Section 13).
25 Disconnect the electrical connector to the release switch.
26 Remove the two switch molding screws.
27 Depress the tabs on the switch and remove the switch from the molding.
28 Installation is the reverse of removal. Perform the de-initialization/initialization procedures (see Section 16).

Manual release switch

29 Remove the trim panel (see Section 13).
30 Disconnect the electrical connector to the release switch molding from inside of the liftgate.
31 Remove the six switch molding nuts from the inside of the liftgate and pull the molding off of the outside of the liftgate.
32 Depress the tabs on the switch and remove the switch from the molding.
33 Disconnect the electrical connector to the release switch.
34 Installation is the reverse of removal. Perform the de-initialization/initialization procedures (see Section 16).

Keyless entry pad

35 Remove the upper trim cap for the outside end of the door.
36 Working through the hole, use a screwdriver to release the clip holding the key pad to the reinforcement handle.
37 Pull the key pad out and disconnect the electrical connector.
38 Installation is the reverse of removal.

16 Door window glass - removal and installation

1 Remove the door trim panel (see Section 13).
2 Remove the front door speaker (see Chapter 12).
3 Remove the access hole cover plugs from the door module.
4 Lower the window glass until the glass clamp bolts are accessible through the access holes (see illustration). Loosen (but don't remove) both glass clamp bolts, then secure the window in place with tape so that it won't go down.
Note: *If the door window regulator motor is broken or disconnected, you'll have to remove it (see Section 17) and move the window glass manually.*
5 Lower the window regulator, then guide the window up and tilt it outward to remove it through the window opening.
6 Installation is the reverse of removal. Perform the de-initialization/initialization procedures in the following order.

De-initialization procedure

7 Turn the ignition key to ON.
8 Activate the power window control switch in the one-touch up-mode and simultaneously remove power from the window motor while the window is moving, by one of the following methods:

 a) *Disconnect the battery negative cable while the window is moving.*
 b) *Disconnect the power window motor electrical connector while the window is moving.*
 c) *Remove the left or right front power window motor fuse while the window is moving.*

9 The power window motor is now de-initialized and the motor is reset to its original factory settings.
10 Reconnect the battery negative cable, power window motor electrical connector or power window motor fuse before proceeding to the initialization procedure.

Initialization procedure

11 Turn the ignition key to ON.
12 Press and hold the window control switch until the window glass stalls for two seconds into the glass top run, then release the switch.
13 Pull and hold the power window control switch until the window glass stalls for two seconds at the bottom of its travel, then release the switch.
14 Test the operation of the power window by activating the power window switch in the one-touch up-mode. If the window doesn't operate correctly, repeat this procedure.

17 Door window regulator and motor - removal and installation

Regulator motor

1 Remove the door trim panel (see Section 13).
2 Disconnect the electrical connector from the regulator motor.
3 Secure the window with tape or use a set of window suction cups to hold the window up and out of the way of the regulator (see illustration). Remove the window switch and battery connection before continuing.

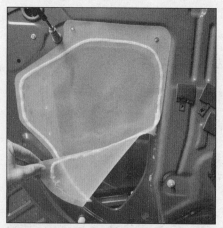

17.8 Carefully peel the watershield from the door

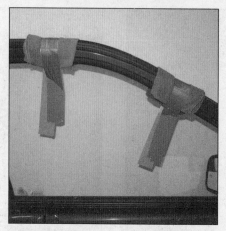

17.10 Some strong tape can suffice if you do not have any window suction cups to hold the glass up

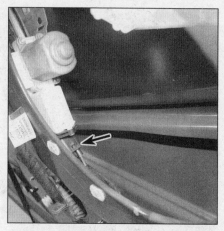

17.11 Use a small flat screwdriver to lift the locking tab to the electrical connector, then pull the connector free

17.12 You may find the regulator mounting bolt openings covered with a piece of padded tape

18.1 Pry out the sail panel

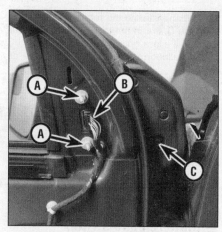

18.2 Mirror details

A   *Mirror retaining nuts*
B   *Mirror electrical connector*
C   *Mirror retaining bolt*

4   Remove the three regulator motor mounting screws, then remove the motor.
5   Installation is the reverse of removal. Perform the de-initialization/initialization procedures (see Section 16).

Regulator module assembly

6   Remove the door trim panel (see Section 13).
7   Remove the front door speaker (see Chapter 12).
8   Remove the water shield. Be careful not to tear the water shield so it can be reused (see illustration).
9   Temporarily reattach the battery connection and window switch and lower the window glass until the glass clamp bolts are accessible through the access holes. Loosen (but don't remove) both glass clamp bolts (see illustration 16.4), then pull the glass from the brackets and raise it to the highest possible position.

10   Secure the window with tape or use a set of window suction cups to hold the window up and out of the way of the regulator (see illustration). Remove the window switch and battery connection before continuing.
11   Disconnect the electrical connection from the window motor (see illustration).
12   Remove the top regulator mounting bolt then remove the two lower mounting nuts (see illustration).
13   Use the speaker opening to reach in and hold the regulator to help guide it out of the access opening.
Note: *If the door window regulator motor is broken or disconnected, you'll have to cut the cables to the regulator to move the window glass manually.*
14   Installation is the reverse of removal. Torque all the fasteners to the proper specifications listed in (this Chapter's Specifications). Perform the de-initialization/initialization procedures (see Section 16).

18   Mirrors - removal and installation

Outside mirror

1   Disengage the sail panel clips and remove the panel (see illustration).
2   Disconnect the electrical connector from the mirror, then remove the two nuts and the single bolt (see illustration). Be prepared to hold onto the mirror once the last nut or bolt has been removed. Now remove the mirror.
3   Installation is the reverse of removal.

Interior mirror

4   There are four different styles of interior mirrors used.
5   Each type of mirror are equipped with a different configuration of features or in some cases no added feature at all. Those features are the rain sensor, Compass, and Lane Departure Camera.

Chapter 11 Body

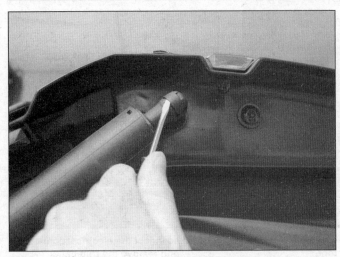

19.4 Detach the support strut retaining clips

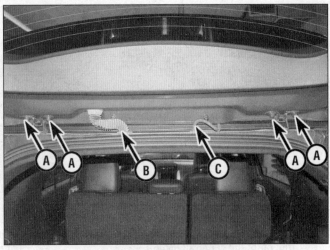

19.5 Liftgate upper components
 A Liftgate hinge bolts
 B Wiring harness boot
 C Washer hose connection

6 The interior mirror requires the use of a special mirror removal and installation tools.
Note: *The windshield should be above 65 or 70 degrees before attempting to remove the inside mirror.*
7 To remove the mirror, first disconnect the electrical connectors from the mirror. Then, tilt the mirror up to gain a better access to the mirror base clamp. Using the mirror removal tool, compress the retaining spring securing the mirror to the mirror base. Now, gently bump the tool and the mirror up slightly to release the retaining spring. Once the spring is free gently slide the mirror off of the base.
Note: *With some luck, you can remove the mirror without the special tool. This requires a small allen tool that is able to slide up between the mirror base and the mirror. Gently nudge the allen tool up to apply pressure to the retaining clip as you pull upwards on the mirror.*
8 One other type of mirror mount requires you to remove the electrical connectors and trim, then pull slightly outwards in order to release the mirror stop clip. Now rotate the mirror counter-clockwise to remove it from the base.
Note: *Using a small flat pocket screwdriver can help in lifting the stop clip out of the mirror base. While the clip has been lifted gently rotate the mirror counterclockwise.*
9 Installation for either style mirror is reverse of removal.

19 Liftgate - removal and installation

Note: *Any time the battery is disconnected, or the liftgate hinges, striker, LTM motor or pinch strip switch are removed or replaced, the liftgate power system must go through the initialization procedure (see Section 20).*

1 Have an assistant support the liftgate in its fully open position.
2 Pull the weatherstrip from the liftgate opening.

19.6 Remove the boot and the wire harness from the body before removing the shock/motor unit

3 Pry down the edge of the headliner to gain access to the electrical and wiper washer hose connections. Disconnect the high mount stop lamp, license plate lighting, rear windshield wiper motor, and rear latch electrical connectors. Pry the electrical lead wiring harness boot out of the body and pull the harness from the headliner area and let it dangle freely.
Note: *Do not pull the headliner too far down or it can become creased.*
Note: *On power liftgate models, disconnect the lift mechanism actuator rod nut and remove the rod from the liftgate.*
4 While an assistant supports the liftgate, detach the support struts (see illustration).
5 Mark or scribe around the hinges, then remove the hinge bolts and detach the liftgate from the vehicle (see illustration).
6 On power liftgates, disconnect the electrical connector and weather boot from the body, then disconnect the shock/motor assembly (see illustration).
7 Installation is the reverse of removal. Make sure to perform the power liftgate initialization procedure (see Section 20).
8 Close the liftgate and make sure it is in proper alignment with the surrounding body panels. Adjustments are made by changing the position of the hinge bolts in the slots. Loosen the hinge bolts and reposition the hinges (either side-to-side or front and back) the desired amount and retighten the bolts.
9 The engagement of the liftgate can be adjusted by loosening the lock striker bolts, repositioning the striker and retightening the bolts.

20 Liftgate/Trunk Module (LTM) and power liftgate motor - removal and installation

Note: *Any time the battery is disconnected, or the liftgate hinges, striker, LTM, motor or pinch strip switch are removed or replaced, the liftgate power system must go through the initialization procedure in this Section.*

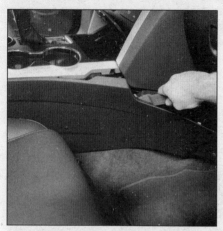

21.3 Use a flat trim tool to work the side trim off of the center console

22.4a Gently pry out the trim panel with a flat trim tool

22.4b Disconnect the electrical connection and remove the trim panel

Liftgate/Trunk Module (LTM)

1 Remove the driver's side rear quarter panel trim (see Section 26).
2 Disconnect the module electrical connectors.
3 Remove the two mounting bolts and remove the LTM.
4 Installation is the reverse of removal. Make sure to perform the power liftgate initialization procedure (see Steps 26 through 30).

Power liftgate motor

5 Support the liftgate in the open position before disconnecting the power liftgate motor.

Liftgate motor

6 Remove the driver's side rear quarter panel trim access cover.
7 Disconnect the power liftgate motor from the mount.
8 Disconnect the liftgate motor electrical connector and remove the motor assembly.
9 Installation is the reverse of removal. Make sure to perform the power liftgate initialization procedure.
10 Replacing the right hand liftgate arm does not require liftgate initialization. There is no electrical connection to the right-side liftgate arm.

Pinch strip switch

11 Remove the liftgate trim panel (see Section 13).
12 Disconnect the switch electrical connector.
13 Remove the plastic pin-type fasteners from the liftgate, then pull the grommet out along with the harness and switch assembly.
14 Installation is the reverse of removal. Make sure to perform the power liftgate initialization procedure (see Steps 26 through 30).

Power liftgate initialization

Note: *Make sure the power liftgate system is turned to the On position in the message center before starting.*
15 Disconnect the battery (see Chapter 5) or remove the LTM fuse for 20-seconds (see the wiring diagrams at the end of Chapter 12).
16 With the battery disconnected or LTM fuse removed, wait 20 seconds, then manually close the liftgate (if it's not already closed).
Note: *Make sure the liftgate is turned on in the message center before continuing.*
17 Open the liftgate using the control switch or key fob.
18 Allow the liftgate to fully open, then close the liftgate using the control switch or the key fob.
19 The system should be initialized; if the liftgate does not open or close properly, readjust the liftgate. If a trouble code sets during this time, take the vehicle to your local dealer service department or other qualified repair shop for diagnosis.

Liftgate height adjustment

Note: *The liftgate height is stored if the battery has been disconnected.*
20 To adjust the height, manually move the liftgate to the desired height.
21 Press and hold the lift gate control switch until a chime is heard.
22 The liftgate is now reprogrammed.
23 If you need to change the height again, follow the same procedure again.

21 Center console - removal and installation

1 Apply the parking brake and place the shift lever in the neutral position.
2 Slide the front seats all the way rearward, then place them in the fully reclined position.
3 Using a trim tool, carefully pry off the console edge trim for both sides of the console (see illustration).
4 Remove the shift cable from the shift lever and bracket (see Chapter 7A).
5 Disconnect the electrical bulkhead connectors, USB connectors, and 110 power outlet connectors (if equipped) from the console.
6 Remove the carpet pushpins on either side of the console.
7 Remove the eight console mounting bolts from the base of the console.
8 Lift up the console from the rear and remove the center console.
9 Installation is the reverse of removal.

Rear center console

10 Fold the second row seats forward as far as possible.
11 Remove the four center console bolts.
12 Installation is the reverse of removal.

22 Dashboard trim panels - removal and installation

Warning: *Models covered by this manual are equipped with a Supplemental Restraint System (SRS), more commonly known as airbags. Always disable the airbag system before working in the vicinity of any airbag system component to avoid the possibility of accidental deployment of the airbag, which could cause personal injury (see Chapter 12).*
Note: *Refer to Section 6 for fastener and trim removal.*
1 Removing various dashboard trim panels provides access to electrical/electronic components such as the instrument cluster, the audio unit, the heater and air conditioning control unit and various instrument panel-mounted switches. If you're going to remove the entire instrument panel, you'll need to remove various trim panels to access the instrument panel mounting bolts.

Instrument Panel Cluster (IPC) trim cover

2 Lower the steering wheel to the lowest position.
3 Lift off the gap trim panel between the steering column upper trim and the IPC trim.
4 Release the four clips securing the IPC trim (see illustrations). Disconnect the electrical connections and remove the IPC trim.
5 Installation is the reverse of removal.

22.6 Apply even pressure around the end panel to remove it

22.12a Knee bolster trim panel retaining screw locations

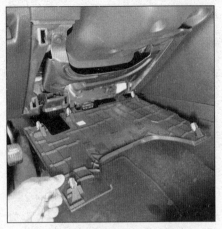

22.12b Pull outward to release the pressure clips to the knee bolster trim panel

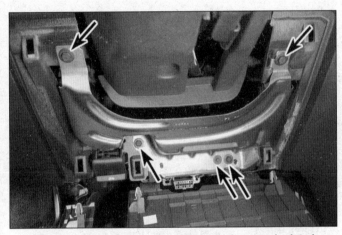

22.12c To gain access to the steering column and related components, remove the knee bolster brace as well

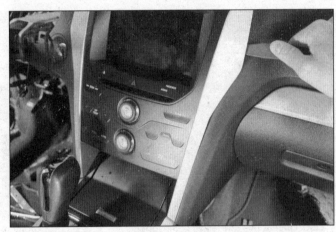

22.15 It's easier to start at the outer edge and work your way around the trim panel

Instrument panel end trim

6   Using a flat trim tool, pry the instrument panel end trim off (see illustration). Keep the tool as close as possible to the clips before applying pressure.

Cowling side trim

7   Remove the instrument panel end trim.
8   Grasp the cowling trim by the edges and pull straight out to remove.

Instrument panel defroster grille

9   With a flat trim tool, pry the edge of the defroster grille up and work around the outer edge until all the clips are free. Then lift the defroster grille off.

Lower left hand trim panel

10   Grasp the panel by the edges and pull straight out. Disconnect the electrical connections and remove the panel.
11   Installation is the reverse of removal.

Knee bolster trim panel and knee bolster

12   Remove the two knee bolster trim panel screws, then carefully pry off the trim panel (see illustrations).
13   Installation is the reverse of removal.
14   Tighten the mounting bolts securely.

Instrument cluster center finish trim, FCIM trim removal, and center speaker grille

Note: *The FCIM trim is basically the HVAC and radio face plate trim.*

15   Remove the right and left side center trim panels from around the FCIM (Front Control Interface Module). Work up or down along the length of the trim to loosen the retaining clips. Once the clips are loose gently remove the left and right center finish trim sections. (see illustration).
16   Using a flat bladed trim tool, pry near the edge of the FCIM trim panel to start the removal process. Work the trim tool around the outer edges of the FCIM trim panel until it comes free (see illustration). With the FCIM

22.16a The FCIM trim panel shown with the right and left center finish trims removed.

11-20   Chapter 11 Body

22.16b With the FCIM trim panel removed, the center speaker grille can be removed

22.18 Remove the dampener string by pulling down to clearing the locking tab, then remove it

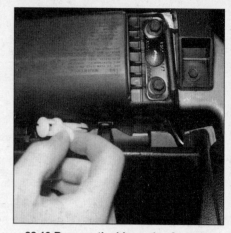

22.19 Remove the hinge pins from the bottom of the glove box

22.27 Once it has started to come loose work your way to the opposite end.

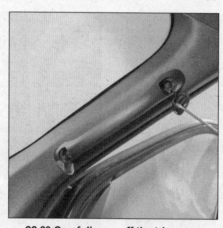

22.29 Carefully pry off the trim caps

Door threshold trim (sill panel) removal

27 Pry up the threshold trim with a flat trim tool. Work along the edge to free the entire length of the trim (see illustration).
28 Installation is the reverse of removal.

A pillar trim removal

29 Lift the trim cap from the bolts securing the A pillar trim to the A pillar (see illustration).
30 Remove the bolts from the A pillar and remove it (see illustration). The driver's side A pillar on some models does not have a grab handle. Those panels are held in place with pressure clips. Pry from the outside edge a few inches from each end to loosen the pressure clips. Then remove it.

22.30 Once the bolts are removed from the passenger's grab handle, pry the trim panel off

trim panel removed you can now lift off the center speaker grille (see illustration).

Glove box door

17 Remove the right hand instrument panel end trim.
18 Unhook the glove dampener string from the glove box door (see illustration).
19 Remove the hinge pins from the bottom of the glove box (see illustration).
20 Firmly grasp the corners of the glove box and push them inwards until the stops clear the sides, and remove the glove box.
21 Installation is the reverse of removal.

Glove box compartment trim panel

22 Remove glove box door.
23 Remove the upper glove box trim panel by pulling it down to release the clips.
24 Disconnect the electrical connector.
25 Remove the four upper and the four lower screws securing the glove box compartment trim to the instrument panel. Then remove the panel.
26 Installation is the reverse of removal

23 Instrument panel - removal and installation

Warning: *Models covered by this manual are equipped with a Supplemental Restraint System (SRS), more commonly known as airbags. Always disable the airbag system before working in the vicinity of any airbag system component to avoid the possibility of accidental deployment of the airbag, which could cause personal injury (see Chapter 12).*
Warning: *Wait until the engine is completely cool before beginning this procedure.*
Note: *This is a difficult procedure for the home mechanic. There are many hidden fasteners, difficult angles to work in and many electrical connectors to tag and disconnect/connect. We recommend that this procedure be done only by an experienced do-it-yourselfer.*
Note: *During removal of the instrument panel, make careful notes of how each piece comes off, where it fits in relation to other pieces and what holds it in place. If you note how each part is installed before removing it, getting the instrument panel back together again will be much easier.*

1 Have the refrigerant recovered by a

Chapter 11 Body   11-21

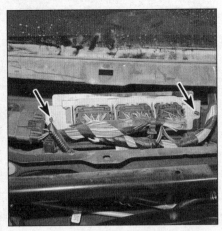

23.9 PCM mounting nuts

23.12 The two driver's side cowling to dash bolts shown. One single bolt is on the passenger side

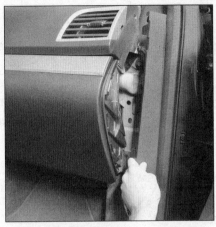

23.17 Remove the cowling trim to gain access to the instrument panel bolts

23.21a Disconnect the electrical connections between the instrument panel and the firewall

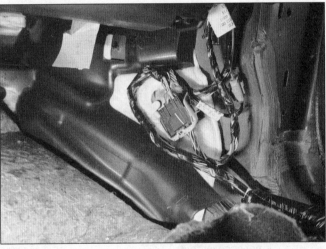

23.21b Remove the connectors from their clips, then disconnect them. Be sure to put them back in the same places

licensed air conditioning technician.
2   Drain the cooling system (see Chapter 1).
3   Remove the center console and rear center console (if equipped) (See Section 21).
4   Remove the FCIM and the instrument panel speakers (see Chapter 12).
5   Remove the instrument panel speakers.
6   Remove the cowling grille (See Section 12).
7   Remove both front seats for adding working area (see Section 25).
8   Remove the steering wheel (see Chapter 10).
9   Disconnect the PCM and remove the PCM (see illustration).
10 Remove the strut tower cross brace.
11 Disconnect the heater hoses at the heater core tubes at the firewall and the refrigerant lines from the TXV (Thermostatic Expansion Valve for the air conditioning system).
12 From the engine compartment side, remove the three bolts securing the instrument panel (see illustration).
13 Remove the left and right threshold (sill panel) trim (see Section 22).
14 Pull the weatherstripping out of the way of the dash. Position the weatherstripping so it will not hinder the dash removal.
15 Remove the A pillar trim (see Section 22).
16 Remove the left and instrument panel end trim pieces (see Section 22).
17 Remove the left and right cowling trim located in front of the instrument panel end trim panels. Pry these off with a flat trim tool (see illustration).
18 Remove the knee bolster trim and knee bolster brace (see Section 22).
19 Remove the inside hood release handle.
20 Remove the steering column pinch bolt (see Chapter 10).
21 Disconnect the left and right hand electrical connectors and antenna cable connector (see illustrations).
22 Remove the driver's foot rest pad (see illustration).

23.22 Pry the pad off of the carpet

11-22   Chapter 11   Body

23.23 Use the trim tool to remove the pressure clips

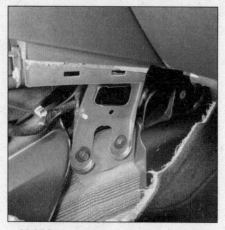

23.27 Remove the two bolts from each side of the center section

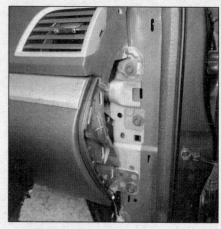

23.28 Right-side instrument panel bolts

24.2a Pry the tabs back as you lift the upper cover free

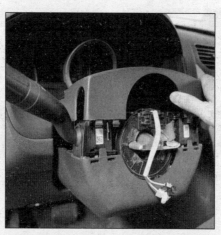

24.2b Lift the upper cover up away from the steering column

24.2c To remove the upper cover completely, pinch the clips together to detach the upper flexible trim from the steering wheel upper cover

24.2d Remove the three screws from the lower cover to remove it

23 Remove the front carpet sections (see illustration).
24 Remove the center duct work (if equipped).
25 Remove the two ground leads positioned between the center instrument panel support braces and below where the center duct work was located.
26 Disconnect the restraint control module.
27 Remove the center instrument panel support bolts (two on each side) (see illustration).
28 Remove the three bolts from the left and right hand side of the instrument panel (see illustration).
29 Pull the dash board out a few inches and check for any remaining items that were not disconnected or are going to hang up or get caught as you remove the dash board. Slowly tilt the dash board face down and work your way out the right side of the vehicle.
30 Installation is the reverse of removal.

24 Steering column covers - removal and installation

Warning: *Models covered by this manual are equipped with a Supplemental Restraint System (SRS), more commonly known as airbags. Always disable the airbag system before working in the vicinity of any airbag system component to avoid the possibility of accidental deployment of the airbag, which could cause personal injury (see Chapter 12).*

1 Disconnect the cable from the negative terminal of the battery (see Chapter 5).
2 To remove the steering column covers follow the accompanying illustrations (see illustrations). The steering wheel has been removed for clarity, however you can remove the covers without removing the steering wheel. Just rotate the steering wheel 90-degrees to gain access to the tabs connecting the upper and lower cover halves.

Chapter 11 Body

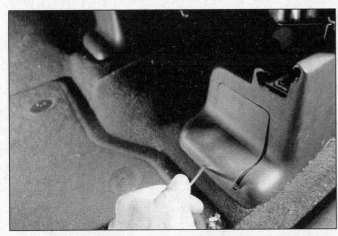

25.2a Remove the trim covers…

25.2b … then remove the bolts

25.2c Move the seat forward enough to expose the rear bolts, then remove them

25.3 On some models there will be more than one connector to disconnect

Note: *On column-shift vehicles, work the shift lever grommet off of the upper and lower covers and pull out of the way of the covers. To install, guide the edge of the grommet around the edge of the upper and lower covers.*

3   Installation is the reverse of removal.

25 Front seats - removal and installation

Warning: *All models covered by this manual are equipped with a Supplemental Restraint System (SRS), more commonly known as airbags. Always disable the airbag system before working in the vicinity of any airbag system component to avoid the possibility of accidental deployment of the airbag, which could cause personal injury (see Chapter 12).*

Warning: *During the removal and installation procedures, if a seat is dropped, or sat in while removed, or if pressure/load is applied to the seat tracks, the Occupant Safety System (OCS) may become damaged and the system may fail in a crash.*

Warning: *The seats are heavy, so have an assistant handy to help you lift the seat from the vehicle.*

Note: *This procedure applies to the driver and passenger seats.*

1   If you're removing the passenger seat, remove the cover from the seatbelt anchor, remove the anchor bolt, and disconnect electrical connector the seatbelt from the seat.
2   Remove the covers from the seat mounting bolts and remove the bolts (see illustrations).
3   Disconnect all electrical connectors from underneath the seat (see illustration).
4   Using a helper, carefully lift the seat out of the vehicle.

Warning: *The seat is heavy, so trying to remove it by yourself could cause injury.*

5   Installation is the reverse of removal.
6   Tighten all fasteners to the torque listed in this Chapter's Specifications.
7   The remainder of the seats come out in a similar manner. Sliding the seat forward and backward will gain access to the bolts. Disconnect any electrical connections, then remove the seat (see illustration).

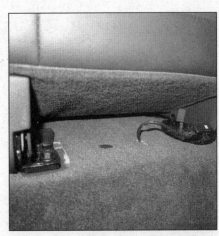

25.7 Second row seat bolt locations and electrical connector locations shown. The remainder of the seats are similar

26.4a Pry off the trim cover…

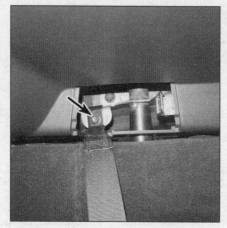

26.4b … and remove the seat belt anchor bolt

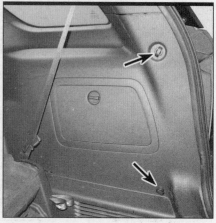

26.5 Unscrew the cargo net hooks

26.6 Pry off the liftgate scuff plate

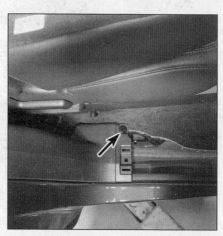

26.9 Pry out the plastic retainer

26.10a Pull the quarter trim panel off…

26.10b … then disconnect the electrical connectors

26 Quarter trim panel - removal and installation

1 On models equipped with a power liftgate, pry the liftgate switch cover off at the bottom of the liftgate rod and disconnect the electrical connector.
2 Remove the power liftgate rod-to-liftgate nut cover and nut (see Section 20).
3 Remove the rear mat and storage tray from the floor, then remove the foam blocks in the spare tire well.
4 Remove the lower seat belt anchor bolt cover, then remove the bolt (see illustrations).
Warning: *Pay close attention to how the seatbelts are routed. They must be installed exactly the same way. In the event of an accident, failure to do so could result in serious injury to someone secured by an incorrectly routed seatbelt.*
5 Unscrew the cargo net hooks (see illustration).
6 Using a plastic trim tool, disengage the mounting clips and pry out the liftgate scuff plate (see illustration).
7 Lay the second row seats flat, then tilt them forward.
8 Lay the third row seats flat.
9 Open the rear door and remove the scuff plate (sill panel), then remove the plastic retainer at the front of the quarter trim panel (see illustration).
10 Remove the quarter trim panel and disconnect any electrical connectors (see illustrations).
11 Installation is the reverse of removal.

Chapter 12
Chassis electrical system

Contents

	Section		Section
Airbags - general information	27	Instrument cluster - removal and installation	11
Antenna - removal and installation	13	Instrument panel switches - replacement	10
Bulb replacement	18	Multi-function switch - replacement	7
Circuit breakers - general information	4	Power door lock and keyless entry system - description check and battery replacement	25
Cruise control system - description and check	23	Power mirror control system - description and check	22
Daytime Running Lights (DRL) - general information	26	Power window system - description and check	24
Electrical connectors - general information	6	Radio and speakers - removal and installation	12
Electrical troubleshooting - general information	2	Rear window defogger - check and repair	21
Fuses and fusible links - general information	3	Relays - general information and testing	5
General information	1	Steering column control module	8
Headlight bulb - replacement	14	Taillight housing - removal and installation	17
Headlight housing - removal and installation	15	Wiper system	19
Headlights - adjustment	16	Wiring diagrams - general information	28
Horn	20		
Ignition switch and key lock cylinder - replacement	9		

Specifications

Torque specifications

Note: *One foot-pound (ft-lb) of torque is equivalent to 12 inch-pounds (in-lbs) of torque. Torque values below approximately 15 ft-lbs are expressed in inch-pounds, because most foot-pound torque wrenches are not accurate at these smaller values.*

	Ft-lbs (unless otherwise indicated)	Nm
Windshield mounting arm and pivot shaft bolts	177 in-lbs	20
Windshield pivot nuts	22	30
Rear wiper pivot nut	106 in-lbs	12
Rear wiper mounting bolts	80 in-lbs	9

1 General information

1   The electrical system is a 12-volt, negative ground type. Power for the lights and all electrical accessories is supplied by a lead/acid-type battery that is charged by the alternator.
2   This Chapter covers repair and service procedures for the various electrical components not associated with the engine. Information on the battery, alternator, ignition system and starter motor can be found in Chapter 5.
3   It should be noted that when portions of the electrical system are serviced, the negative cable should be disconnected from the battery to prevent electrical shorts and/or fires.

2 Electrical troubleshooting - general information

1   A typical electrical circuit consists of an electrical component, any switches, relays, motors, fuses, fusible links or circuit breakers related to that component and the wiring and connectors that link the component to both the battery and the chassis. To help you pinpoint an electrical circuit problem, wiring diagrams are included at the end of this Chapter.
2   Before tackling any troublesome electrical circuit, first study the appropriate wiring diagrams to get a complete understanding of what makes up that individual circuit. Trouble spots, for instance, can often be narrowed down by noting if other components related to the circuit are operating properly. If several components or circuits fail at one time, chances are the problem is in a fuse or ground connection, because several circuits are often routed through the same fuse and ground connections.
3   Electrical problems usually stem from simple causes, such as loose or corroded connections, a blown fuse, a melted fusible link or a failed relay. Visually inspect the condition of all fuses, wires and connections in a problem circuit before troubleshooting the circuit.
4   If test equipment and instruments are going to be utilized, use the diagrams to plan ahead of time where you will make the necessary connections in order to accurately pinpoint the trouble spot.

12-2   Chapter 12 Chassis electrical system

5   The basic tools needed for electrical troubleshooting include a circuit tester or voltmeter (a 12-volt bulb with a set of test leads can also be used), a continuity tester, which includes a bulb, battery and set of test leads, and a jumper wire, preferably with a circuit breaker incorporated, which can be used to bypass electrical components (see illustrations). Before attempting to locate a problem with test instruments, use the wiring diagram(s) to decide where to make the connections.

Voltage checks

6   Voltage checks should be performed if a circuit is not functioning properly. Connect one lead of a circuit tester to either the negative battery terminal or a known good ground. Connect the other lead to a connector in the circuit being tested, preferably nearest to the battery or fuse (see illustration). If the bulb of the tester lights, voltage is present, which means that the part of the circuit between the connector and the battery is problem free. Continue checking the rest of the circuit in the same fashion. When you reach a point at which no voltage is present, the problem lies between that point and the last test point with voltage. Most of the time the problem can be traced to a loose connection.

Note: *Keep in mind that some circuits receive voltage only when the ignition key is in the Accessory or Run position.*

Finding a short

7   One method of finding shorts in a circuit is to remove the fuse and connect a test light or voltmeter in place of the fuse terminals. There should be no voltage present in the circuit. Move the wiring harness from side-to-side while watching the test light. If the bulb goes on, there is a short to ground somewhere in that area, probably where the insulation has rubbed through. The same test can be performed on each component in the circuit, even a switch.

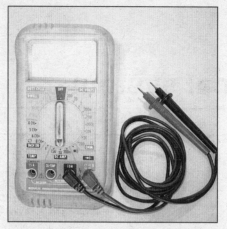

2.5a The most useful tool for electrical troubleshooting is a digital multimeter that can check volts, amps, and test continuity

Ground check

8   Perform a ground test to check whether a component is properly grounded. Disconnect the battery and connect one lead of a continuity tester or multimeter (set to the ohms scale), to a known good ground. Connect the other lead to the wire or ground connection being tested. If the resistance is low (less than 5 ohms), the ground is good. If the bulb on a self-powered test light does not go on, the ground is not good.

Continuity check

9   A continuity check is done to determine if there are any breaks in a circuit - if it is passing electricity properly. With the circuit off (no power in the circuit), a self-powered continuity tester or multimeter can be used to check the circuit. Connect the test leads to both ends of the circuit (or to the power end and a good ground), and if the test light comes on the circuit is passing current properly (see illustration). If the resistance is low (less than

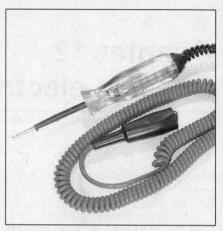

2.5b A test light is a very handy tool for checking voltage

5 ohms), there is continuity; if the reading is 10,000 ohms or higher, there is a break somewhere in the circuit. The same procedure can be used to test a switch, by connecting the continuity tester to the switch terminals. With the switch turned On, the test light should come on (or low resistance should be indicated on a meter).

Finding an open circuit

10   When diagnosing for possible open circuits, it is often difficult to locate them by sight because the connectors hide oxidation or terminal misalignment. Merely wiggling a connector on a sensor or in the wiring harness may correct the open circuit condition. Remember this when an open circuit is indicated when troubleshooting a circuit. Intermittent problems may also be caused by oxidized or loose connections.

11   Electrical troubleshooting is simple if you keep in mind that all electrical circuits are basically electricity running from the battery, through the wires, switches, relays, fuses

2.6 In use, a basic test light's lead is clipped to a known good ground, then the pointed probe can test connectors, wires or electrical sockets - if the bulb lights, the part being tested has battery voltage

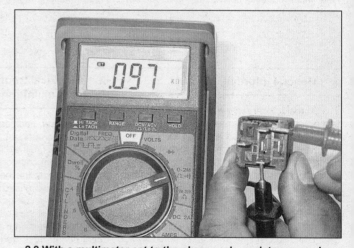

2.9 With a multimeter set to the ohms scale, resistance can be checked across two terminals - when checking for continuity, a low reading indicates continuity, a high reading indicates lack of continuity

Chapter 12 Chassis electrical system

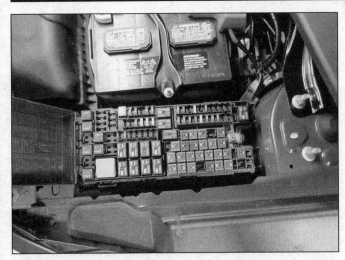

3.1a The engine compartment fuse and relay panel is located at the left side of the engine compartment

3.1b The passenger compartment fuse box is located under the left end of the instrument panel

and fusible links to each electrical component (light bulb, motor, etc.) and to ground, from which it is passed back to the battery. Any electrical problem is an interruption in the flow of electricity to and from the battery.

3 Fuses and fusible links - general information

Fuses

1 The electrical circuits of the vehicle are protected by a combination of fuses, circuit breakers and fusible links. The main fuse/relay panel is in the engine compartment (see illustration), while the interior fuse/relay panel is located inside the passenger compartment (see illustration). Each of the fuses is designed to protect a specific circuit, and the various circuits are identified on the fuse panel itself.
2 Several sizes of fuses are employed in the fuse blocks. There are small, medium and large sizes of the same design, all with the same blade terminal design. The medium and large fuses can be removed with your fingers, but the small fuses require the use of pliers or the small plastic fuse-puller tool found in most fuse boxes.
3 If an electrical component fails, always check the fuse first. The best way to check the fuses is with a test light. Check for power at the exposed terminal tips of each fuse. If power is present at one side of the fuse but not the other, the fuse is blown. A blown fuse can also be identified by visually inspecting it (see illustration).
4 Be sure to replace blown fuses with the correct type. Fuses (of the same physical size) of different ratings may be physically interchangeable, but only fuses of the proper rating should be used. Replacing a fuse with

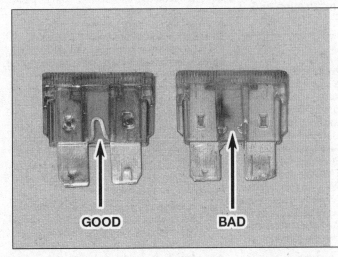

3.3 When a fuse blows, the element between the terminals melts

one of a higher or lower value than specified is not recommended. Each electrical circuit needs a specific amount of protection. The amperage value of each fuse is molded into the top of the fuse body.
5 If the replacement fuse immediately fails, donít replace it again until the cause of the problem is isolated and corrected. In most cases, this will be a short circuit in the wiring caused by a broken or deteriorated wire.

Fusible links

6 On these models, large, high-amperage fuses are used instead of traditional fusible links. They are located in the underhood fuse/relay box.

4 Circuit breakers - general information

1 Circuit breakers protect certain circuits, such as the power windows or heated seats. Depending on the vehicle's accessories, there may be one or two circuit breakers, located in the fuse/relay box in the engine compartment.
2 Because the circuit breakers reset automatically, an electrical overload in a circuit breaker-protected system will cause the circuit to fail momentarily, then come back on. If the circuit does not come back on, check it immediately.
3 For a basic check, pull the circuit breaker up out of its socket on the fuse panel, but just far enough to probe with a voltmeter. The breaker should still contact the sockets. With the voltmeter negative lead on a good chassis ground, touch each end prong of the circuit breaker with the positive meter probe. There should be battery voltage at each end. If there is battery voltage only at one end, the circuit breaker must be replaced.
4 Some circuit breakers must be reset manually.

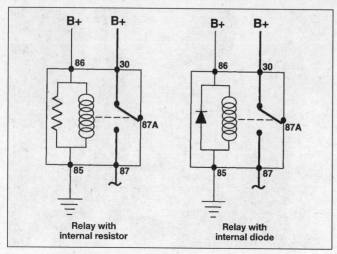

5.2a Typical ISO relay designs, terminal numbering and circuit connections

5.2b Most relays are marked on the outside to easily identify the control circuits and the power circuits (four-terminal type shown)

5 Relays - general information and testing

General information

1 Several electrical circuits in the vehicle use relays to transmit the electrical signal to the component. Relays use a low-current circuit (the control circuit) to open and close a high-current circuit (the power circuit). If the relay is defective, that component will not operate properly. Most relays are mounted in the engine compartment fuse/relay box, with some specialized relays located above the interior fuse box in the dash (see illustrations 3.1a and 3.1b). If a faulty relay is suspected, it can be removed and tested using the procedure below or by a dealer service department or a repair shop. Defective relays must be replaced as a unit.

Testing

2 Most of the relays used in these vehicles are of a type often called "ISO" relays, which refers to the International Standards Organization. The terminals of ISO relays are numbered to indicate their usual circuit connections and functions. There are two basic layouts of terminals on the relays used in these vehicles (see illustrations).
3 Refer to the wiring diagram for the circuit to determine the proper connections for the relay you're testing. If you can't determine the correct connection from the wiring diagrams, however, you may be able to determine the test connections from the information that follows.
4 Two of the terminals are the relay control circuit and connect to the relay coil. The other relay terminals are the power circuit. When the relay is energized, the coil creates a magnetic field that closes the larger contacts of the power circuit to provide power to the circuit loads.

5 Terminals 85 and 86 are normally the control circuit. If the relay contains a diode, terminal 86 must be connected to battery positive (B+) voltage and terminal 85 to ground. If the relay contains a resistor, terminals 85 and 86 can be connected in either direction with respect to B+ and ground.
6 Terminal 30 is normally connected to the battery voltage (B+) source for the circuit loads. Terminal 87 is connected to the circuit leading to the component being powered. If the relay has several alternate terminals for load or ground connections, they usually are numbered 87A, 87B, 87C, and so on.
7 Use an ohmmeter to check continuity through the relay control coil.
   a) Connect the meter according to the polarity shown in illustration 5.2a for one check; then reverse the ohmmeter leads and check continuity in the other direction.
   b) If the relay contains a resistor, resistance will be indicated on the meter, and should be the same value with the ohmmeter in either direction.
   c) If the relay contains a diode, resistance should be higher with the ohmmeter in the forward polarity direction than with the meter leads reversed.
   d) If the ohmmeter shows infinite resistance in both directions, replace the relay.
8 Remove the relay from the vehicle and use the ohmmeter to check for continuity between the relay power circuit terminals. There should be no continuity between terminal 30 and 87 with the relay de-energized.
9 Connect a fused jumper wire to terminal 86 and the positive battery terminal. Connect another jumper wire between terminal 85 and ground. When the connections are made, the relay should click.
10 With the jumper wires connected, check for continuity between the power circuit terminals. Now, there should be continuity between terminals 30 and 87.

11 If the relay fails any of the above tests, replace it.

6 Electrical connectors - general information

1 Most electrical connections on these vehicles are made with multiwire plastic connectors. The mating halves of many connectors are secured with locking clips molded into the plastic connector shells. The mating halves of some large connectors, such as some of those under the instrument panel, are held together by a bolt through the center of the connector.
2 To separate a connector with locking clips, use a small screwdriver to pry the clips apart carefully, then separate the connector halves. Pull only on the shell, never pull on the wiring harness, as you may damage the individual wires and terminals inside the connectors. Look at the connector closely before trying to separate the halves. Often the locking clips are engaged in a way that is not immediately clear. Additionally, many connectors have more than one set of clips.
3 Each pair of connector terminals has a male half and a female half. When you look at the end view of a connector in a diagram, be sure to understand whether the view shows the harness side or the component side of the connector. Connector halves are mirror images of each other, and a terminal shown on the right side end-view of one half will be on the left side end-view of the other half.
4 It is often necessary to take circuit voltage measurements with a connector connected. Whenever possible, carefully insert a small straight pin (not your meter probe) into the rear of the connector shell to contact the terminal inside, then clip your meter lead to the pin. This kind of connection is called "backprobing." When inserting a test probe into a terminal, be careful not to distort

Electrical connectors

Most electrical connectors have a single release tab that you depress to release the connector

Some electrical connectors have a retaining tab which must be pried up to free the connector

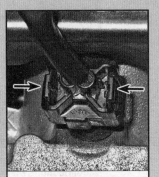

Some connectors have two release tabs that you must squeeze to release the connector

Some connectors use wire retainers that you squeeze to release the connector

Critical connectors often employ a sliding lock (1) that you must pull out before you can depress the release tab (2)

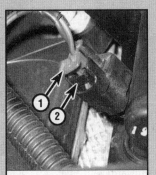

Here's another sliding-lock style connector, with the lock (1) and the release tab (2) on the side of the connector

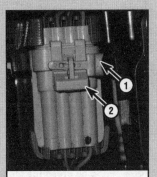

On some connectors the lock (1) must be pulled out to the side and removed before you can lift the release tab (2)

Some critical connectors, like the multi-pin connectors at the Powertrain Control Module employ pivoting locks that must be flipped open

the terminal opening. Doing so can lead to a poor connection and corrosion at that terminal later. Using the small straight pin instead of a meter probe results in less chance of deforming the terminal connector.

7 Multi-function switch - replacement

Warning: *The models covered by this manual are equipped with a Supplemental Restraint System (SRS), more commonly known as airbags. Always disable the airbag system before working in the vicinity of any airbag system components to avoid the possibility of accidental deployment of the airbags, which could cause personal injury (see Section 27).*

1 Disconnect the cable from the negative terminal of the battery (see Chapter 5). Remove the steering column covers (see Chapter 11).
2 Turn the steering wheel 90 degrees from straight forward to gain access to the multi-function switch screws.
3 Remove the multi-function switch retaining screws and remove the switch (see illustration).

Note: *The multi-function switch is mounted on the steering column control module.*

4 Installation is the reverse of removal.

7.3 Remove the switch retaining screws (steering wheel removed for clarity)

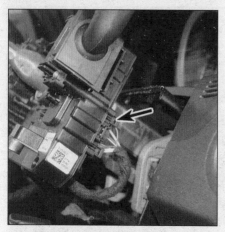

8.5a Disconnect the electrical connection

8.5b There are a total of three electrical connections that need to be removed

8.5c Remove the upper two screws securing the control module to the steering column

8.5d Remove the wire harness clips for extra clearance

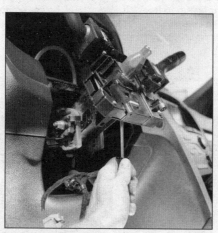

8.5e Press the tab in (upward) with a flat screwdriver to release the lower fastener

8.5f Carefully remove the control module

8 Steering column control module

1 The steering column control module is attached to the upper part of the steering column. The module receives information from the multi-function switch, steering angle sensor, pedal adjustment controls, and windshield washer system and carries that information to the needed areas of the vehicle by way of the HS-CAN (High Speed Controller Area Network) or the LIN (Local Interconnect Network). The steering column control module also provides an electrical path for the horn, cruise control, air bag clock spring, the steering wheel switches, and steering wheel illumination. The steering column control module can also detect faults in the data lines from the various inputs connected to it. If any faults occur a code will set in the steering column control module.

Note: *The multi-function switch and the wiper switch can be removed without removing the control module.*

Removal

2 Disconnect the cable from the negative terminal of the battery (see Chapter 5).
3 Remove the steering wheel covers (see Chapter 11).
4 Remove the steering wheel and the clockspring (see Chapter 10).
5 Follow along with the illustrations for the steering column control module removal (see illustrations).
6 Installation is the reverse of removal.

9 Ignition switch and key lock cylinder - replacement

Warning: *The models covered by this manual are equipped with a Supplemental Restraint System (SRS), more commonly known as airbags. Always disable the airbag system before working in the vicinity of any airbag system components to avoid the possibility of accidental deployment of the airbags, which could cause personal injury (see Section 27).*

1 Disconnect the cable from the negative terminal of the battery (see Chapter 5).
2 Remove the instrument cluster upper and lower trim panels (see Chapter 11).

Key lock housing

3 Remove the instrument cluster center trim panels and FCIM unit (radio) to gain better access to the ignition switch.
4 Use a small flat screwdriver and carefully remove the bezel trim from around the ignition switch without marring the finish or the trim (see illustration).
5 To disconnect the ignition lock housing push in on the tabs on either side of the switch.
6 Disconnect the electrical connectors.
7 To install the key lock cylinder, insert it into the lock cylinder housing and push it in until it clicks into place.
8 Verify that the ignition switch operates correctly in the OFF, ACC, RUN and START positions.

Chapter 12 Chassis electrical system

9.4 Carefully pry the bezel trim off of the ignition switch.

9.12 Once the tool has pressed in on the tab, pull the lock tumbler out by the key (lock tumbler has been partially pulled out already in this photo)

10.3 Use a trim removal tool to pry the headlight switch/instrument panel dimmer switch assembly out of the instrument panel

11.4 Instrument cluster mounting screws

9 Installation is otherwise the reverse of removal.

Ignition lock cylinder with intelligent access

10 Insert a non-marring tool to remove the ignition switch bezel trim. Gently pry from the outer edges to remove the trim plate.
11 Insert the key and turn to the ACC position.
12 Use a thin pick tool with a 90 degree bend near the tip of it to press in on the release tab on the top side of the ignition switch housing through the opening where the bezel trim covered up the release tab (see illustration). Press in on the release tab while pulling the ignition lock cylinder outward.
13 Gently slide the lock cylinder from the lock housing.
14 Installation is the reverse of removal. The lock cylinder will click into place once it has been inserted into the lock housing correctly.

10 Instrument panel switches - replacement

Warning: *The models covered by this manual are equipped with a Supplemental Restraint System (SRS), more commonly known as airbags. Always disable the airbag system before working in the vicinity of any airbag system components to avoid the possibility of accidental deployment of the airbags, which could cause personal injury (see Section 27).*

1 Disconnect the cable from the negative terminal of the battery (see Chapter 5).
2 Remove the left end trim panel from the instrument panel (see Chapter 11).
3 Reach through the opening in the end of the instrument panel and push on the back of the switch while simultaneously prying the switch out of the dash with a trim removal tool (see illustration). Disconnect the electrical connectors and remove the switch assembly.
4 No further disassembly is possible. If you're replacing the headlight switch, you must replace this assembly.
5 Pop a new instrument panel dimmer switch into the bezel and make sure that it snaps into place. Installation is otherwise the reverse of removal.

11 Instrument cluster - removal and installation

Warning: *The models covered by this manual are equipped with a Supplemental Restraint System (SRS), more commonly known as airbags. Always disable the airbag system before working in the vicinity of any airbag system components to avoid the possibility of accidental deployment of the airbags, which could cause personal injury (see Section 27).*

1 Disconnect the cable from the negative battery terminal (see Chapter 5).
2 Release the tilt wheel lever and lower the steering wheel to its lowest position.
3 Remove the instrument panel cluster trim (see Chapter 11).
4 Remove the instrument cluster mounting screws (see illustration).

11.5 Disconnect the electrical connection and remove the cluster

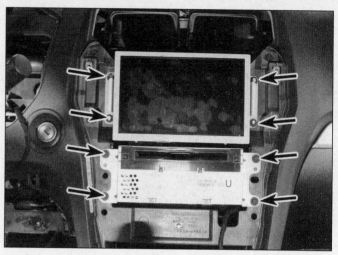

12.3 FCIM fastener locations

12.4 ACM mounting screws (A) and the FCIM mounting screws (B)

12.5 FCIM electrical connector locations

5 Carefully pull out the instrument cluster from the instrument panel and disconnect the electrical connector (see illustration).
6 Installation is the reverse of removal.

12 Radio and speakers - removal and installation

Warning: *The models covered by this manual are equipped with a Supplemental Restraint System (SRS), more commonly known as airbags. Always disable the airbag system before working in the vicinity of any airbag system components to avoid the possibility of accidental deployment of the airbags, which could cause personal injury (see Section 27).*

Radio

Note: *The radio consists of the Front Controls Interface Module (FCIM) and the Audio Control Module (ACM). The FCIM (the controls) is what you see in the center trim panel; the ACM (the radio) is a separate component in the dash behind the FCIM. You can replace the FCIM at home, but not the ACM. You can remove the ACM to access something else, or to remove the instrument panel, but if the ACM must be replaced, it will have to be done by a dealer service department because module configuration must be programmed into the new ACM unit. Without the correct module configuration, the new ACM will not work.*

1 Disconnect the cable from the negative battery terminal (see Chapter 5).
2 Remove the center console side trim moldings and FCIM trim panel (see Chapter 11).
3 Remove the fasteners securing the Front Controls Interface Module (FCIM), remove the FCIM (see illustration), then disconnect the electrical connector
4 If you're removing the ACM, remove the ACM mounting screws (see illustration), then pull out the ACM and disconnect the antenna cable and all electrical connectors.
5 If you're removing the FCIM, remove the FCIM mounting screws, pull out the FCIM and disconnect the electrical connectors (see illustration).
6 Installation is the reverse of removal.

Speakers

Door speakers

Note: *This procedure applies to front and rear door speakers.*

7 Remove the door trim panel (see Chapter 11).
8 Remove the speaker mounting screws (see illustration).
9 Pull the speaker out of its enclosure and disconnect the speaker electrical connectors.

Chapter 12 Chassis electrical system

12.8 Door speaker mounting screws

12.12 Speaker connection location

14.2 Pull the bulb access cover off the housing

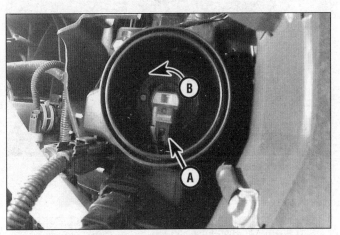

14.3 Unplug the electrical connector (A), then rotate the bulb counterclockwise (B) and pull it out of the housing

Center speaker
10 Remove the left and right center trim panels, and remove the FCIM trim panel (see Chapter 11).
11 Lift the center speaker grille up (see Chapter 11).
12 With grille lifted, disconnect the speaker electrical connector (see illustration) then unscrew the fasteners securing the speaker to the grille.
13 Installation is the reverse of removal.

13 Antenna - removal and installation

1 Remove the C and D pillar trim panels (see Chapter 11).
2 Gently lower the rear portion of the headliner.
3 Remove the antenn mounting bolt and disconnect the cable from the antenna base.

4 To install, check that the insulator and seal are in proper position, then reinstall in reverse of order of removal.

14 Headlight bulb - replacement

Halogen headlights
Warning: *Halogen bulbs are gas-filled and under pressure and might shatter if the surface is scratched or the bulb is dropped. Wear eye protection and handle the bulbs carefully, grasping only the base whenever possible. Don't touch the surface of the bulb with your fingers because the oil from your skin could cause it to overheat and fail prematurely. If you do touch the bulb surface, clean it with rubbing alcohol.*
Note: *The manufacturer recommends that you do not turn on the headlights if the headlight bulb is disconnected.*
1 Disconnect the cable from the negative battery terminal (see Chapter 5).
2 Remove the bulb access cover (see illustration).
3 Disconnect the electrical connector, then rotate the bulb counterclockwise and pull it out of the housing (see illustration).
4 Installation is the reverse of removal.

High Intensity Discharge (HID) headlights
Warning: *Some models use High Intensity Discharge (HID) bulbs instead of conventional halogen bulbs. According to the manufacturer, the high voltages produced by this system can be fatal in the event of a shock. Also, the voltage can remain in the circuit even after the headlight switch has been turned to OFF and the ignition key has been removed. Therefore, for your safety, we don't recommend that you replace these bulbs yourself. Instead, have this service performed by a dealer service department or other qualified repair shop.*

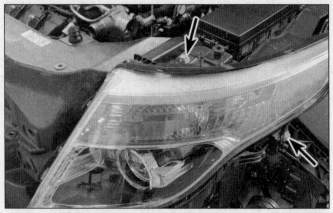

15.2a Headlight housing mounting fasteners (one not visible here; the last one is to the rear section of the headlamp assembly)

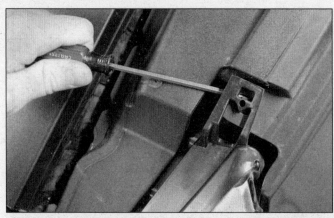

15.2b Pry the rear retainer up to remove the headlamp housing

15.2c Pull out the headlamp housing and disconnect the electrical connetions

16.1 Vertical adjustment screw location

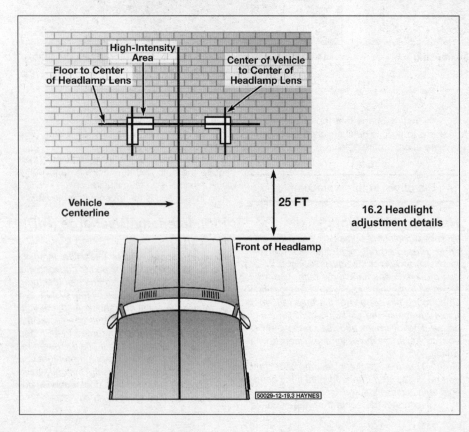

16.2 Headlight adjustment details

15 Headlight housing - removal and installation

1 Remove the front bumper cover (see Chapter 11).
2 Remove the three headlight housing mounting fasteners (see illustrations).
3 Installation is the reverse of removal.

16 Headlights - adjustment

Warning: *The headlights must be aimed correctly. If adjusted incorrectly, they could temporarily blind the driver of an oncoming vehicle and cause an accident or seriously reduce your ability to see the road. The headlights should be checked for proper aim every 12 months and any time a new headlight is installed or front-end bodywork is performed. The following procedure is only an interim step to provide temporary adjustment until the headlights can be adjusted by a properly equipped shop.*

1 The headlight adjustment screws (see illustration) control up-and-down movement. Left-and-right movement is not adjustable.
2 There are several methods of adjusting the headlights. The simplest method requires a blank wall 25 feet in front of the vehicle and a level floor (see illustration).

Chapter 12 Chassis electrical system

17.2a Remove the mounting fasteners . . .

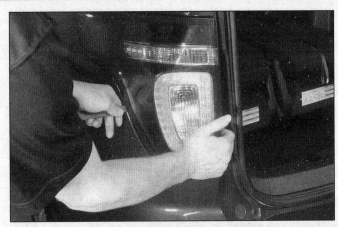

17.2b . . . then pull the housing towards you to disengage it from the vehicle

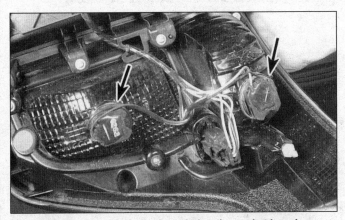

17.2c Remove the bulbs by twisting the socket housing counterclockwise, then pull the bulb from the socket

18.13 To remove an overhead light lens, carefully pry it off with a trim removal tool or screwdriver. If you're only removing the bulbs, that can be done now without going any further

3   Position masking tape on the wall in reference to the vehicle centerline and the centerlines of both headlights.
4   Measure the height of the headlight reference marks (in the centers of the headlight lenses) from the ground. Position a horizontal tape line on the wall at the same height as the headlight reference marks.
Note: *It may be easier to position the tape on the wall with the vehicle parked only a few inches away.*
5   Adjustment should be made with the vehicle sitting level, the gas tank half-full and no unusually heavy load in the vehicle.
6   Turn on the low beams. Turn the adjusting screw to position the high intensity zone so it is two inches below the horizontal line.
7   Have the headlights adjusted by a dealer service department at the earliest opportunity.

17 Taillight housing - removal and installation

1   Open the liftgate.
2   Remove the taillight housing mounting fasteners and remove the taillight housing (see illustrations).
3   Installation is the reverse of removal.

18 Bulb replacement

Exterior light bulbs

Front turn signal/parking light bulbs

1   Remove the headlight housing (see Section 15). Turn the bulb holder counterclockwise and remove it from the headlight housing. It's not necessary to disconnect the electrical connector to replace the bulb.
2   To remove the bulb, pull it straight out of the holder.
3   Installation is the reverse of removal.

Taillight bulbs

4   Remove the taillight housing (see Section 17).
5   To remove a taillight bulb socket, rotate it counterclockwise and pull it out of the taillight housing. To remove the bulb from the socket, pull it straight out.
6   Installation is the reverse of removal.

Puddle light bulbs

Note: *The puddle lights, which illuminate the ground below the front doors, are located in the outside mirrors.*
7   Release the puddle light housing tab and remove the puddle light housing from the bottom of the mirror housing.
8   Remove the puddle light bulb from its socket and install a new bulb in the socket.
9   To install the bulb socket in the puddle light housing, insert the socket into the housing. You will hear an audible click when the socket is fully seated and locked into place.

License plate light bulbs

10   Using a small screwdriver, unclip the tab at the side of the lens that secures the license plate light housing to the door. Remove the housing.
11   Remove the license plate light socket by rotating it counterclockwise and pulling it out of the license plate light housing. To remove the bulb from the socket, pull it straight out.
12   Installation is the reverse of removal.

Interior front overhead light bulb and fixture removal

13   Carefully pry off the overhead lens (see illustration).

Chapter 12 Chassis electrical system

18.14 Carefully work your way around the outer edges to release the pressure clips

18.15 Disconnect the electrical connections

18.17 Reuse the Torx screw

18.18 Pull the fixture down and disconnect the electrical connection

18.19 Rotate the bulb socket and remove the bulb. Pull the bulb out of the bulb socket and replace the bulb

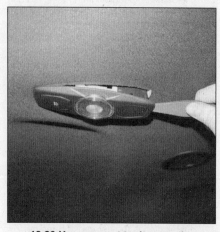

18.20 Use care not to damage the headliner as you work the fixture free

18.21 Disconnect the electrical connection, then remove the fixture

14 Pry the housing from the headliner (see illustration).
15 With the unit pulled down, disconnect the electrical connections and remove the unit (see illustration).
16 Installation is the reverse of removal.

Middle row overhead dome light bulb

17 Remove the single torx screw (see illustration).
18 Work the fixture from the headliner (see illustration).
19 Remove the bulb (see illustration).

Rear overhead lighting fixture removal

20 Pry the fixture from the headliner (see illustration).
21 Pull the fixture down and disconnect the electrical connector. Bulb replacement is similar to the middle row lighting fixtures (see illustration).

Chapter 12 Chassis electrical system

Bulb removal

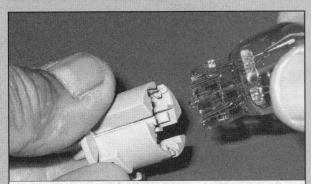

To remove many modern exterior bulbs from their holders, simply pull them out

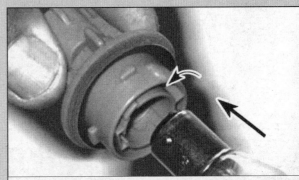

On bulbs with a cylindrical base ("bayonet" bulbs), the socket is spring-loaded; a pair of small posts on the side of the base hold the bulb in place against spring pressure. To remove this type of bulb, push it into the holder, rotate it 1/4-turn counterclockwise, then pull it out

If a bayonet bulb has dual filaments, the posts are staggered, so the bulb can only be installed one way

To remove most overhead interior light bulbs, simply unclip them

19 Wiper system

1 The wiper system is activated through the wiper switch which is incorporated into the multi-function switch mounted on the right side of the SCCM (Steering Column Control Module). On police package equipped vehicles the wiper switch is part of the multi-function switch.

2 Some systems will have a rain sensor control rather than intermittent wipers. The rain sensor system uses a rain sensor mounted to the front windshield, between the windshield and the inside rear view mirror. The sensitivity of the rain sensor has five different ranges that are controlled by the wiper switch.

3 The rain sensor system can be activated or deactivated through the message center.

4 If the rain sensor system has been deactivated the SCCM sends a signal to the wiper motor to switch to a default mode and will retain an intermittent wiper mode similar to a vehicle not equipped with a rain sensor system. The wipers will react as if they are operating as a intermittent wiper system with no regard to the intensity of the rain fall.

Wiper switch

5 Tilt the steering wheel to its highest position and remove the steering column covers (see Chapter 11).

6 Rotate steering wheel to gain access to the wiper switch screws.

7 Remove the wiper switch screws (see illustration).

8 Remove the wiper switch

9 Installation is the reverse of removal.

19.7 Wiper switch screws (steering wheel removed for clarity)

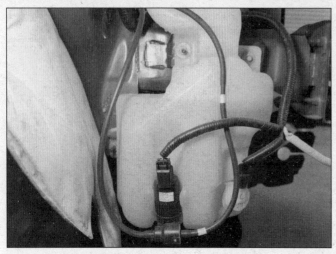
19.16 Washer pump and reservoir location

19.18a Remove the protective cap and the wiper arm retaining nut

19.18b Mark the relationship of the wiper arms to their shafts

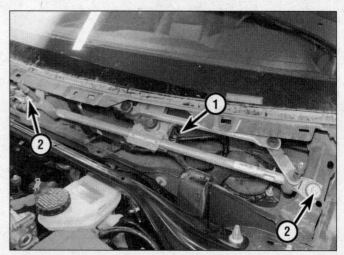
19.20 Front wiper motor assembly details
1   Electrical connector
2   Wiper motor assembly mounting bolts

Rain sensor and rain sensor bracket

10 Remove the inside mirror.
11 Raise the edge of the retaining clips and pull the rain sensor up and off of the rain sensor bracket.
12 Rain sensor bracket is not serviced and is only replaced if the windshield is damaged. The glass must be cleaned with alcohol and a new bracket installed.
13 Rain sensor bracket has to properly measured and adjusted to the new windshield and can take over 6 hours for the adhesive to setup.
Note: *Poor workmanship and faulty installation of the rain sensor can also damage either the windshield or the forward windshield defrost camera. Due to the differences in updated adhesives and revised procedures, the replacement procedures are included with the new replacement rain sensor bracket. Follow the direction to ensure proper installation.*

Washer fluid reservoir and washer pump

14 The washer fluid reservoir and washer pump are located in the right front corner of the front fender area.
15 Loosen the right-front wheel lug nuts. Raise the front of the vehicle and support it securely on jackstands, then remove the wheel. Also remove the inner fender splash shield (see Chapter 11).
16 Reservoir and washer pump are located in the right front corner of the front fender area. Remove the reservoir bolts and nuts and pull the washer pump hoses off as well as the washer pump electrical connections. Some fluid will leak out as the hoses are removed. Use a catch pan to contain the washer fluid. Then guide the neck of the reservoir out through the bottom side of the fender. Move any electrical or brackets out of the way to remove it. The washer pump is held in place with a large rubber grommet. Pull the pump straight out to remove (see illustration).
17 Installation is the reverse of removal.

Front wiper motor

18 Remove the wiper arm nuts and mark the relationship of the wiper arms to their shafts (see illustrations). Remove both wiper arms.
19 Remove the plastic cowl cover (see Chapter 11).
20 Disconnect the electrical connector from the wiper motor (see illustration).

Chapter 12 Chassis electrical system

20.3 The single bolt holds both horns in place. Each horn has its own electrical connection

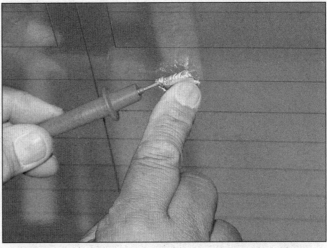

21.4 When measuring the voltage at the rear window defogger grid, wrap a piece of aluminum foil around the positive probe of the voltmeter and press the foil against the wire with your finger

21 Remove the two wiper motor and link assembly mounting bolts and remove the wiper motor and link assembly.
22 Using a trim removal tool or a screwdriver, carefully separate the linkage from the motor's crank arm.
23 Remove the nut that secures the crank arm to the motor shaft.
24 Mark the relationship of the crank arm to the motor shaft and remove the crank arm from the shaft.
25 Remove the four motor mounting bolts and remove the motor from its mounting bracket.
26 Installation is the reverse of removal.

Rear wiper motor

27 Remove the wiper arm nut and mark the relationship of the wiper arm to the shaft (see illustration 18.1). Remove the wiper arms.
28 Remove the liftgate trim panel (see Chapter 11).
29 Disconnect the electrical connector from the wiper motor.
30 Remove the three wiper motor mounting bolts and remove the wiper motor.
31 Installation is the reverse of removal.

20 Horn - replacement

Note: *The horn is located at the lower right corner of the vehicle, in the void ahead of the right front wheel well.*

1 Loosen the left-front wheel lug nuts. Raise the vehicle and place it securely on jackstands, then remove the wheel.
2 To access the horn, remove the left front inner fender splash shield fasteners and pull down the splash shield (see Chapter 11).
3 Disconnect the horn electrical connector (see illustration).
4 Remove the horn mounting bolt, then remove the horn.
5 Installation is the reverse of removal. Tighten the mounting bolts to 115 in-lbs (13 Nm).

21 Rear window defogger - check and repair

1 The rear window defogger consists of a number of horizontal elements baked onto the glass surface.
2 Small breaks in the element can be repaired without removing the rear window.

Check

3 Turn the ignition switch and defogger system switches to the ON position. Using a voltmeter, place the positive probe against the defogger grid positive terminal and the negative probe against the ground terminal. If battery voltage is not indicated, check the fuse, defogger switch and related wiring. If voltage

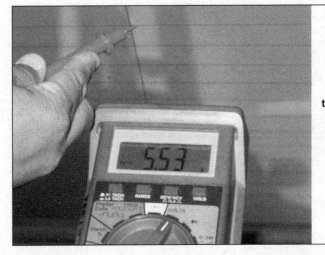

21.5 To determine if a heating element has broken, check the voltage at the center of each element; if the voltage is 5 or 6 volts, the element is unbroken, but if the voltage is 10 or 12 volts, the element is broken between the center and the ground side. If there is no voltage, the element is broken between the center and the positive side

is indicated, but all or part of the defogger doesn't heat, proceed with the following tests.
4 When measuring voltage during the next two tests, wrap a piece of aluminum foil around the tip of the voltmeter positive probe and press the foil against the heating element with your finger (see illustration). Place the negative probe on the defogger grid ground terminal.
5 Check the voltage at the center of each heating element (see illustration). If the voltage is 5 or 6-volts, the element is okay (there is no break). If the voltage is zero, the element is broken between the center of the element and the positive end. If the voltage is 10 to 12-volts, the element is broken between the center of the element and ground. Check each heating element.
6 Connect the negative lead to a good body ground. The reading should stay the same. If it doesn't, the ground connection is bad.
7 To find the break, place the voltmeter negative probe against the defogger ground

12-16   Chapter 12   Chassis electrical system

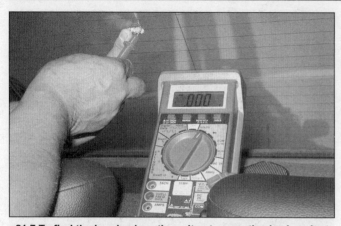

21.7 To find the break, place the voltmeter negative lead against the defogger ground terminal, place the voltmeter positive lead with the foil strip against the heating element at the positive terminal end and slide it toward the negative terminal end. The point at which the voltmeter reading changes abruptly is the point at which the element is broken

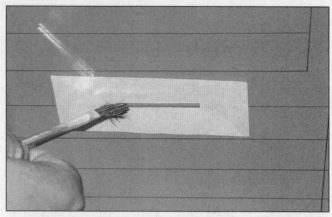

21.13 To use a defogger repair kit, apply masking tape to the inside of the window at the damaged area, then brush on the special conductive coating

terminal. Place the voltmeter positive probe with the foil strip against the heating element at the positive terminal end and slide it toward the negative terminal end. The point at which the voltmeter deflects from several volts to zero is the point at which the heating element is broken (see illustration).

Repair

8 Repair the break in the element using a repair kit specifically recommended for this purpose, available at most auto parts stores. Included in this kit is plastic conductive epoxy.
9 Prior to repairing a break, turn off the system and allow it to cool off for a few minutes.
10 Lightly buff the element area with fine steel wool, then clean it thoroughly with rubbing alcohol.
11 Use masking tape to mask off the area being repaired.
12 Thoroughly mix the epoxy, following the instructions provided with the repair kit.
13 Apply the epoxy material to the slit in the masking tape, overlapping the undamaged area about 3/4-inch on either end (see illustration).
14 Allow the repair to cure for 24 hours before removing the tape and using the system.

22 Power mirror control system - description and check

1 Electric rear view mirrors use two motors to move the glass; one for up and down adjustments and one for left-right adjustments.
2 The control switch has a selector portion which sends voltage to the left or right side mirror. With the ignition ON but the engine OFF, roll down the windows and operate the mirror control switch through all functions (left-right and up-down) for both the left and right side mirrors.

3 Listen carefully for the sound of the electric motors running in the mirrors.
4 If the motors can be heard but the mirror glass doesn't move, there's probably a problem with the drive mechanism inside the mirror.
5 If the mirrors do not operate and no sound comes from the mirrors, check the fuse (see Section 3).
6 If the fuse is OK, remove the mirror control switch from its mounting without disconnecting the wires attached to it. Turn the ignition ON and check for voltage at the switch. There should be voltage at one terminal. If there's no voltage at the switch, check for an open or short in the circuit between the fuse panel and the switch.
7 If the mirror motor fails to operate as described, replace the mirror assembly (see Chapter 11).

23 Cruise control system - description and check

1 All models have an electronically-controlled throttle body - there is no accelerator cable (or cruise control cable). When you select the speed that you want to maintain, the PCM controls vehicle speed by opening and closing the throttle plate by means of a computer-controlled solenoid (motor) inside the throttle body.
2 The diagnostic procedures for troubleshooting the cruise control system are beyond the scope of this manual, but if the system can't be set, or the set speed doesn't cancel when the brake pedal is depressed, check the fuses. Start with the fuses in the engine compartment fuse and relay box, then check the fuses in the under-dash fuse and relay box. If the set speed doesn't cancel when the CANCEL button is depressed, check the fuse for that circuit.
3 Other than checking the fuses, the diagnostic procedures for troubleshooting the cruise control system on these models are beyond the scope of this manual. A dealer service department should handle any further testing.

24 Power window system - description and check

Note: *These models are equipped with a Smart Junction Box (SJB) (manufacturer terminology), otherwise known as a Body Control Module (BCM). Several systems are linked to this centralized control module, which allows simple and accurate troubleshooting, but only with a professional-grade scan tool. The SJB or BCM governs the door locks, the power windows, the ignition lock and security system, the interior lights, the Daytime Running Lights system, the horn, the windshield wipers, the heating/air conditioning system and the power mirrors. In the event of malfunction with this system, have the vehicle diagnosed by a dealership service department or other qualified automotive repair facility.*
1 The power window system operates electric motors, mounted in the doors, which lower and raise the windows. The system consists of the control switches, the motors, regulators, glass mechanisms, the Smart Junction Box (SJB) and associated wiring.
2 The power windows can be lowered and raised from the master control switch by the driver or by remote switches located at the individual windows. Each window has a separate motor that is reversible. The position of the control switch determines the polarity and therefore the direction of operation.
3 The circuit is protected by a fuse and a circuit breaker. Each motor is also equipped with an internal circuit breaker; this prevents one stuck window from disabling the whole system.
4 The power window system will only oper-

Chapter 12 Chassis electrical system

25.14 Using a coin, carefully pry the halves of the transmitter apart

25.15 Carefully pry out the old battery

ate when the ignition switch is ON, and for a period of time after the ignition key has been turned Off (unless one of the doors is opened). In addition, many models have a window lockout switch at the master control switch which, when activated, disables the switches at the rear windows and, sometimes, the switch at the passenger's window also. Always check these items before troubleshooting a window problem.

5 These procedures are general in nature, so if you can't find the problem using them, take the vehicle to a dealer service department or other properly equipped repair facility.

6 If the power windows won't operate, always check the fuse and circuit breaker first.

7 If only the rear windows are inoperative, or if the windows only operate from the master control switch, check the rear window lockout switch for continuity in the unlocked position. Replace it if it doesn't have continuity.

8 Check the wiring between the switches and fuse panel for continuity. Repair the wiring, if necessary.

9 If only one window is inoperative from the master control switch, try the other control switch at the window.

Note: *This doesn't apply to the driver's door window.*

10 If the same window works from one switch, but not the other, check the switch for continuity.

11 If the switch tests OK, check for a short or open in the circuit between the affected switch and the window motor.

12 If one window is inoperative from both switches, remove the switch panel from the affected door. Check for voltage at the switch and at the motor (refer to Chapter 11 for door panel removal) while the switch is operated.

13 If voltage is reaching the motor, disconnect the glass from the regulator (see Chapter 11). Move the window up and down by hand while checking for binding and damage. Also check for binding and damage to the regulator. If the regulator is not damaged and the window moves up and down smoothly, replace the motor. If there's binding or damage, lubricate, repair or replace parts, as necessary.

14 If voltage isn't reaching the motor, check the wiring in the circuit for continuity between the switches and the body control module, and between the body control module and the motors. You'll need to consult the wiring diagram at the end of this Chapter. If the circuit is equipped with a relay, check that the relay is grounded properly and receiving voltage.

15 Test the windows after you are done to confirm proper repairs.

25 Power door lock and keyless entry system - description, check and battery replacement

Note: *These models are equipped with a Smart Junction Box (SJB) (manufacturer terminology), otherwise known as a Body Control Module (BCM). Several systems are linked to this centralized control module, which allows simple and accurate troubleshooting, but only with a professional-grade scan tool. The SJB or BCM governs the door locks, the power windows, the ignition lock and security system, the interior lights, the Daytime Running Lights system, the horn, the windshield wipers, the heating/air conditioning system and the power mirrors. In the event of malfunction with this system, have the vehicle diagnosed by a dealership service department or other qualified automotive repair facility.*

Description and check

1 The power door lock system operates the door lock actuators mounted in each door. The system consists of the switches, actuators, Smart Junction Box (SJB) and associated wiring. Diagnosis can usually be limited to simple checks of the wiring connections and actuators for minor faults that can be easily repaired.

2 Power door lock systems are operated by bi-directional solenoids located in the doors. The lock switches have two operating positions: Lock and Unlock. These switches send a signal to the SJB, which in turn sends a signal to the door lock solenoids.

3 If you are unable to locate the trouble using the following general steps, consult your dealer service department.

4 Always check the circuit protection first. Some vehicles use a combination of circuit breakers and fuses. Refer to the wiring diagrams at the end of this Chapter.

5 Check for voltage at the switches. If no voltage is present, check the wiring between the fuse panel and the switches for shorts and opens.

6 If voltage is present, test the switch for continuity. Replace it if there's not continuity in both switch positions. To remove the switch, use a flat-bladed trim tool to pry out the door/window switch assembly (see Chapter 11).

7 If the switch has continuity, check the wiring between the switch and door lock solenoid.

8 If all but one lock solenoids operate, remove the trim panel from the affected door (see Chapter 11) and check for voltage at the solenoid while the lock switch is operated. One of the wires should have voltage in the Lock position; the other should have voltage in the Unlock position.

9 If the inoperative solenoid is receiving voltage, replace the solenoid.

10 If the inoperative solenoid isn't receiving voltage, check for an open or short in the wire between the lock solenoid and the relay.

11 On the models covered by this manual, power door lock system communication goes through the Smart Junction Box. If the above tests do not pinpoint a problem, take the vehicle to a dealer or qualified shop with the proper scan tool to retrieve trouble codes from the SJB.

Keyless entry system

12 The keyless entry system consists of a remote control transmitter that sends a coded infrared signal to a receiver, which then operates the door lock system.

13 Replace the battery when the transmitter doesn't operate the locks at a distance of ten feet. Normal range should be about 30 feet.

Key remote control battery replacement

14 Use a coin to carefully separate the case halves (see illustration).
15 Replace the battery (see illustration).
16 Snap the case halves together.

Transmitter programming

17 Programming replacement transmitters requires the use of a specialized scan tool. Take the vehicle and the transmitter(s) to a dealer service department or other qualified repair shop equipped with the necessary tool to have the transmitter(s) programmed to the vehicle.

26 Daytime Running Lights (DRL) - general information

1 The Daytime Running Lights (DRL) system used on some models illuminates the headlights whenever the engine is running. The only exception is with the engine running and the parking brake engaged. Once the parking brake is released, the lights will remain on as long as the ignition switch is on, even if the parking brake is later applied.

2 The DRL system supplies reduced power to the headlights so they won't be too bright for daytime use, while prolonging headlight life.

27 Airbags - general information

1 These models are equipped with a Supplemental Restraint System (SRS), more commonly known as airbags. This system is designed to protect the driver and the front seat passenger from serious injury in the event of a head-on or frontal collision. It consists of an airbag module in the center of the steering wheel and another airbag module on the right side of the instrument panel plus, on some and later models, side airbags and curtain shield airbags designed to protect the occupants in a side impact and a sensing/diagnostic module which is mounted in the center of the vehicle below the instrument panel. These models are also equipped with a pair of impact sensors that are located at the front of the vehicle.

2 Some later models are equipped with seatbelt pre-tensioners, also part of the airbag system. The pre-tensioners are pyrotechnic (explosive) devices designed to retract the seat belts in the event of a collision.

3 On models equipped with pre-tensioners, do not remove the front seat belt retractor assemblies. Problems with the pre-tensioners will turn on the SRS (airbag) warning light on the dash. If any pre-tensioner problems are suspected, take the vehicle to a dealer service department.

Airbag module

Steering wheel-mounted

4 The airbag inflator module contains a housing incorporating the cushion (airbag) and inflator unit, mounted in the center of the steering wheel. The inflator assembly is mounted on the back of the housing over a hole through which gas is expelled, inflating the bag almost instantaneously when an electrical signal is sent from the system. A spiral cable assembly on the steering column under the module carries this signal to the module. This spiral cable assembly can transmit an electrical signal regardless of steering wheel position.

Instrument panel-mounted

5 The passenger side airbag is mounted above the glove compartment and designated by the letters SRS (Supplemental Restraint System). It consists of an inflator containing an igniter, a bag assembly, a reaction housing and a trim cover.

6 The passenger airbag is considerably larger than the steering wheel-mounted unit and is supported by the steel reaction housing. The trim cover has a molded seam which splits when the bag inflates.

Side and curtain airbags

7 The side airbag and inflator modules are mounted on the sides of the front seats and contain an inflator containing an igniter and bag assembly. The curtain shield airbag assemblies run along the interior of the roof from the front A-pillar to the rear of the passenger compartment. In the event of a side impact, both airbag assemblies are activated by the sensors mounted at the base of the center pillar behind the seats.

Sensing and diagnostic module

8 The sensing and diagnostic module supplies the current to the airbag system in the event of the collision, even if battery power is cut off. It checks this system every time the vehicle is started, causing the "AIR BAG" light to go on then off, if the system is operating properly. If there is a fault in the system, the light will go on and stay on, flash, or the dash will make a beeping sound. If this happens, the vehicle should be taken to your dealer immediately for service.

Seat belt pre-tensioners

9 Some models are equipped with pyrotechnic (explosive) units in the front seat belt retracting mechanisms. During an impact that would trigger the airbag system, the airbag control unit also triggers the seat belt retractors. When the pyrotechnic charges go off, they accelerate the retractors to instantly take up any slack in the seat belt system to more fully prepare the driver and front seat passenger for impact.

10 The airbag system should be disabled any time work is done to or around the seats.

Warning: *Never strike the pillars or floorpan with a hammer or use an impact-driver tool in these areas unless the system is disabled.*

Precautions

Disabling the SRS system

Warning: *Failure to follow these precautions could result in accidental deployment of the airbag and personal injury.*

Warning: *Never install a memory-saver device, used to preserve PCM memory and radio station presets, when working on or around any of the airbag system components.*

11 Whenever working in the vicinity of the steering wheel, instrument panel or any of the other SRS system components, the system must be disarmed. To disarm the system:

a) Point the wheels straight ahead and turn the ignition key to the LOCK position.
b) Disconnect the cable from the negative terminal of the battery.
c) Wait at least two minutes for the back-up power supply capacitor to be depleted.

12 Whenever handling an airbag module, always keep the airbag opening (trim side) pointed away from your body. Never place the airbag module on a bench or other surface with the airbag opening facing the surface. Always place the airbag module in a safe location with the airbag opening (trim side) facing up.

13 Never measure the resistance of any SRS component. An ohmmeter has a built-in battery supply that could accidentally deploy the airbag.

14 Never use electrical welding equipment on a vehicle equipped with an airbag without first disconnecting the negative battery cable.

15 Never dispose of a live airbag module. Return it to your dealer for safe deployment, using special equipment, and disposal.

28 Wiring diagrams - general information

1 Since it isn't possible to include all wiring diagrams for every year covered by this manual, the following diagrams are those that are typical and most commonly needed.

2 Prior to troubleshooting any circuits, check the fuse and circuit breakers (if equipped) to make sure they're in good condition. Make sure the battery is properly charged and check the cable connections (see Chapter 1).

3 When checking a circuit, make sure that all connectors are clean, with no broken or loose terminals. When unplugging a connector, do not pull on the wires. Pull only on the connector housings themselves.

Chapter 12 Chassis electrical system

12-19

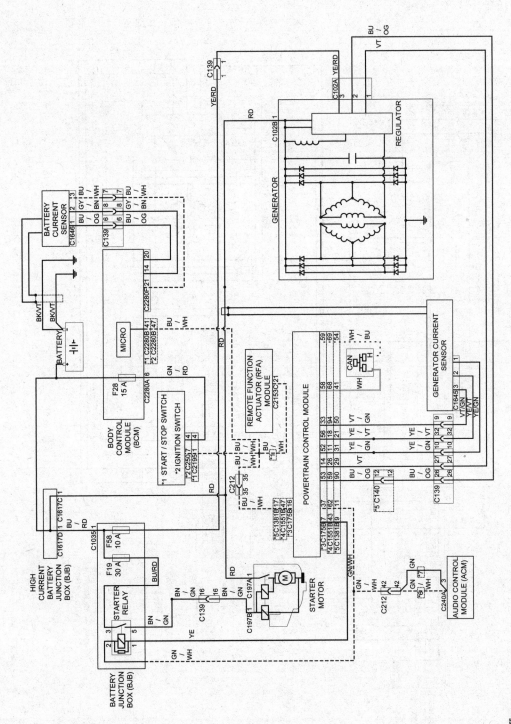

Starting and charging systems - 2015 and earlier models

*1 With intelligent access
*2 Without intelligent access
*3 Engine code: 3.5 L TiVCT
*4 Engine code: 3.5 L GTDI
*5 Engine code: 2.0 L GTDI
*6 From 2011 to 2012
*7 From 2013 to 2015

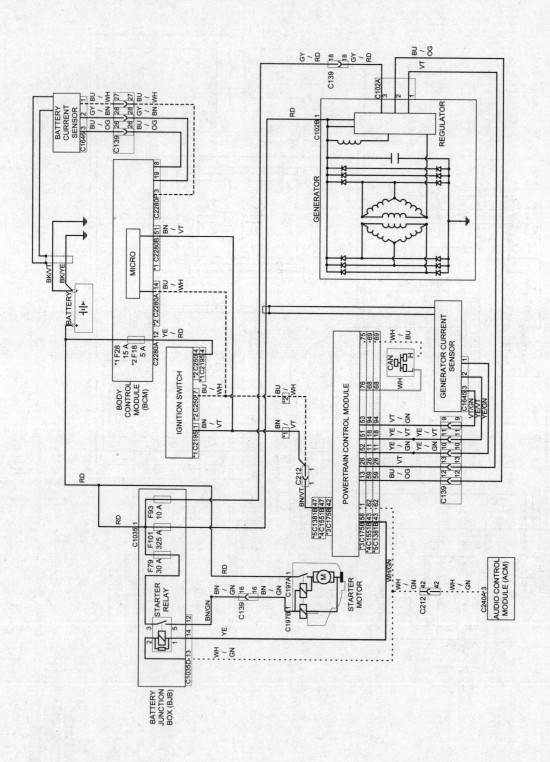

Chapter 12 Chassis electrical system

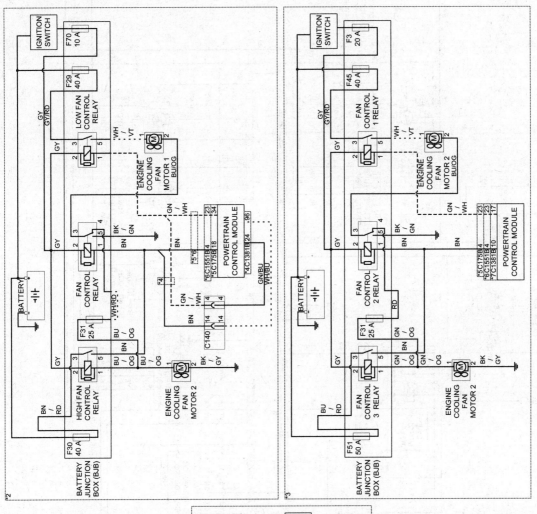

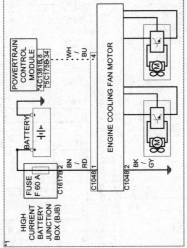

Radiator cooling fan system

*1 For 2011 to 2012
*2 For 2013 to 2015
*3 For 2015 to 2017
*4 Engine code: 2.0 L GTDI
*5 Engine code: 3.5 L V6
*6 Engine code: 3.5 L GTDI
*7 Engine code: 2.3 L GTDI

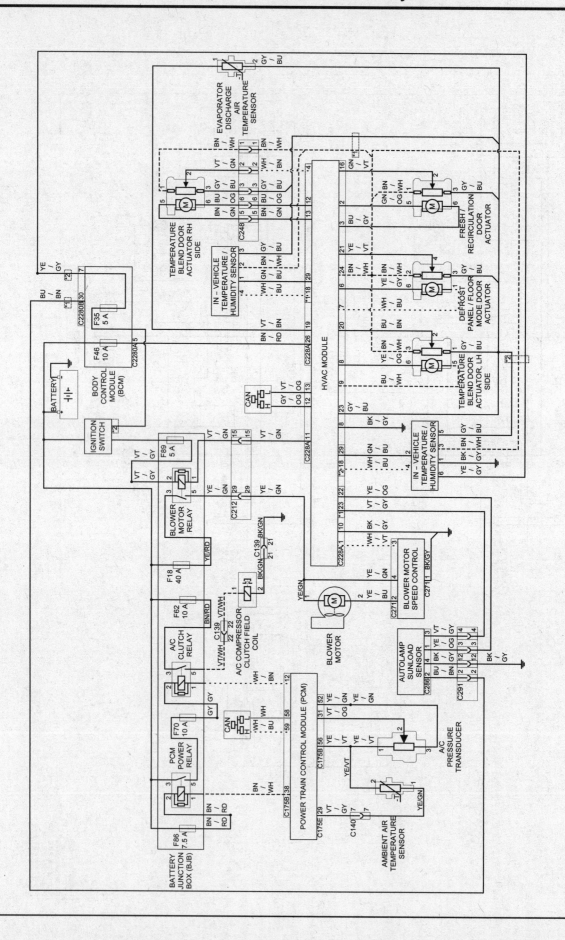

Air conditioning and heating system - 2015 and earlier models

*1 Automatic climate control system
*2 Manual climate control system

Chapter 12 Chassis electrical system

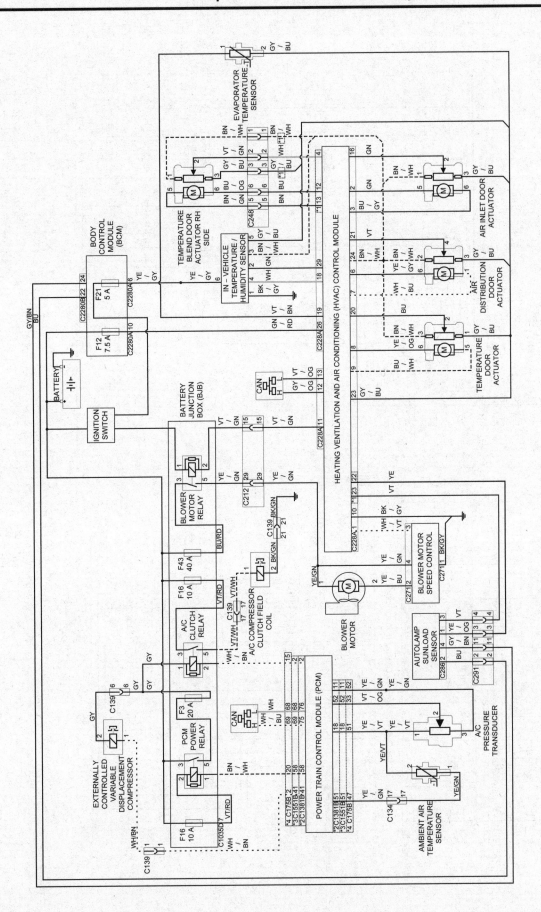

Air conditioning and heating system - 2016 and later models

*1 Automatic climate control system
*2 Engine code: 2.3 L
*3 Engine code: 3.5 L GTDI
*4 Engine code: 3.5 L / 3.7 L TiVCT

12-24 Chapter 12 Chassis electrical system

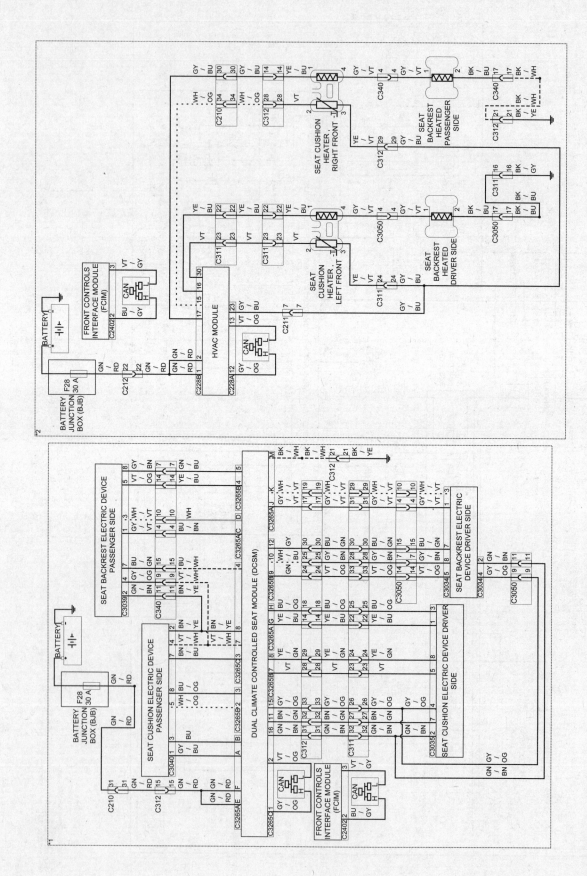

Seat heater system - 2015 and earlier models

*1 With climate control
*2 Without climate control

Chapter 12 Chassis electrical system

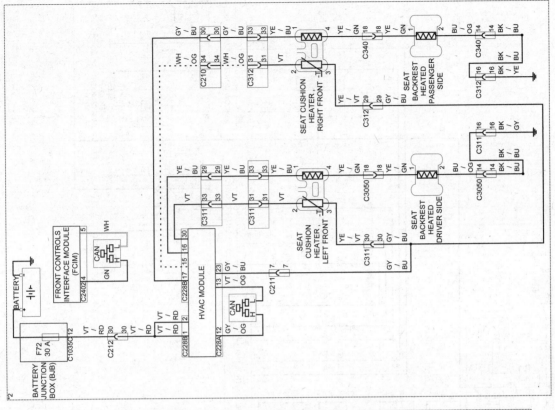

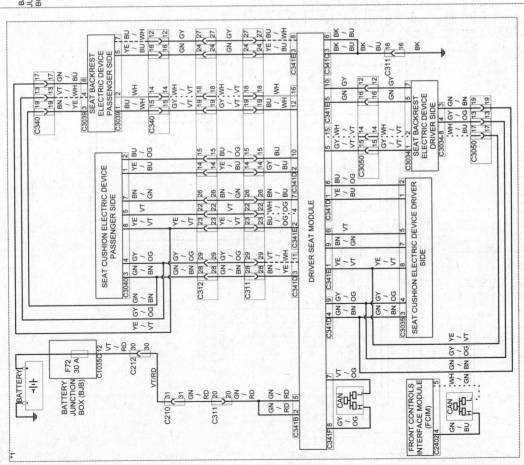

Seat heater system - 2015 and earlier models

*1 With climate control
*2 Without climate control

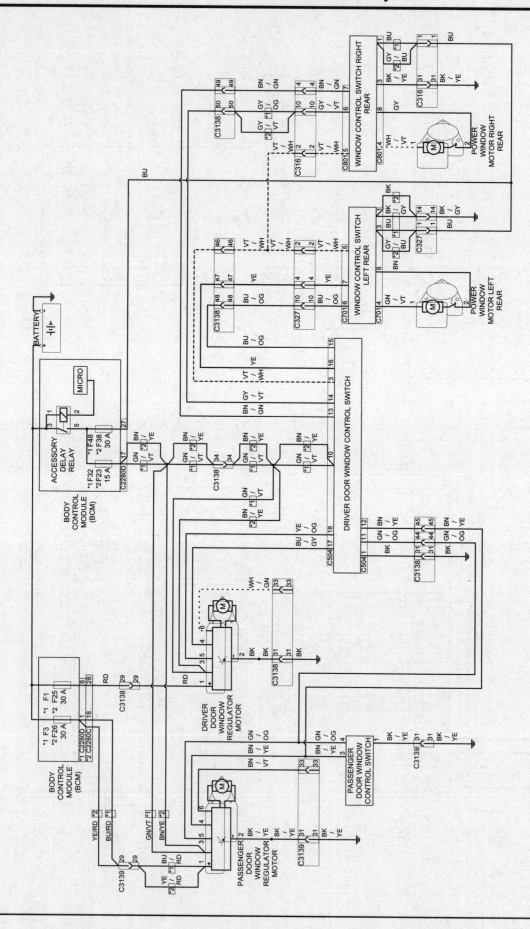

Chapter 12 Chassis electrical system

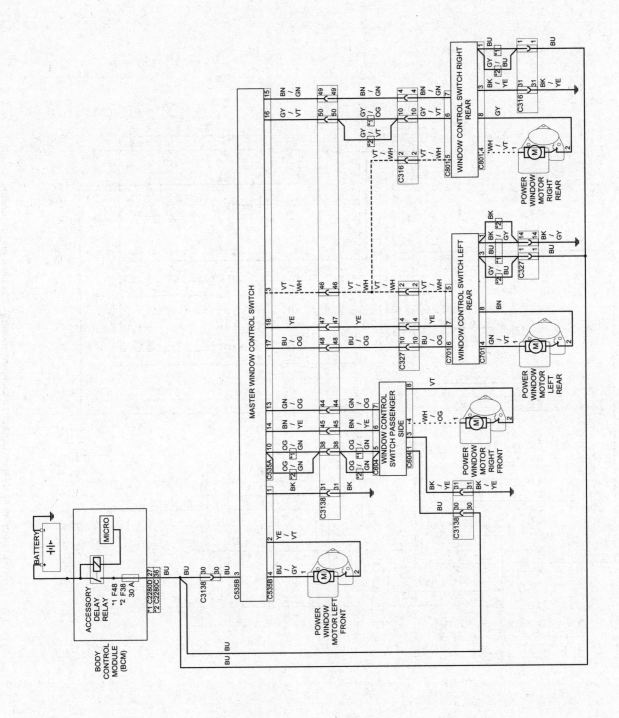

Power window system (standard equipment)

*1 From 2011 to 2015
*2 From 2016 to 2017

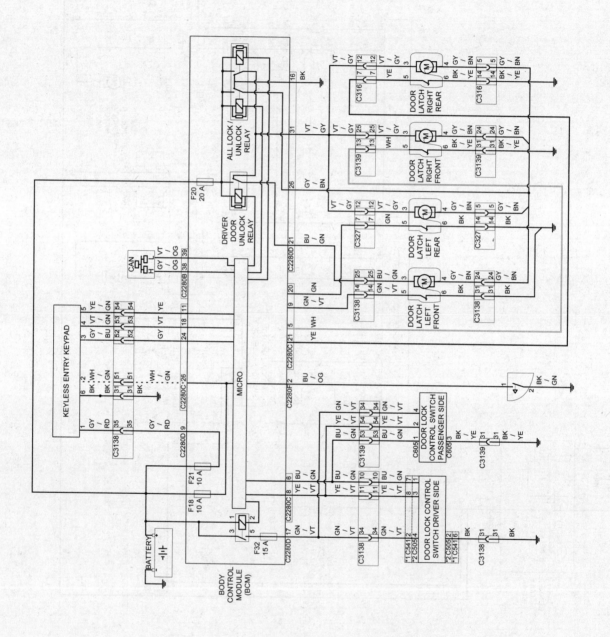

Chapter 12 Chassis electrical system

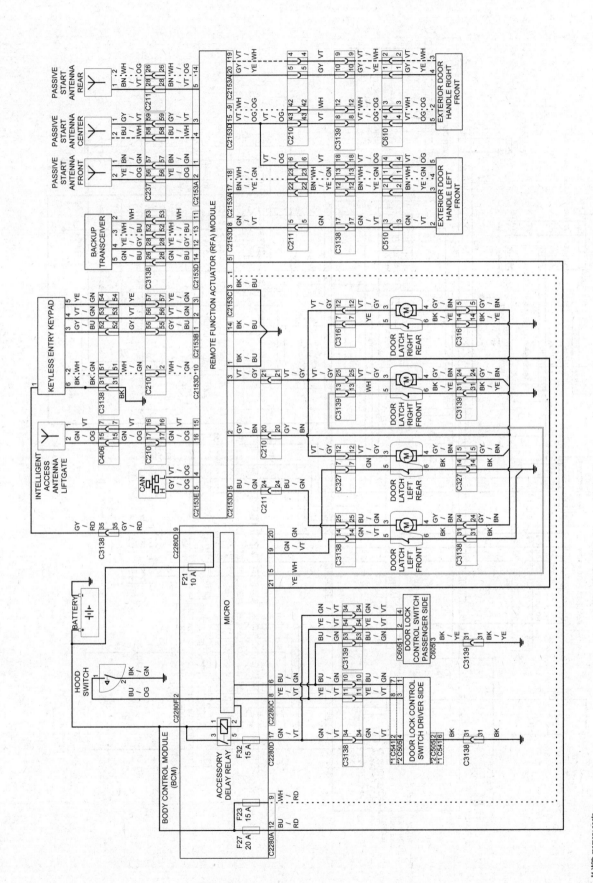

Power door locks system (with intelligent access) - 2015 and earlier models

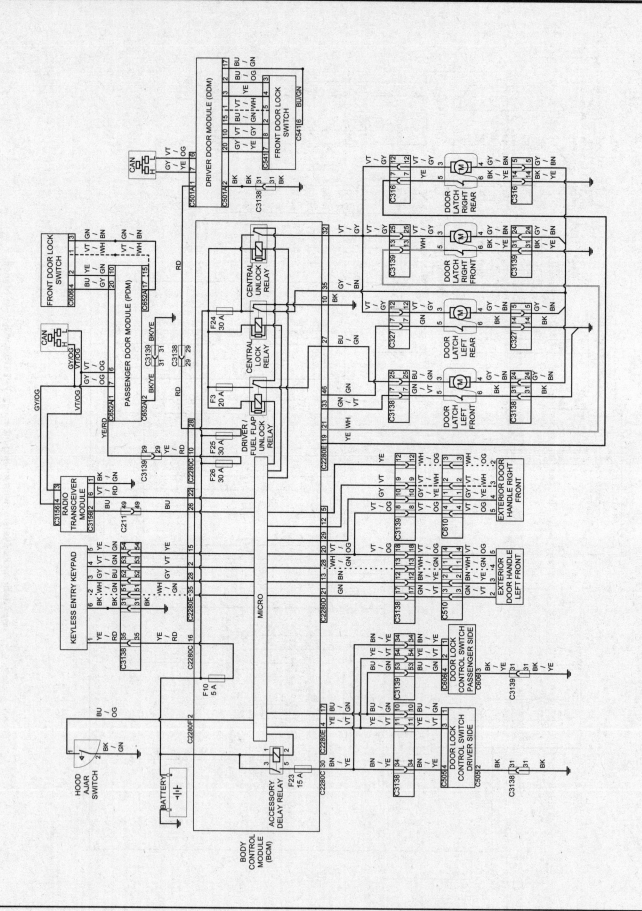

Power door locks system - 2016 and later models

Chapter 12 Chassis electrical system

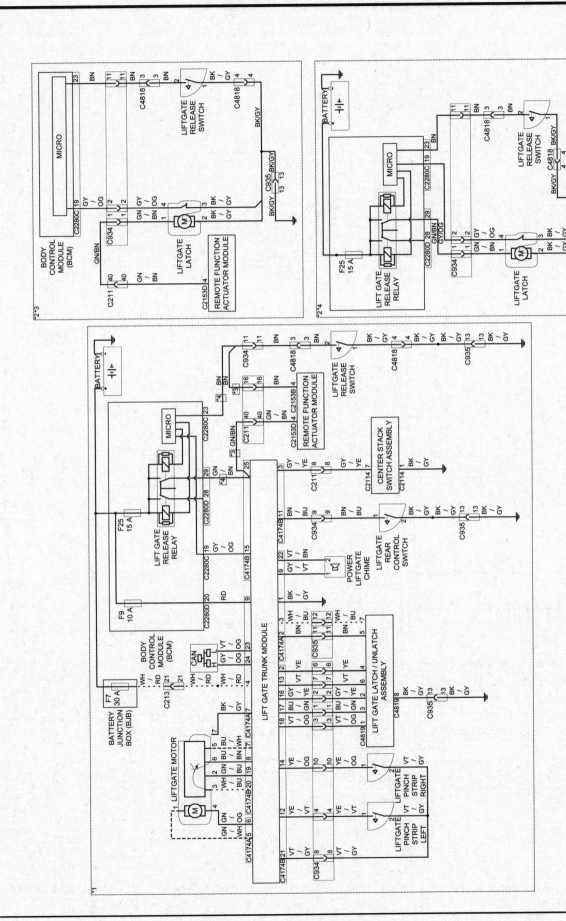

Power liftgate system - 2015 and earlier models

*1 With power liftgate
*2 With manual liftgate
*3 With intelligent access
*4 Without intelligent access

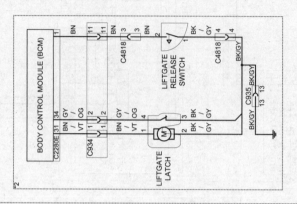
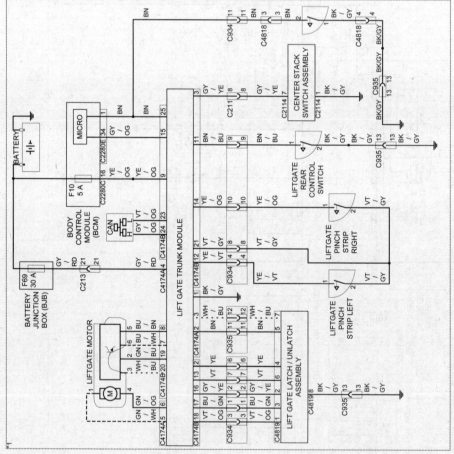

Power liftgate system - 2016 and later models

*1 With power liftgate
*2 With manual liftgate

Chapter 12 Chassis electrical system

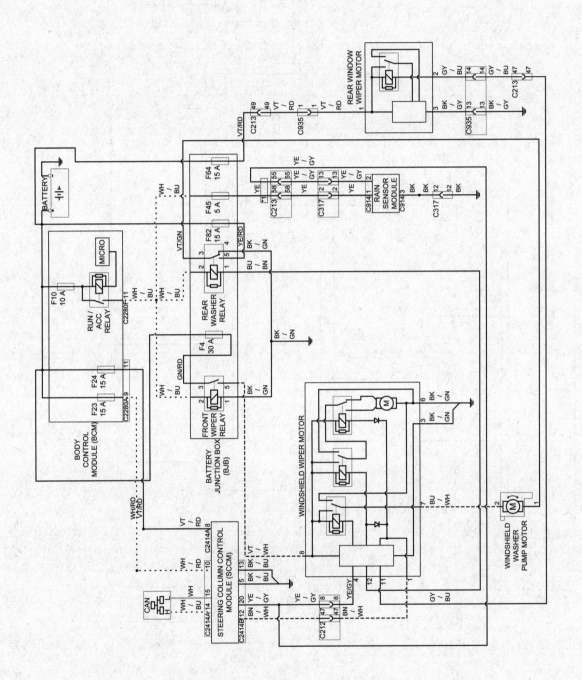

Windshield and rear wiper/washer systems - 2015 and earlier models

*1 With rain sensor

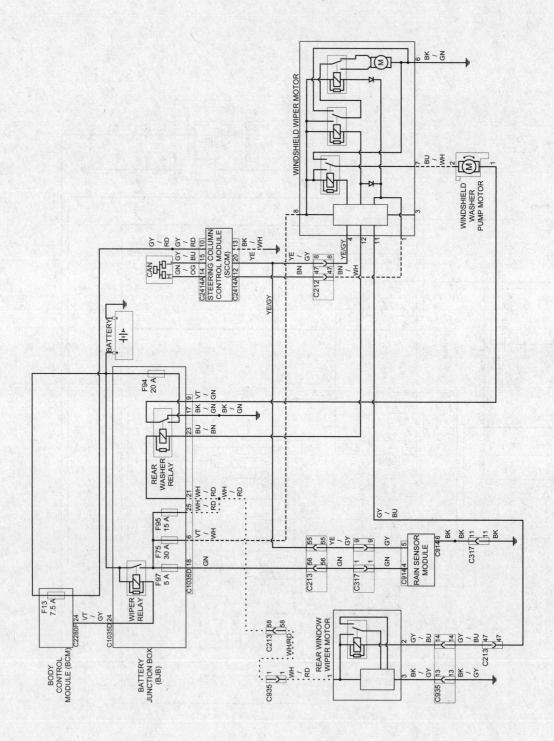

Windshield and rear wiper/washer systems - 2016 and later models

Chapter 12 Chassis electrical system

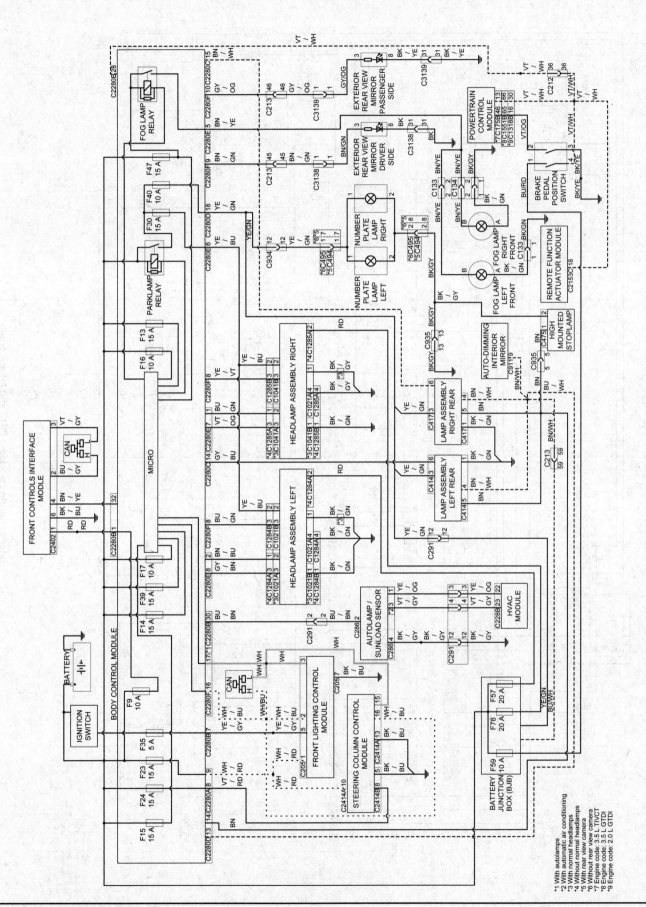

Exterior lighting system - 2015 and earlier models

*1 With autolamps
*2 With automatic air conditioning
*3 With normal headlamps
*4 Without normal headlamps
*5 With rear view camera
*6 Without rear view camera
*7 Engine code: 3.5L TiVCT
*8 Engine code: 3.5L GTDI
*9 Engine code: 2.0L GTDI

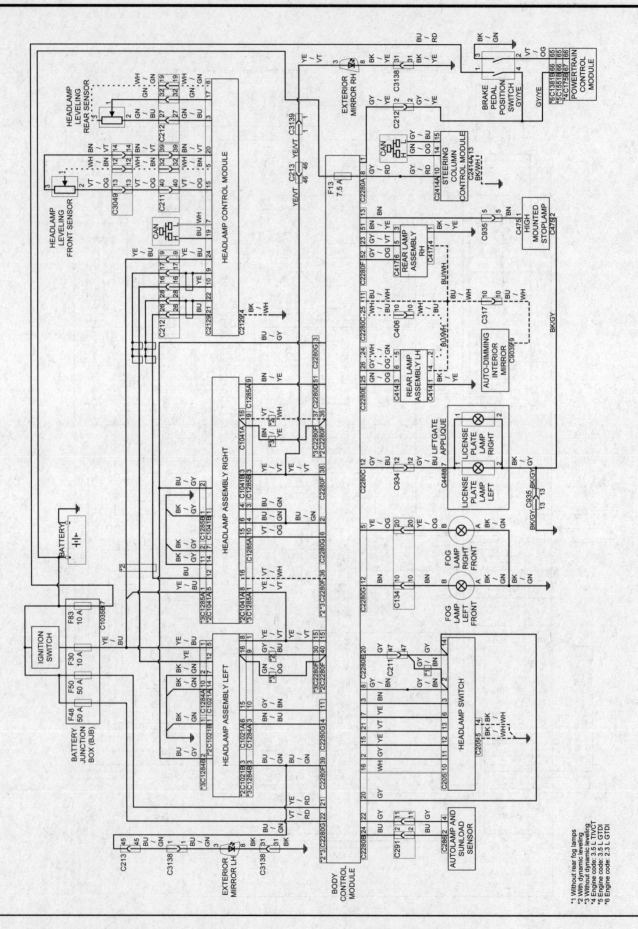

Chapter 12 Chassis electrical system

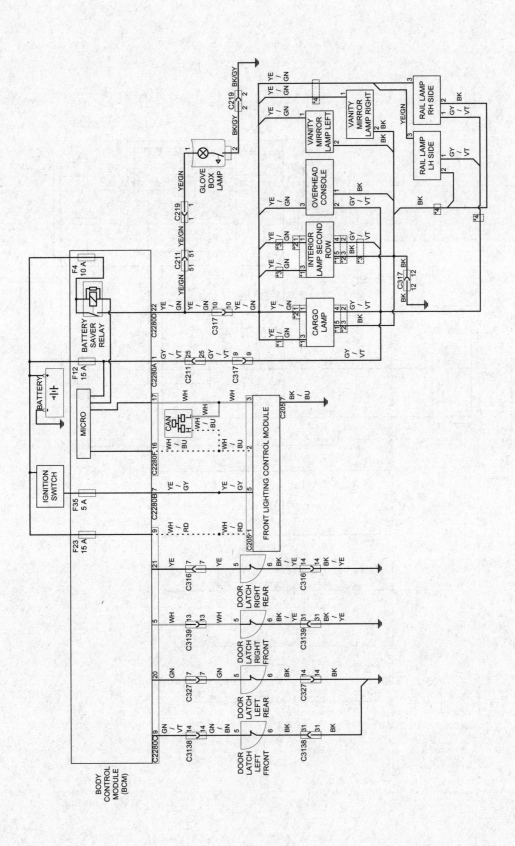

Interior lighting system - 2015 and earlier models

*1 With led lamps
*2 With incandescent lamps
*3 Without moonroof
*4 With moonroof

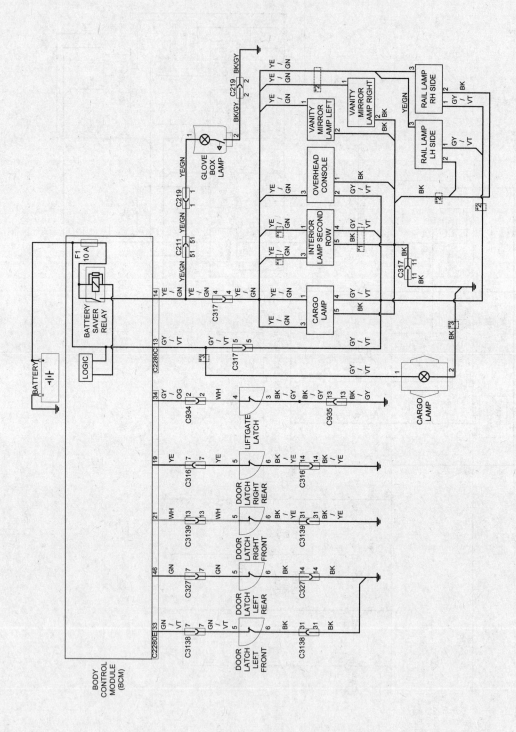

Interior lighting system - 2016 and later models

Chapter 12 Chassis electrical system

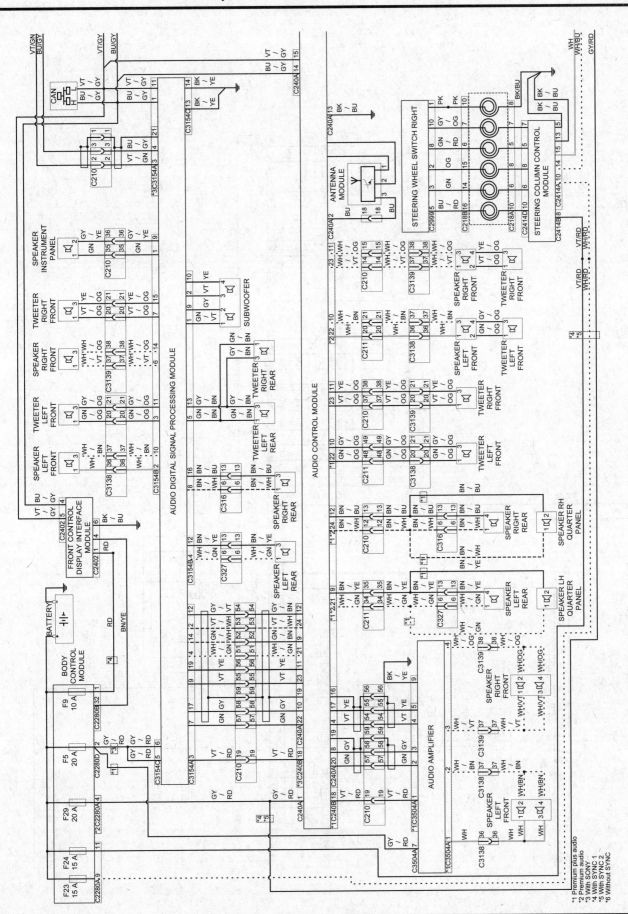

Audio system - 2015 and earlier models (1 of 2)

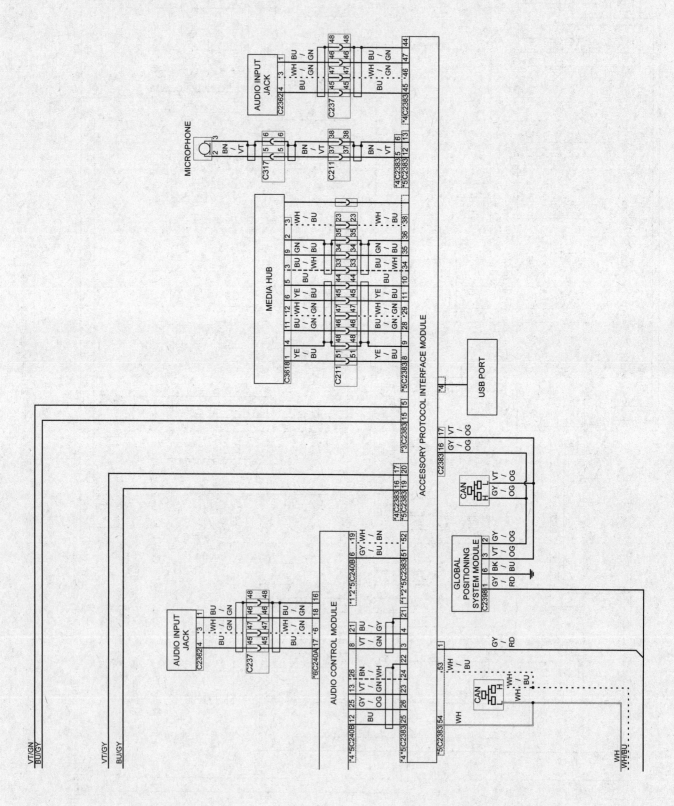

Audio system - 2015 and earlier models (2 of 2)

Chapter 12 Chassis electrical system

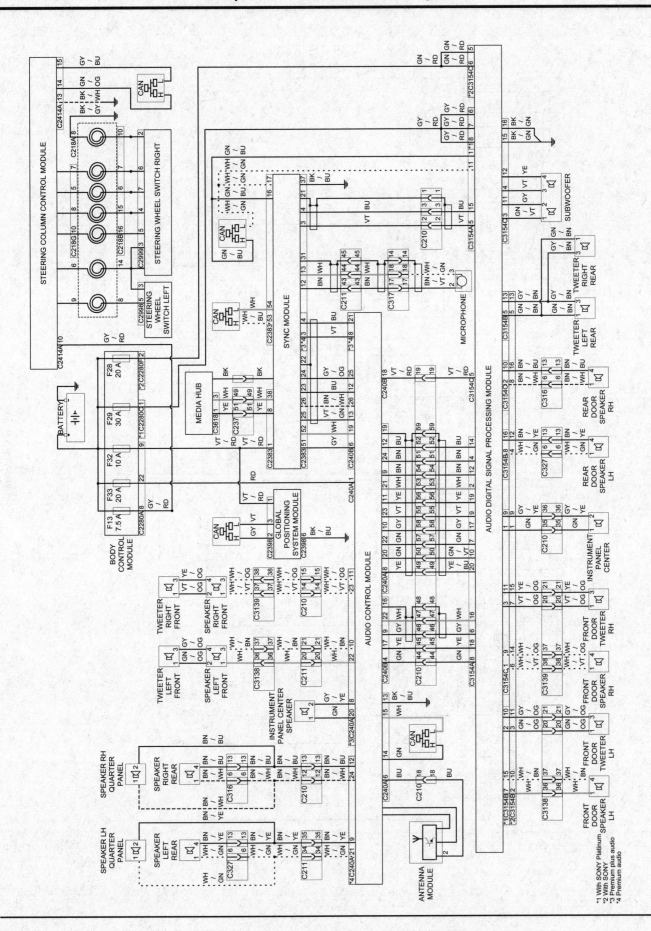

Audio system - 2016 and later models

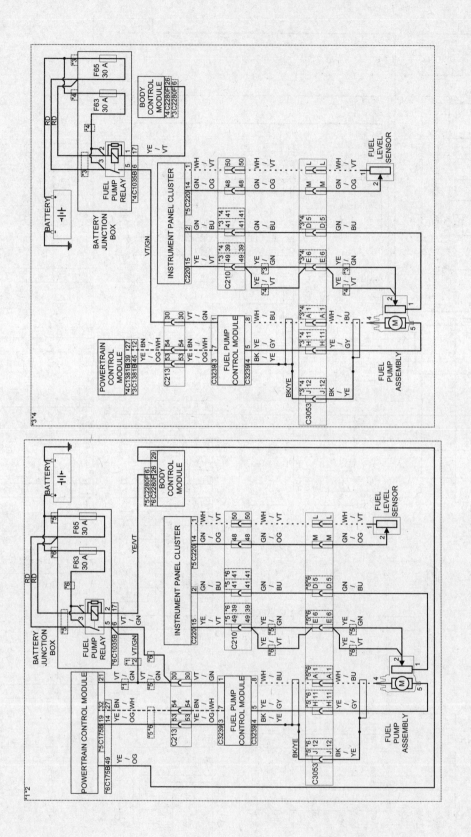

Fuel pump system

Chapter 12 Chassis electrical system

BATTERY JUNCTION BOX (FROM 2011 TO 2015)

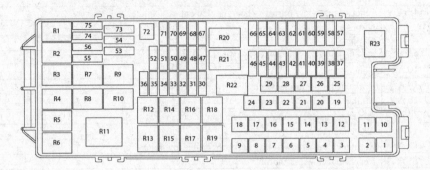

FROM 2011 TO 2012

FUSE/RELAY	VALUE	DESCRIPTION	OEM NAME
1	-	Not used	-
2	-	Not used	-
3	30 A	Trailer brake control module connector	F3
4	30 A	Front wiper relay	F4
5	50 A	Anti-lock brake system	F5
6	-	Not used	-
7	30 A	Liftgate / Trunk module	F7
8	20 A	Roof opening panel module	F8
9	20 A	Power point, console rear	F9
10	-	Not used	-
11	40 A	DC/AC inverter module	F17
12	40 A	Blower motor relay	F18
13	30 A	Starter relay	F19
14	20 A	Power point, instrument panel	F20
15	20 A	Power point, third row	F21
16	30 A	Third row power seat relay	F22
17	30 A	Driver seat module - with memory, Seat control switch, driver side front - without memory	F23
18	30 A	Trailer tow relay, Battery charge	F24
19	40 A	Rear window defrost relay	F26
20	20 A	Power point, console front	F27
21	30 A	HVAC module, Dual climate controlled seat module	F28
22	-	Not used	-
23	-	Not used	-
24	-	Not used	-
25	40 A	Auxiliary blower motor relay	F39
26	-	Not used	-
27	-	Not used	-
28	30 A	Seat control switch, passenger side front	F42
29	40 A	Anti-lock brake system	F43
30	5 A	Rain sensor module	F45
31	-	Not used	-
32	-	Not used	-

33	-	Not used	-
34	-	Not used	-
35	15 A	Exterior rear view mirrors - heating element	F50
36	-	Not used	-
37	-	Not used	-
38	20 A	Headlamp assembly, left with HID headlamps	F57
39	10 A	Generator	F58
40	10 A	Brake pedal position switch	F59
41	10 A	Trailer tow relay, reversing	F60
42	20 A	Power release seat motors, second row	F61
43	10 A	A/C clutch relay	F62
44	15 A	Right turn trailer tow relay, Left turn trailer tow relay	F63
45	15 A	Rear window wiper motor	F64
46	30 A	Fuel pump relay	F65
47	20 A	Heated oxygen sensor, Variable camshaft timing solenoids, Evaporative emission canister purge valve, Mass air flow / Intake air temperature sensor	F67
48	15 A	Coil on plugs	F68
49	15 A	Powertrain control module	F69
50	10 A	A/C clutch relay, All wheel drive relay module	F70
51	-	Not used	-
52	-	Not used	-
53	-	Not used	-
54	-	Not used	-
55	-	Not used	-
56	-	Not used	-
57	20 A	Headlamp assembly, right with HID headlamps	F78
58	5 A	Cruise control module	F79
59	-	Not used	-
60	-	Not used	-
61	15 A	Rear washer relay	F82
62	-	Not used	-
63	20 A	Trailer tow relay, parking lamp	F84
64	-	Not used	-
65	7.5 A	EVAP canister vent control solenoid, PCM power relay, Powertrain control module	F86
66	5 A	Run/start relay	F87
67	5 A	Power steering control module, Blower motor relay	F89
68	10 A	Powertrain control module	F90
69	10 A	Cruise control module	F91
70	10 A	Anti-lock brake system	F92
71	5 A	Auxiliary blower motor relay, Rear window defrost relay, Battery charge trailer tow relay	F93
72	30 A	Body control module F33-F37, F41,F42	F94
73	-	Not used	-
74	-	Not used	-
75	-	Not used	-
R1	-	A/C clutch relay	R98
R2	-	Trailer tow relay, parking lamp	R77
R3	-	Not used	-
R4	-	Trailer tow relay, reversing	R38
R5	-	Not used	-

R6	-	Fuel pump relay	R15
R7	-	Trailer tow relay left turn	R53
R8	-	Trailer tow relay right turn	R37
R9	-	Not used	-
R10	-	Not used	-
R11	-	Not used	-
R12	-	Not used	-
R13	-	Starter relay	R13
R14	-	Blower motor relay	R34
R15	-	Trailer tow relay battery charge	R12
R16	-	Not used	-
R17	-	Rear window defrost relay	R11
R18	-	Auxiliary blower motor relay	R32
R19	-	Third row power seat relay	R10
R20	-	Run/start relay	R88
R21	-	PCM power relay	R66
R22	-	Rear washer relay	R44
R23	-	Front wiper relay	R55

FROM 2013 TO 2015

FUSE/RELAY	VALUE	DESCRIPTION	OEM NAME
1	-	Not used	-
2	-	Not used	-
3	30 A	Trailer brake control module connector	F3
4	30 A	Front wiper relay	F4
5	50 A	Anti-lock brake system	F5
6	-	Not used	-
7	30 A	Liftgate / Trunk module	F7
8	20 A	Roof opening panel module	F8
9	20 A	Power point, console rear	F9
10	-	Not used	-
11	40 A	DC/AC inverter module	F17
12	40 A	Blower motor relay	F18
13	30 A	Starter relay	F19
14	20 A	Power point, instrument panel	F20
15	20 A	Power point, third row	F21
16	30 A	Third row power seat relay	F22
17	30 A	Driver seat module - with memory, Seat control switch, driver side front - without memory	F23
18	30 A	Battery charge trailer tow relay	F24
19	40 A	Rear window defrost relay	F26
20	20 A	Power point, console front	F27
21	30 A	HVAC module, Dual climate controlled seat module	F28
22	40 A	Low fan control relay	F29
23	40 A	High fan control relay	F30
24	25 A	Engine cooling fan motors	F31
25	40 A	Auxiliary blower motor relay	F39

26	-	Not used	-
27	-	Not used	-
28	30 A	Seat control switch, passenger side front	F42
29	40 A	Anti-lock brake system	F43
30	5 A	Rain sensor module	F45
31	-	Not used	-
32	-	Not used	-
33	-	Not used	-
34	-	Not used	-
35	15 A	Exterior rear view mirrors - heating element	F50
36	-	Not used	-
37	15 A	Transmission control module	F56
38	20 A	Headlamp assembly, left	F57
39	10 A	Generator	F58
40	10 A	Brake pedal position switch	F59
41	10 A	Trailer tow relay, reversing	F60
42	20 A	Power release seat motors, second row	F61
43	10 A	A/C clutch relay	F62
44	15 A	Right turn trailer tow relay, Left turn trailer tow relay	F63
45	15 A	Rear window wiper motor	F64
46	30 A	Fuel pump relay	F65
47	20 A	Heated oxygen sensors, Variable camshaft timing solenoids, Evaporative emission purge valve, Mass air flow / Intake air temperature sensor	F67
48	20 A	Coil on plugs	F68
49	20 A	Powertrain control module	F69
50	10 A	A/C clutch relay, Fan control relay, High fan control relay, Low fan control relay, Externally controlled variable, Displacement compressor, Turbocharger wastegate regulating valve solenoid, Turbocharger bypass valve (2.0 L and 3.5 L GTDI), Turbocharger bypass valve 2 (3.5 L GTDI), All wheel drive AWD relay module (3.5 L TIVCT)	F70
51	-	Not used	-
52	-	Not used	-
53	-	Not used	-
54	-	Not used	-
55	-	Not used	-
56	20 A	Headlamp assembly, right with HID headlamps	F78
57	5 A	Cruise control module	F79
58	-	Not used	-
59	-	Not used	-
60	15 A	Rear washer relay	F82
61	-	Not used	-
62	20 A	Trailer tow relay, parking lamp	F84
63	-	Not used	-
64	7.5 A	Evaporative emission canister vent valve, PCM power relay, Powertrain control module (3.5 L TIVCT)	F86
65	5 A	Run / Start relay	F87
66	5 A	Power steering control module, Blower motor relay	F89
67	10 A	Powertrain control module, Transmission control module	F90
68	10 A	Cruise control module	F91
69	10 A	Anti-lock brake system	F92

Chapter 12 Chassis electrical system

70	5 A	Auxiliary blower motor relay, Rear window defrost relay, Battery charge trailer tow relay	F93
71	30 A	Body control module - F33-F37, F41, F42	F94
72	-	Not used	-
73	-	Not used	-
74	-	Not used	-
R1	-	A/C clutch relay	R98
R2	-	Trailer tow relay, parking lamp	R77
R3	-	Not used	-
R4	-	Trailer tow relay, reversing	R38
R5	-	Not used	-
R6	-	Fuel pump relay	R15
R7	-	Trailer tow relay left turn	R53
R8	-	Trailer tow relay right turn	R37
R9	-	Not used	-
R10	-	Not used	-
R11	-	High fan control relay	R14
R12	-	Low fan control relay	R35
R13	-	Starter relay	R13
R14	-	Blower motor relay	R34
R15	-	Trailer tow relay battery charge	R12
R16	-	Fan control relay	R33
R17	-	Rear window defrost relay	R11
R18	-	Auxiliary blower motor relay	R32
R19	-	Third row power seat relay	R10
R20	-	Run/start relay	R88
R21	-	PCM power relay	R66
R22	-	Rear washer relay	R44
R23	-	Front wiper relay	R55

BATTERY JUNCTION BOX TOP VIEW FROM 2016 TO 2017

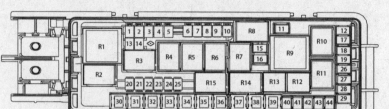

FUSE/RELAY	VALUE	DESCRIPTION	OEM NAME
1	20 A	Powertrain control module	F1
2	20 A	Heated oxygen sensors, Variable camshaft timing solenoids, Evaporative Emission purge valve, Evaporative Emission vent valve	F2
3	20 A	A/C clutch relay, Fan control relays, Externally controlled variable, Grille shutter actuator, Oil pressure control solenoid, and Displacement compressor, Turbocharger wastegate regulating valve solenoid, Turbocharger Bypass valve for 2.3 L / 3.5 L GTDI or Turbocharger Bypass valve 2, Third row power seat relay for 3.5 L GTDI or All wheel drive (AWD) relay module for 3.5 L TIVCT	F3
4	20 A	Ignition coil on plugs	F4
5	-	Not used	-
6	-	Not used	-
7	-	Not used	-
8	-	Not used	-
9	-	Not used	-
10	15 A	Exterior rear view mirrors - heating element	F10
11	40 A	Rear window defrost grid	F12
12	-	Not used	-
13	20 A	Horn relay	F15
14	10 A	A/C clutch relay	F16
15	25 A	Engine cooling fan	F22
16	-	Not used	-
17	30 A	Anti-lock brake system (ABS) module	F26
18	30 A	Trailer tow relay, battery charge	F27
19	40 A	Heated windshield relay, left	F28
20	10 A	Headlamp leveling	F30
21	10 A	Power steering control module	F31
22	10 A	Anti-lock brake system (ABS) module	F32
23	10 A	Powertrain control module	F33
24	10 A	Image processing module B, Video cameras, Side obstacle detection control module left, Side obstacle detection control module right, Cruise control module	F34
25	-	Not used	-
26	40 A	Auxiliary blower motor relay	F41
27	40 A	Heated windshield relay, right	F42
28	40 A	Blower motor relay	F43
29	-	Not used	-

30	50 A	Body control module	F44
31	40 A	Low fan control relay	F45
32	50 A	Power point, right rear	F46
33	-	Not used	-
34	50 A	Body control module	F48
35	-	Not used	-
36	50 A	Body control module	F50
37	50 A	Fan control 3 relay	F51
38	60 A	Anti-lock brake system (ABS) module	F52
39	50 A	Power point, right rear	F53
40	-	Not used	-
41	-	Not used	-
42	-	Not used	-
43	-	Not used	-
44	-	Not used	-
45	20 A	Power point, console front, Power point 2, instrument panel	F60
46	-	Not used	-
47	20 A	Power point, instrument panel with console	F62
48	30 A	Fuel pump relay	F63
49	-	Not used	-
50	20 A	Power point, console rear	F65
51	-	Not used	-
52	20 A	Power point, third row	F67
53	-	Not used	-
54	20 A	Liftgate / Trunk module	F69
55	20 A	Trailer tow relays, stop/turn	F70
56	-	Not used	-
57	-	Not used	-
58	20 A	Driver seat module - with memory, Seat control switch, driver side front - without memory	F73
59	20 A	Seat control switch, passenger side front	F74
60	30 A	Windshield wiper motor	F75
61	-	Not used	-
62	-	Not used	-
63	30 A	Third row seat release relay	F78
64	30 A	Starter relay	F79
65	-	Not used	-
66	-	Not used	-
67	20 A	Steering column lock relay	F82
68	30 A	Brake pedal position switch	F83
69	-	Not used	-
70	5 A	2nd row USB charger	F85
71	-	Not used	-
72	-	Not used	-
73	-	Not used	-
74	-	Not used	-
75	-	Not used	-
76	-	Not used	-
77	15 A	Adjustable pedal relay, multi-contour seat module relay	F92

78	10 A	Alternator sense	F93
79	20 A	Rear washer relay	F94
80	15 A	Rear window wiper motor	F95
81	10 A	PCM power relay	F96
82	5 A	Rain sensor module	F97
83	20 A	Power release seat motors, second row	F98
84	20 A	Trailer tow relay, parking lamp	F99
R1	-	PCM power relay	-
R2	-	Run/start relay	-
R3	-	Rear window defrost relay	-
R4	-	Auxiliary blower motor relay	-
R5	-	Not used	-
R6	-	Fan control 1 relay	-
R7	-	Fan control 2 relay	-
R8	-	Fan control 3 relay	-
R9	-	Steering column lock relay	-
R10	-	Heated windshield relay, left	-
R11	-	Heated windshield relay, right	-
R12	-	Horn relay	-
R13	-	A/C clutch relay	-
R14	-	Trailer tow relay, battery charge	-
R15	-	Blower motor relay	-

BATTERY JUNCTION BOX BOTTOM VIEW FROM 2016 TO 2017

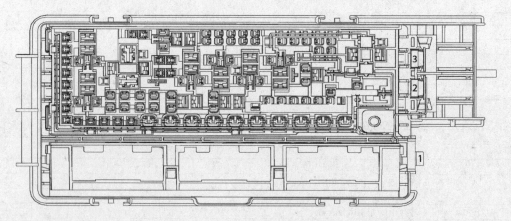

FUSE/RELAY	VALUE	DESCRIPTION	OEM NAME
1	125 A	Body control module	F100
2	100 A	Generator	F101
3	325 A	Power steering control module	F102

BODY CONTROL MODULE FROM 2011 TO 2015

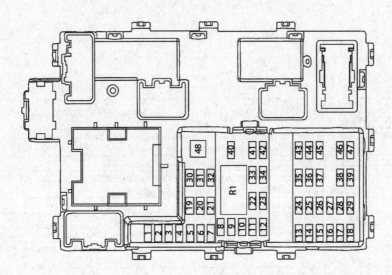

FUSE/RELAY	VALUE	DESCRIPTION	OEM NAME
1	30 A	Power window motor left front	F1
2	15 A	Spare fuse	-
3	30 A	Power window motor right front	F3
4	10 A	Third row power seat relay, Second row seat control switches, Glove box lamp, Overhead console, Vanity mirror lamps, Rail lamps, Second row interior lamp, Cargo lamp	F4
5	20 A	Audio amplifier or audio digital signal processing module	F5
6	5 A	Spare fuse	-
7	7.5 A	Driver seat module	F7
8	10 A	Spare fuse	-
9	10 A	Front controls interface module, Liftgate / Trunk module, Front control / Display interface module	F9
10	10 A	Front wiper relay, Rear washer relay, Battery junction box	F10
11	10 A	Instrument panel cluster, Head up display	F11
12	15 A	Exterior rear view mirrors, puddle lamps, Backlighting, Overhead console, Rail lamps, Second row interior lamp, Cargo lamp	F12
13	15 A	Headlamp assembly right - turn lamp, Lamp assembly, right rear - turn lamp, Right turn trailer tow relay	F13
14	15 A	Headlamp assembly left - turn lamp, Lamp assembly left rear - turn lamp, Left turn trailer tow relay	F14
15	15 A	High mounted stop lamp, Rear lamp assemblies - Stop lamp, Reversing trailer tow relay, Rear lamp assemblies - Reversing lamp, Auto-dimming interior mirror	F15
16	10 A	Headlamp assembly right, low beam	F16
17	10 A	Headlamp assembly left, low beam	F17
18	10 A	Power fold seat module, Brake shift interlock, Powertrain control module and Keyless entry keypad, Passive anti-theft transceiver without intelligent access or Start / Stop switch with intelligent access	F18
19	20 A	Seat control switch, driver side front with memory	F19
20	20 A	Door latch actuators	F20

21	10 A	Keyless entry keypad	F21
22	20 A	Horn	F22
23	15 A	Remote function actuator module with intelligent access, Front lighting control module, Steering column control module	F23
24	15 A	Data link connector, Steering column control module	F24
25	15 A	Liftgate latch without power liftgate and without intelligent access or Liftgate / Trunk module with power liftgate without intelligent access	F25
26	5 A	Tire pressure monitor module	F26
27	20 A	Remote function actuator module with intelligent access	F27
28	15 A	Ignition switch without intelligent access or Start / Stop switch with intelligent access	F28
29	20 A	Audio control module, Accessory protocol interface module, Global position system module	F29
30	15 A	Headlamp assemblies park lamp	F30
31	5 A	Trailer brake control module connector	F31
32	15 A	Master window control switch, Passenger window control switch, Front window motors, Door lock control switches, DC/AC inverter module	F32
33	10 A	Occupant classification system module	F33
34	10 A	Parking aid module, Video camera, Left side obstacle detection control module, Right side obstacle detection control module	F34
35	5 A	Head up display module, Front lighting control module, In-vehicle temperature / humidity sensor, All terrain control module	F35
36	10 A	Spare fuse	-
37	10 A	Restraints control module	F37
38	10 A	Auto-dimming interior mirror, Roof opening panel control switch, Roof opening panel module	F38
39	15 A	Headlamp high beam	F39
40	10 A	Parking lamp trailer tow relay, Rear lamp assemblies park lamp, License plate lamps	F40
41	7.5 A	Center stack switch assembly, brake shift interlock	F41
42	5 A	Spare fuse	-
43	10 A	Spare fuse	-
44	10 A	Spare fuse	-
45	5 A	Spare fuse	-
46	10 A	HVAC module	F46
47	15 A	Fog lamps, Exterior rear view mirrors cornering lamp	F47
48	30 A	Master window control switch, Passenger window control switch, Rear window control switches	F48
R1	-	Accessory delay relay	-

Chapter 12 Chassis electrical system

BODY CONTROL MODULE FROM 2016 TO 2017

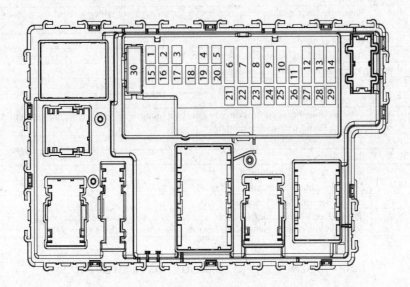

FUSE/RELAY	VALUE	DESCRIPTION	OEM NAME
1	10 A	Battery saver relay	F1
2	7.5 A	Memory seat switch	F2
3	20 A	Driver unlock relay	F3
4	5 A	Aftermarket electronic brake controller	F4
5	20 A	Heated seat module rear	F5
6	10 A	Not used	-
	10 A	Not used	-
7	10 A	Not used	-
	10 A	Not used	-
8	5 A	Keyless entry keypad, Liftgate / Trunk Module, Hands free liftgate actuation module	F10
	5 A	Rear heating, ventilation and air conditioning control module	F11
9	7.5 A	Rear heating, ventilation and air conditioning control module	F12
	7.5 A	Steering column control module, Instrument panel cluster, Gateway module A	F13
10	10 A	Spare fuse	-
	10 A	Data link connector	F15
11	15 A	Spare fuse	-
12	5 A	Front controls interface module, Adjustable pedal switch	F17
	5 A	Ignition switch	F18
13	5 A	Center stack switch assembly	F19
	-	Not used	-
14	5 A	In-vehicle temperature and humidity sensor, Head up display module, All terrain control module	F21
	5 A	Occupant classification system module	F22
15	10 A	Roof opening panel module, DC / AC inverter module, Driver door window regulator motor / switch, Passenger door window regulator motor / switch, Front door lock switches	F23
16	30 A	Lock relays	F24

17	30 A	Driver door module with memory or Driver door window regulator motor without memory	F25
18	30 A	Passenger door module with memory or Passenger door window regulator motor without memory	F26
19	30 A	Roof opening panel module	F27
20	20 A	Audio digital signal processing module	F28
21	30 A	Audio digital signal processing module	F29
22	30 A	Spare fuse	-
23	15 A	Spare fuse	-
24	10 A	Sync module, Front control display interface module, Global positioning system module, Radio transceiver module	F32
25	20 A	Audio control module	F33
26	30 A	Run / Start relay	F34
27	5 A	Restraints control module	F35
28	15 A	Auto-dimming interior mirror, Heated Steering Wheel module	F36
29	15 A	Heated Steering wheel module	F37
30	30 A	Driver door window control switch, Passenger door window control switch, Rear door window control switch LH, Rear door window control switch RH	F38

HIGH CURRENT BATTERY JUNCTION BOX FROM 2011 TO 2015

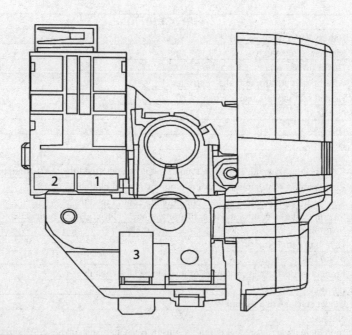

FUSE/RELAY	VALUE	DESCRIPTION	OEM NAME
1	100 A	Body control module	FUSE B
2	60 A	Engine cooling fan motor or Not used	FUSE F
3	100 A	Power steering control module	FUSE E

Index

A

About this manual, 0-5
Accelerator Pedal Position (APP) sensor - replacement, 6-15
Air conditioning
  and heating system - check and maintenance, 3-5
  compressor - removal and installation, 3-12
  condenser - removal and installation, 3-13
  pressure cycling switch - replacement, 3-14
  receiver-drier - removal and installation, 3-13
  thermostatic expansion valve (TXV) - general information, 3-14
Air filter
  check and replacement, 1-21
  housing - removal and installation, 4-8
Airbags - general information, 12-18
All-Terrain Control Switch/Module (ACTM) - replacement, 7B-2
All-Wheel Drive (AWD) module - replacement, 7B-2
Alternator - removal and installation, 5-5
Antenna - removal and installation, 12-9
Anti-lock Brake System (ABS) - general information, 9-6
Automatic transaxle fluid change, 1-24
Automatic transaxle, 7A-1
  Automatic transaxle - removal and installation, 7A-5
  Automatic transaxle overhaul - general information, 7A-6
  Diagnosis - general, 7A-2
  Driveaxle oil seals - replacement, 7A-6
  General information, 7A-2
  Shift cable - replacement and adjustment, 7A-3
  Shift lever - replacement, 7A-2
  Tow/haul switch - replacement, 7A-4
  Transaxle mount - replacement, 7A-6
Automotive chemicals and lubricants, 0-18

B

Balance shaft assembly (four-cylinder engine) - removal and installation, 2C-16
Battery
  and battery tray - removal and installation, 5-4
  cables - replacement, 5-4
  check, maintenance and charging, 1-16
  disconnection, 5-3
Blower motor resistor and blower motor - replacement, 3-11
Body, 11-1
  Body repair
    major damage, 11-3
    minor damage, 11-2
  Bumper covers - removal and installation, 11-7
  Center console - removal and installation, 11-18
  Cowl panels - removal and installation, 11-10
  Dashboard trim panels - removal and installation, 11-18
  Door
    latch, lock cylinder and handles - removal and installation, 11-14
    removal, installation and adjustment, 11-13
    trim panel - removal and installation, 11-11
    window glass - removal and installation, 11-15
    window regulator and motor - removal and installation, 11-15
  Fastener and trim removal, 11-5
  Front fender - removal and installation, 11-9
  Front seats - removal and installation, 11-23
  General information, 11-1
  Hood - removal, installation and adjustment, 11-6
  Hood latch and release cable - removal and installation, 11-6
  Instrument panel - removal and installation, 11-20
  Liftgate - removal and installation, 11-17
  Liftgate/Trunk Module (LTM) and power liftgate motor - removal and installation, 11-17
  Mirrors - removal and installation, 11-16
  Quarter trim panel - removal and installation, 11-24

Radiator grille and shutter - removal and installation, 11-10
Repair of minor paint scratches, 11-2
Steering column covers - removal and installation, 11-22
Upholstery, carpets and vinyl trim - maintenance, 11-6

Booster battery (jump) starting, 0-17

Brake
check, 1-18
fluid change, 1-22

Brakes, 9-1
Anti-lock Brake System (ABS) - general information, 9-6
Brake
booster vacuum pump (four-cylinder models) - removal and installation, 9-15
disc - inspection, removal and installation, 9-10
hoses and lines - inspection and replacement, 9-13
hydraulic system - bleeding, 9-14
light switch - replacement, 9-15
Disc brake caliper - removal and installation, 9-10
Disc brake pads - replacement, 9-7
General information, 9-2
Master cylinder - removal and installation, 9-12
Power brake booster - check removal and installation, 9-14
Troubleshooting, 9-2

Bulb replacement, 12-11
Bumper covers - removal and installation, 11-7
Buying parts, 0-9

C

Cabin air filter replacement, 1-21
Camshaft Position (CMP) sensor - replacement, 6-16
Camshafts and lifters - removal, inspection and installation, 2A-9
Camshafts and tappets/roller followers and lash adjusters - removal, inspection and installation, 2B-13
Catalytic converter - replacement, 6-21
Center console - removal and installation, 11-18
Charge air cooler - removal and installation, 4-12
Chassis electrical system, 12-1
Airbags - general information, 12-18
Antenna - removal and installation, 12-9
Bulb replacement, 12-11
Circuit breakers - general information, 12-3
Cruise control system - description and check, 12-16
Daytime Running Lights (DRL) - general information, 12-18
Electrical connectors - general information, 12-4
Electrical troubleshooting - general information, 12-1
Fuses and fusible links - general information, 12-3
General information, 12-1
Headlight bulb - replacement, 12-9
Headlight housing - removal and installation, 12-10
Headlights - adjustment, 12-10
Horn, 12-15
Ignition switch and key lock cylinder - replacement, 12-6
Instrument cluster - removal and installation, 12-7
Instrument panel switches - replacement, 12-7
Multi-function switch - replacement, 12-5
Power door lock and keyless entry system - description check and battery replacement, 12-17
Power mirror control system - description and check, 12-16
Power window system - description and check, 12-16
Radio and speakers - removal and installation, 12-8
Rear window defogger - check and repair, 12-15
Relays - general information and testing, 12-4
Steering column control module, 12-6
Taillight housing - removal and installation, 12-11
Wiper system, 12-13
Wiring diagrams - general information, 12-18

Circuit breakers - general information, 12-3
Coil spring (rear) - removal and installation, 10-12
Control arm (front) - removal, inspection and installation, 10-8
Conversion factors, 0-19
Cooling system
check, 1-17
servicing, 1-22

Cooling, heating and air conditioning systems, 3-1
Air conditioning
and heating system - check and maintenance, 3-5
compressor - removal and installation, 3-12
condenser - removal and installation, 3-13
pressure cycling switch - replacement, 3-14
receiver-drier - removal and installation, 3-13
thermostatic expansion valve (TXV) - general information, 3-14
Blower motor resistor and blower motor - replacement, 3-11

Coolant expansion tank - removal and installation, 3-8
Coolant temperature sending unit - check and replacement, 3-11
Engine cooling fans - removal and installation, 3-7
General information, 3-2
Heater core - replacement, 3-12
Heater/air conditioner control assembly and HVAC module - removal and installation, 3-11
Radiator - removal and installation, 3-8
Thermostat - replacement, 3-6
Troubleshooting, 3-2
Water pump - replacement, 3-10

Cowl panels - removal and installation, 11-10

Crankshaft
front oil seal - replacement, 2A-8
removal and installation, 2C-14

Crankshaft Position (CKP) sensor - replacement, 6-16

Crankshaft pulley
and crankshaft front oil seal - replacement, 2B-9
removal and installation, 2A-6

Cruise control system - description and check, 12-16

Cylinder compression check, 2C-4

Cylinder head - removal and installation
four-cylinder engines, 2A-11
V6 engines, 2B-9

Cylinder Head Temperature (CHT) sensor - replacement, 6-17

D

Dashboard trim panels - removal and installation, 11-18

Daytime Running Lights (DRL) - general information, 12-18

Diagnosis - general, 7A-2

Differential
(AWD models) - removal and installation, 8-4
lubricant change (AWD models), 1-25
pinion oil seal - replacement, 8-4

Disc brake
caliper - removal and installation, 9-10
pads - replacement, 9-7

Door
latch, lock cylinder and handles - removal and installation, 11-14
removal, installation and adjustment, 11-13
trim panel - removal and installation, 11-11
window glass - removal and installation, 11-15

window regulator and motor - removal and installation, 11-15

Driveaxle oil seals - replacement, 7A-6

Drivebelt check and replacement/tensioner replacement, 1-22

Driveline, 8-1
Differential (AWD models) - removal and installation, 8-4
Differential pinion oil seal - replacement, 8-4
Driveaxles - removal and installation, 8-2
Driveshaft (AWD models) - removal and installation, 8-4
General information, 8-2
Rear driveaxle oil seals (AWD models) - replacement, 8-4

Driveplate - removal and installation, 2B-16

Driveplate - removal, inspection and installation, 2A-13

Driveshaft (AWD models) - removal and installation, 8-4

E

Electrical connectors - general information, 12-4

Electrical troubleshooting - general information, 12-1

Emissions and engine control systems, 6-1
Accelerator Pedal Position (APP) sensor - replacement, 6-15
Camshaft Position (CMP) sensor - replacement, 6-16
Catalytic converter - replacement, 6-21
Crankshaft Position (CKP) sensor - replacement, 6-16
Cylinder Head Temperature (CHT) sensor - replacement, 6-17
Engine Coolant Temperature (ECT) sensor - replacement, 6-18
Engine Oil Pressure (EOP) sensor/oil pressure switch - replacement, 6-18
Evaporative Emissions Control (EVAP) system - component replacement, 6-22
Fuel Rail Pressure (FRP) sensor - replacement, 6-23
General information, 6-2
Intake Air Temperature (IAT) sensor, 6-19
Knock sensors - replacement, 6-18
Manifold Absolute Pressure Temperature (MAPT) sensor - replacement, 6-18
Mass Air Flow/Intake Air Temperature (MAF/IAT) sensor - replacement, 6-18

Obtaining and clearing Diagnostic Trouble Codes (DTCs), 6-4
On Board Diagnosis (OBD) system, 6-2
Oxygen sensors - replacement, 6-19
Positive Crankcase Ventilation (PCV) valve - replacement, 6-24
Powertrain Control Module (PCM) - removal and installation, 6-21
Throttle Position (TP) sensor - replacement, 6-20
Transmission Range (TR) sensor - removal and installation, 6-20
Turbine Shaft Speed (TSS) sensor - replacement, 6-20
Turbocharger Boost Pressure (TCBP)/Charge Air Cooler Temperature (CACT) sensor - replacement, 6-20
Turbocharger bypass valve - replacement, 6-20
Variable Valve Timing (VVT) system - component replacement, 6-24

Engine - removal and installation, 2C-7

Engine, in-vehicle repair procedures

Four-cylinder engines, 2A-1
- Camshafts and lifters - removal, inspection and installation, 2A-9
- Crankshaft front oil seal - replacement, 2A-8
- Crankshaft pulley - removal and installation, 2A-6
- Cylinder head - removal and installation, 2A-11
- Driveplate - removal, inspection and installation, 2A-13
- Engine mount - check and replacement, 2A-14
- General information, 2A-3
- Intake manifold - removal and installation, 2A-5
- Oil pan - removal and installation, 2A-12
- Oil pump - removal and installation, 2A-13
- Rear main oil seal - replacement, 2A-14
- Repair operations possible with the engine in the vehicle, 2A-3
- Timing chain cover, timing chain and tensioner - removal and installation, 2A-7
- Top Dead Center (TDC) for number 1 piston - locating, 2A-3
- Valve clearances - check and adjustment, 2A-5
- Valve cover - removal and installation, 2A-4

V6 engines, 2B-1
- Camshafts and tappets/roller followers and lash adjusters - removal, inspection and installation, 2B-13
- Crankshaft pulley and crankshaft front oil seal - replacement, 2B-9
- Cylinder heads - removal and installation, 2B-9
- Driveplate - removal and installation, 2B-16
- Exhaust manifolds - removal and installation, 2B-8
- General information, 2B-4
- Intake manifold(s) - removal and installation, 2B-6
- Oil pan - removal and installation, 2B-14
- Oil pump - removal and installation, 2B-16
- Rear main oil seal - replacement, 2B-16
- Repair operations possible with the engine in the vehicle, 2B-4
- Timing chain cover - removal and installation, 2B-10
- Timing chains and sprockets - removal and installation, 2B-11
- Valve clearance - check and adjustment, 2B-4
- Valve covers - removal and installation, 2B-4

Engine Coolant Temperature (ECT) sensor - replacement, 6-18

Engine cooling fans - removal and installation, 3-7

Engine electrical systems, 5-1
- Alternator - removal and installation, 5-5
- Battery
  - and battery tray - removal and installation, 5-4
  - cables - replacement, 5-4
  - disconnection, 5-3
- General information and precautions, 5-1
- Ignition coil(s) - removal and installation, 5-4
- Starter motor - removal and installation, 5-6
- Troubleshooting, 5-2

Engine mount - check and replacement
- four-cylinder engines, 2A-14
- V6 engines, 2B-17

Engine oil and filter change, 1-14

Engine Oil Pressure (EOP) sensor/oil pressure switch - replacement, 6-18

Engine overhaul
- dissassembly sequence, 2C-8
- reassembly sequence, 2C-16

Engine rebuilding alternatives, 2C-6

Engine removal - methods and precautions, 2C-6

Evaporative Emissions Control (EVAP) system - component replacement, 6-22

Exhaust manifolds - removal and installation, 2B-8

Exhaust system
- check, 1-21
- servicing - general information, 4-6

F

Fastener and trim removal, 11-5
Fluid level checks, 1-9
Four-cylinder engines, 2A-1
    Camshafts and lifters - removal, inspection and installation, 2A-9
    Crankshaft front oil seal - replacement, 2A-8
    Crankshaft pulley - removal and installation, 2A-6
    Cylinder head - removal and installation, 2A-11
    Driveplate - removal, inspection and installation, 2A-13
    Engine mount - check and replacement, 2A-14
    General information, 2A-3
    Intake manifold - removal and installation, 2A-5
    Oil pan - removal and installation, 2A-12
    Oil pump - removal and installation, 2A-13
    Rear main oil seal - replacement, 2A-14
    Repair operations possible with the engine in the vehicle, 2A-3
    Timing chain cover, timing chain and tensioner - removal and installation, 2A-7
    Top Dead Center (TDC) for number 1 piston - locating, 2A-3
    Valve clearances - check and adjustment, 2A-5
    Valve cover - removal and installation, 2A-4
Franction/decimal/millimeter equivalents, 0-20
Front fender - removal and installation, 11-9
Front seats - removal and installation, 11-23
Fuel and exhaust systems, 4-1
    Air filter housing - removal and installation, 4-8
    Charge air cooler - removal and installation, 4-12
    Exhaust system servicing - general information, 4-6
    Fuel
        lines and fittings - general information and disconnection, 4-4
        pressure - check, 4-4
        pressure relief procedure, 4-3
    Fuel Pump Control Module (FPCM) - replacement, 4-14
    Fuel
        pump module - removal and installation, 4-6
        rail and injectors - removal and installation, 4-8
        tank - removal and installation, 4-7
    General information and precautions, 4-3
    High-pressure fuel pump (turbocharged models) - removal and installation, 4-11
    Throttle body - removal and installation, 4-8
    Troubleshooting, 4-3
    Turbocharger - removal and installation, 4-12
Fuel Rail Pressure (FRP) sensor - replacement, 6-23
Fuel system check, 1-20
Fuses and fusible links - general information, 12-3

G

General engine overhaul procedures, 2C-1
    Balance shaft assembly (four-cylinder engine) - removal and installation, 2C-16
    Crankshaft - removal and installation, 2C-14
    Cylinder compression check, 2C-4
    Engine
        overhaul
            disassembly sequence, 2C-8
            reassembly sequence, 2C-16
        rebuilding alternatives, 2C-6
        removal
            and installation, 2C-7
            methods and precautions, 2C-6
    General information - engine overhaul, 2C-3
    Initial start-up and break-in after overhaul, 2C-16
    Oil pressure check, 2C-4
    Pistons and connecting rods - removal and installation, 2C-8
    Vacuum gauge diagnostic checks, 2C-5
General information - engine overhaul, 2C-3

H

Headlight
    adjustment, 12-10
    bulb - replacement, 12-9
    housing - removal and installation, 12-10
Heater core - replacement, 3-12
Heater/air conditioner control assembly and HVAC module - removal and installation, 3-11
High-pressure fuel pump (turbocharged models) - removal and installation, 4-11
Hood - removal, installation and adjustment, 11-6
Hood latch and release cable - removal and installation, 11-6
Horn, 12-15
Hub and bearing assembly - removal and installation
    front, 10-9
    rear, 10-12

I

Ignition coil(s) - removal and installation, 5-4
Ignition switch and key lock cylinder - replacement, 12-6
Initial start-up and break-in after overhaul, 2C-16
Instrument cluster - removal and installation, 12-7
Instrument panel
   removal and installation, 11-20
   switches - replacement, 12-7
Intake Air Temperature (IAT) sensor, 6-19
Intake manifold - removal and installation
   four-cylinder engines, 2A-5
   V6 engines, 2B-6

J

Jacking and towing, 0-16

K

Knock sensors - replacement, 6-18

L

Liftgate - removal and installation, 11-17
Liftgate/Trunk Module (LTM) and power liftgate motor - removal and installation, 11-17

M

Maintenance schedule, 1-8
Maintenance techniques, tools and working facilities, 0-9
Manifold Absolute Pressure Temperature (MAPT) sensor - replacement, 6-18
Mass Air Flow/Intake Air Temperature (MAF/IAT) sensor - replacement, 6-18
Master cylinder - removal and installation, 9-12
Mirrors - removal and installation, 11-16
Multi-function switch - replacement, 12-5

O

Obtaining and clearing Diagnostic Trouble Codes (DTCs), 6-4
Oil pan - removal and installation
   four-cylinder engines, 2A-12
   V6 engines, 2B-14
Oil pressure check, 2C-4
Oil pump - removal and installation
   four-cylinder engines, 2A-13
   V6 engines, 2B-16
On Board Diagnosis (OBD) system, 6-2
Oxygen sensors - replacement, 6-19

P

Pistons and connecting rods - removal and installation, 2C-8
Positive Crankcase Ventilation (PCV) valve - replacement, 6-24
Power brake booster - check removal and installation, 9-14
Power door lock and keyless entry system - description check and battery replacement, 12-17
Power mirror control system - description and check, 12-16
Power window system - description and check, 12-16
Powertrain Control Module (PCM) - removal and installation, 6-21

Q

Quarter trim panel - removal and installation, 11-24

R

Radiator - removal and installation, 3-8
Radiator grille and shutter - removal and installation, 11-10
Radio and speakers - removal and installation, 12-8
Rear driveaxle oil seals (AWD models) - replacement, 8-4
Rear knuckle - removal and installation, 10-12

Index

Rear main oil seal – replacement
four-cylinder engines, 2A-14
V6 engines, 2B-16
Rear window defogger - check and repair, 12-15
Recall information, 0-7
Relays - general information and testing, 12-4
Repair of minor paint scratches, 11-2
Repair operations possible with the engine in the vehicle
four-cylinder engines, 2A-3
V6 engines, 2B-4

S

Safety first!, 0-21
Seat belt check, 1-18
Shift cable - replacement and adjustment, 7A-3
Shift lever - replacement, 7A-2
Shock absorber (rear) - removal and installation, 10-10
Spark plug check and replacement, 1-26
Stabilizer bar, links and bushings - removal and installation
front, 10-7
rear, 10-13
Starter motor - removal and installation, 5-6
Steering column
control module, 12-6
covers - removal and installation, 11-22
Steering, suspension and driveaxle boot check, 1-19
Suspension and steering systems, 10-1
Coil spring (rear) - removal and installation, 10-12
Control arm (front) - removal, inspection and installation, 10-8
General information and precautions, 10-5
Hub and bearing assembly (front) - removal and installation, 10-9
Hub and bearing assembly (rear) - removal and installation, 10-12
Rear knuckle - removal and installation, 10-12
Shock absorber (rear) - removal and installation, 10-10
Stabilizer bar, links and bushings - removal and installation
front, 10-7
rear, 10-13
Steering
column - removal and installation, 10-14
gear - removal and installation, 10-16
gear boots - replacement, 10-16

knuckle and hub - removal and installation, 10-9
wheel - removal and installation, 10-13
Strut/coil spring assembly (front)
removal, inspection and installation, 10-5
replacement, 10-6
Subframe - removal and installation, 10-17
Suspension arms (rear) - removal and installation, 10-10
Tie-rod ends - removal and installation, 10-15
Wheel alignment - general information, 10-18
Wheels and tires - general information, 10-18

T

Taillight housing - removal and installation, 12-11
Thermostat - replacement, 3-6
Throttle body - removal and installation, 4-8
Throttle Position (TP) sensor - replacement, 6-20
Tie-rod ends - removal and installation, 10-15
Timing chain cover
removal and installation, 2B-10
timing chain and tensioner - removal and installation, 2A-7
Timing chains and sprockets - removal and installation, 2B-11
Tire and tire pressure checks, 1-12
Tire rotation, 1-17
Top Dead Center (TDC) for number 1 piston - locating, 2A-3
Tow/haul switch - replacement, 7A-4
Transaxle mount - replacement, 7A-6
Transfer case lubricant
change (AWD models), 1-26
level check, 1-20
Transfer case, 7B-1
All-Terrain Control Switch/Module (ACTM) - replacement, 7B-2
All-Wheel Drive (AWD) module - replacement, 7B-2
General information, 7B-1
Transfer case
case cooler - replacement, 7B-3
driveaxle oil seal - removal and installation, 7B-2
fluid pump - replacement, 7B-3
rear output shaft oil seal - removal and installation, 7B-2
removal and installation, 7B-3

Transmission Range (TR) sensor - removal and installation, 6-20
Tune-up and routine maintenance, 1-1
  Air filter check and replacement, 1-21
  Automatic transaxle fluid change, 1-24
  Battery check, maintenance and charging, 1-16
  Brake check, 1-18
  Brake fluid change, 1-22
  Cabin air filter replacement, 1-21
  Cooling system check, 1-17
  Cooling system servicing, 1-22
  Differential lubricant change (AWD models), 1-25
  Drivebelt check and replacement/tensioner replacement, 1-22
  Engine oil and filter change, 1-14
  Exhaust system check, 1-21
  Fluid level checks, 1-9
  Fuel system check, 1-20
  Introduction, 1-9
  Maintenance schedule, 1-8
  Seat belt check, 1-18
  Spark plug check and replacement, 1-26
  Steering, suspension and driveaxle boot check, 1-19
  Tire and tire pressure checks, 1-12
  Tire rotation, 1-17
  Transfer case lubricant
    change (AWD models), 1-26
    level check, 1-20
  Tune-up general information, 1-9
  Underhood hose check and replacement, 1-18
  Wiper blade inspection and replacement, 1-15
Tune-up general information, 1-9
Turbine Shaft Speed (TSS) sensor - replacement, 6-20
Turbocharger - removal and installation, 4-12
Turbocharger Boost Pressure (TCBP)/Charge Air Cooler Temperature (CACT) sensor - replacement, 6-20
Turbocharger bypass valve - replacement, 6-20

U

Underhood hose check and replacement, 1-18
Upholstery, carpets and vinyl trim - maintenance, 11-6

V

V6 engines, 2B-1
  Camshafts and tappets/roller followers and lash adjusters - removal, inspection and installation, 2B-13
  Crankshaft pulley and crankshaft front oil seal - replacement, 2B-9
  Cylinder heads - removal and installation, 2B-9
  Driveplate - removal and installation, 2B-16
  Engine mounts - check and replacement, 2B-17
  Exhaust manifolds - removal and installation, 2B-8
  General information, 2B-4
  Intake manifold(s) - removal and installation, 2B-6
  Oil pan - removal and installation, 2B-14
  Oil pump - removal and installation, 2B-16
  Rear main oil seal - replacement, 2B-16
  Repair operations possible with the engine in the vehicle, 2B-4
  Timing chain cover - removal and installation, 2B-10
  Timing chains and sprockets - removal and installation, 2B-11
  Valve clearance - check and adjustment, 2B-4
  Valve covers - removal and installation, 2B-4
Vacuum gauge diagnostic checks, 2C-5
Valve clearance - check and adjustment
  four-cylinder engines, 2A-5
  V6 engines, 2B-4
Valve cover - removal and installation
  four-cylinder engines, 2A-4
  V6 engines, 2B-4
Variable Valve Timing (VVT) system - component replacement, 6-24
Vehicle identification numbers, 0-6

W

Water pump - replacement, 3-10
Wheel alignment - general information, 10-18
Wheels and tires - general information, 10-18
Wiper blade inspection and replacement, 1-15
Wiper system, 12-13
Wiring diagrams - general information, 12-18

Haynes Automotive Manuals

NOTE: If you do not see a listing for your vehicle, consult your local Haynes dealer for the latest product information.

ACURA
- **12020** Integra '86 thru '89 & Legend '86 thru '90
- **12021** Integra '90 thru '93 & Legend '91 thru '95
- Integra '94 thru '00 - see HONDA Civic (42025)
- **MDX** '01 thru '07 - see HONDA Pilot (42037)
- **12050** Acura TL all models '99 thru '08

AMC
- Jeep CJ - see JEEP (50020)
- **14020** Mid-size models '70 thru '83
- **14025** (Renault) Alliance & Encore '83 thru '87

AUDI
- **15020** 4000 all models '80 thru '87
- **15025** 5000 all models '77 thru '83
- **15026** 5000 all models '84 thru '88
- Audi A4 '96 thru '01 - see VW Passat (96023)
- **15030** Audi A4 '02 thru '08

AUSTIN-HEALEY
- Sprite - see MG Midget (66015)

BMW
- **18020** 3/5 Series '82 thru '92
- **18021** 3-Series incl. Z3 models '92 thru '98
- **18022** 3-Series incl. Z4 models '99 thru '05
- **18023** 3-Series '06 thru '10
- **18025** 320i all 4 cyl models '75 thru '83
- **18050** 1500 thru 2002 except Turbo '59 thru '77

BUICK
- **19010** Buick Century '97 thru '05
- Century (front-wheel drive) - see GM (38005)
- **19020** Buick, Oldsmobile & Pontiac Full-size (Front-wheel drive) '85 thru '05
- Buick Electra, LeSabre and Park Avenue; Oldsmobile Delta 88 Royale, Ninety Eight and Regency; Pontiac Bonneville
- **19025** Buick, Oldsmobile & Pontiac Full-size (Rear wheel drive) '70 thru '90
- Buick Estate, Electra, LeSabre, Limited, Oldsmobile Custom Cruiser, Delta 88, Ninety-eight, Pontiac Bonneville, Catalina, Grandville, Parisiene
- **19030** Mid-size Regal & Century all rear-drive models with V6, V8 and Turbo '74 thru '87
- Regal - see GENERAL MOTORS (38010)
- Riviera - see GENERAL MOTORS (38030)
- Roadmaster - see CHEVROLET (24046)
- Skyhawk - see GENERAL MOTORS (38015)
- Skylark - see GM (38020, 38025)
- Somerset - see GENERAL MOTORS (38025)

CADILLAC
- **21015** CTS & CTS-V '03 thru '12
- **21030** Cadillac Rear Wheel Drive '70 thru '93
- Cimarron - see GENERAL MOTORS (38015)
- DeVille - see GM (38031 & 38032)
- Eldorado - see GM (38030 & 38031)
- Fleetwood - see GM (38031)
- Seville - see GM (38030, 38031 & 38032)

CHEVROLET
- **10305** Chevrolet Engine Overhaul Manual
- **24010** Astro & GMC Safari Mini-vans '85 thru '05
- **24015** Camaro V8 all models '70 thru '81
- **24016** Camaro all models '82 thru '92
- **24017** Camaro & Firebird '93 thru '02
- Cavalier - see GENERAL MOTORS (38016)
- Celebrity - see GENERAL MOTORS (38005)
- **24020** Chevelle, Malibu & El Camino '69 thru '87
- **24024** Chevette & Pontiac T1000 '76 thru '87
- Citation - see GENERAL MOTORS (38020)
- **24027** Colorado & GMC Canyon '04 thru '10
- **24032** Corsica/Beretta all models '87 thru '96
- **24040** Corvette all V8 models '68 thru '82
- **24041** Corvette all models '84 thru '96
- **24045** Full-size Sedans Caprice, Impala, Biscayne, Bel Air & Wagons '69 thru '90
- **24046** Impala SS & Caprice and Buick Roadmaster '91 thru '96
- Impala '00 thru '05 - see LUMINA (24048)
- **24047** Impala & Monte Carlo all models '06 thru '11
- Lumina '90 thru '94 - see GM (38010)
- **24048** Lumina & Monte Carlo '95 thru '05
- Lumina APV - see GM (38035)
- **24050** Luv Pick-up all 2WD & 4WD '72 thru '82
- Malibu '97 thru '00 - see GM (38026)
- **24055** Monte Carlo all models '70 thru '88
- Monte Carlo '95 thru '01 - see LUMINA (24048)
- **24059** Nova all V8 models '69 thru '79
- **24060** Nova and Geo Prizm '85 thru '92
- **24064** Pick-ups '67 thru '87 - Chevrolet & GMC
- **24065** Pick-ups '88 thru '98 - Chevrolet & GMC
- **24066** Pick-ups '99 thru '06 - Chevrolet & GMC
- **24067** Chevrolet Silverado & GMC Sierra '07 thru '12
- **24070** S-10 & S-15 Pick-ups '82 thru '93, Blazer & Jimmy '83 thru '94
- **24071** S-10 & Sonoma Pick-ups '94 thru '04, including Blazer, Jimmy & Hombre
- **24072** Chevrolet TrailBlazer, GMC Envoy & Oldsmobile Bravada '02 thru '09
- **24075** Sprint '85 thru '88 & Geo Metro '89 thru '01
- **24080** Vans - Chevrolet & GMC '68 thru '96
- **24081** Chevrolet Express & GMC Savana Full-size Vans '96 thru '10

CHRYSLER
- **10310** Chrysler Engine Overhaul Manual
- **25015** Chrysler Cirrus, Dodge Stratus, Plymouth Breeze '95 thru '00
- **25020** Full-size Front-Wheel Drive '88 thru '93
- K-Cars - see DODGE Aries (30008)
- Laser - see DODGE Daytona (30030)
- **25025** Chrysler LHS, Concorde, New Yorker, Dodge Intrepid, Eagle Vision, '93 thru '97
- **25026** Chrysler LHS, Concorde, 300M, Dodge Intrepid, '98 thru '04
- **25027** Chrysler 300, Dodge Charger & Magnum '05 thru '09
- **25030** Chrysler & Plymouth Mid-size front wheel drive '82 thru '95
- Rear-wheel Drive - see Dodge (30050)
- **25035** PT Cruiser all models '01 thru '10
- **25040** Chrysler Sebring '95 thru '06, Dodge Stratus '01 thru '06, Dodge Avenger '95 thru '00

DATSUN
- **28005** 200SX all models '80 thru '83
- **28007** B-210 all models '73 thru '78
- **28009** 210 all models '79 thru '82
- **28012** 240Z, 260Z & 280Z Coupe '70 thru '78
- **28014** 280ZX Coupe & 2+2 '79 thru '83
- 300ZX - see NISSAN (72010)
- **28018** 510 & PL521 Pick-up '68 thru '73
- **28020** 510 all models '78 thru '81
- **28022** 620 Series Pick-up all models '73 thru '79
- 720 Series Pick-up - see NISSAN (72030)
- **28025** 810/Maxima all gasoline models '77 thru '84

DODGE
- 400 & 600 - see CHRYSLER (25030)
- **30008** Aries & Plymouth Reliant '81 thru '89
- **30010** Caravan & Plymouth Voyager '84 thru '95
- **30011** Caravan & Plymouth Voyager '96 thru '02
- **30012** Challenger/Plymouth Saporro '78 thru '83
- **30013** Caravan, Chrysler Voyager, Town & Country '03 thru '07
- **30016** Colt & Plymouth Champ '78 thru '87
- **30020** Dakota Pick-ups all models '87 thru '96
- **30021** Durango '98 & '99, Dakota '97 thru '99
- **30022** Durango '00 thru '03 Dakota '00 thru '04
- **30023** Durango '04 thru '09, Dakota '05 thru '11
- **30025** Dart, Demon, Plymouth Barracuda, Duster & Valiant 6 cyl models '67 thru '76
- **30030** Daytona & Chrysler Laser '84 thru '89
- Intrepid - see CHRYSLER (25025, 25026)
- **30034** Neon all models '95 thru '99
- **30035** Omni & Plymouth Horizon '78 thru '90
- **30036** Dodge and Plymouth Neon '00 thru '05
- **30040** Pick-ups all full-size models '74 thru '93
- **30041** Pick-ups all full-size models '94 thru '01
- **30042** Pick-ups full-size models '02 thru '08
- **30045** Ram 50/D50 Pick-ups & Raider and Plymouth Arrow Pick-ups '79 thru '93
- **30050** Dodge/Plymouth/Chrysler RWD '71 thru '89
- **30055** Shadow & Plymouth Sundance '87 thru '94
- **30060** Spirit & Plymouth Acclaim '89 thru '95
- **30065** Vans - Dodge & Plymouth '71 thru '03

EAGLE
- Talon - see MITSUBISHI (68030, 68031)
- Vision - see CHRYSLER (25025)

FIAT
- **34010** 124 Sport Coupe & Spider '68 thru '78
- **34025** X1/9 all models '74 thru '80

FORD
- **10320** Ford Engine Overhaul Manual
- **10355** Ford Automatic Transmission Overhaul
- **11500** Mustang '64-1/2 thru '70 Restoration Guide
- **36004** Aerostar Mini-vans all models '86 thru '97
- **36006** Contour & Mercury Mystique '95 thru '00
- **36008** Courier Pick-up all models '72 thru '82
- **36012** Crown Victoria & Mercury Grand Marquis '88 thru '10
- **36016** Escort/Mercury Lynx all models '81 thru '90
- **36020** Escort/Mercury Tracer '91 thru '02
- **36022** Escape & Mazda Tribute '01 thru '11
- **36024** Explorer & Mazda Navajo '91 thru '01
- **36025** Explorer/Mercury Mountaineer '02 thru '10
- **36028** Fairmont & Mercury Zephyr '78 thru '83
- **36030** Festiva & Aspire '88 thru '97
- **36032** Fiesta all models '77 thru '80
- **36034** Focus all models '00 thru '11
- **36036** Ford & Mercury Full-size '75 thru '87
- **36044** Ford & Mercury Mid-size '75 thru '86
- **36045** Fusion & Mercury Milan '06 thru '10
- **36048** Mustang V8 all models '64-1/2 thru '73
- **36049** Mustang II 4 cyl, V6 & V8 models '74 thru '78
- **36050** Mustang & Mercury Capri '79 thru '93
- **36051** Mustang all models '94 thru '04
- **36052** Mustang '05 thru '10
- **36054** Pick-ups & Bronco '73 thru '79
- **36058** Pick-ups & Bronco '80 thru '96
- **36059** F-150 & Expedition '97 thru '09, F-250 '97 thru '99 & Lincoln Navigator '98 thru '09
- **36060** Super Duty Pick-ups, Excursion '99 thru '10
- **36061** F-150 full-size '04 thru '10
- **36062** Pinto & Mercury Bobcat '75 thru '80
- **36066** Probe all models '89 thru '92
- Probe '93 thru '97 - see MAZDA 626 (61042)
- **36070** Ranger/Bronco II gasoline models '83 thru '92
- **36071** Ranger '93 thru '10 & Mazda Pick-ups '94 thru '09
- **36074** Taurus & Mercury Sable '86 thru '95
- **36075** Taurus & Mercury Sable '96 thru '05
- **36078** Tempo & Mercury Topaz '84 thru '94
- **36082** Thunderbird/Mercury Cougar '83 thru '88
- **36086** Thunderbird/Mercury Cougar '89 thru '97
- **36090** Vans all V8 Econoline models '69 thru '91
- **36094** Vans full size '92 thru '10
- **36097** Windstar Mini-van '95 thru '07

GENERAL MOTORS
- **10360** GM Automatic Transmission Overhaul
- **38005** Buick Century, Chevrolet Celebrity, Oldsmobile Cutlass Ciera & Pontiac 6000 all models '82 thru '96
- **38010** Buick Regal, Chevrolet Lumina, Oldsmobile Cutlass Supreme & Pontiac Grand Prix (FWD) '88 thru '07
- **38015** Buick Skyhawk, Cadillac Cimarron, Chevrolet Cavalier, Oldsmobile Firenza & Pontiac J-2000 & Sunbird '82 thru '94
- **38016** Chevrolet Cavalier & Pontiac Sunfire '95 thru '05
- **38017** Chevrolet Cobalt & Pontiac G5 '05 thru '11
- **38020** Buick Skylark, Chevrolet Citation, Olds Omega, Pontiac Phoenix '80 thru '85
- **38025** Buick Skylark & Somerset, Oldsmobile Achieva & Calais and Pontiac Grand Am all models '85 thru '98
- **38026** Chevrolet Malibu, Olds Alero & Cutlass, Pontiac Grand Am '97 thru '03
- **38027** Chevrolet Malibu '04 thru '10
- **38030** Cadillac Eldorado, Seville, Oldsmobile Toronado, Buick Riviera '71 thru '85
- **38031** Cadillac Eldorado & Seville, DeVille, Fleetwood & Olds Toronado, Buick Riviera '86 thru '93
- **38032** Cadillac DeVille '94 thru '05 & Seville '92 thru '04 Cadillac DTS '06 thru '10
- **38035** Chevrolet Lumina APV, Olds Silhouette & Pontiac Trans Sport all models '90 thru '96
- **38036** Chevrolet Venture, Olds Silhouette, Pontiac Trans Sport & Montana '97 thru '05
- General Motors Full-size Rear-wheel Drive - see BUICK (19025)
- **38040** Chevrolet Equinox '05 thru '09 Pontiac Torrent '06 thru '09
- **38070** Chevrolet HHR '06 thru '11

GEO
- Metro - see CHEVROLET Sprint (24075)
- Prizm - '85 thru '92 see CHEVY (24060), '93 thru '02 see TOYOTA Corolla (92036)
- **40030** Storm all models '90 thru '93
- Tracker - see SUZUKI Samurai (90010)

GMC
- Vans & Pick-ups - see CHEVROLET

HONDA
- **42010** Accord CVCC all models '76 thru '83
- **42011** Accord all models '84 thru '89
- **42012** Accord all models '90 thru '93
- **42013** Accord all models '94 thru '97
- **42014** Accord all models '98 thru '02
- **42015** Accord '03 thru '07
- **42020** Civic 1200 all models '73 thru '79
- **42021** Civic 1300 & 1500 CVCC '80 thru '83
- **42022** Civic 1500 CVCC all models '75 thru '79

(Continued on other side)

Haynes Automotive Manuals (continued)

NOTE: If you do not see a listing for your vehicle, consult your local Haynes dealer for the latest product information.

HONDA (continued)
- 42023 Civic all models '84 thru '91
- 42024 Civic & del Sol '92 thru '95
- 42025 Civic '96 thru '00, CR-V '97 thru '01, Acura Integra '94 thru '00
- 42026 Civic '01 thru '10, CR-V '02 thru '09
- 42035 Odyssey all models '99 thru '10
- Passport - see ISUZU Rodeo (47017)
- 42037 Honda Pilot '03 thru '07, Acura MDX '01 thru '07
- 42040 Prelude CVCC all models '79 thru '89

HYUNDAI
- 43010 Elantra all models '96 thru '10
- 43015 Excel & Accent all models '86 thru '09
- 43050 Santa Fe all models '01 thru '06
- 43055 Sonata all models '99 thru '08

INFINITI
- G35 '03 thru '08 - see NISSAN 350Z (72011)

ISUZU
- Hombre - see CHEVROLET S-10 (24071)
- 47017 Rodeo, Amigo & Honda Passport '89 thru '02
- 47020 Trooper & Pick-up '81 thru '93

JAGUAR
- 49010 XJ6 all 6 cyl models '68 thru '86
- 49011 XJ6 all models '88 thru '94
- 49015 XJ12 & XJS all 12 cyl models '72 thru '85

JEEP
- 50010 Cherokee, Comanche & Wagoneer Limited all models '84 thru '01
- 50020 CJ all models '49 thru '86
- 50025 Grand Cherokee all models '93 thru '04
- 50026 Grand Cherokee '05 thru '09
- 50029 Grand Wagoneer & Pick-up '72 thru '91 Grand Wagoneer '84 thru '91, Cherokee & Wagoneer '72 thru '83, Pick-up '72 thru '88
- 50030 Wrangler all models '87 thru '11
- 50035 Liberty '02 thru '07

KIA
- 54050 Optima '01 thru '10
- 54070 Sephia '94 thru '01, Spectra '00 thru '09, Sportage '05 thru '10

LEXUS
- ES 300/330 - see TOYOTA Camry (92007) (92008)
- RX 330 - see TOYOTA Highlander (92095)

LINCOLN
- Navigator - see FORD Pick-up (36059)
- 59010 Rear-Wheel Drive all models '70 thru '10

MAZDA
- 61010 GLC Hatchback (rear-wheel drive) '77 thru '83
- 61011 GLC (front-wheel drive) '81 thru '85
- 61012 Mazda3 '04 thru '11
- 61015 323 & Protogé '90 thru '03
- 61016 MX-5 Miata '90 thru '09
- 61020 MPV all models '89 thru '98
- Navajo - see Ford Explorer (36024)
- 61030 Pick-ups '72 thru '93
- Pick-ups '94 thru '00 - see Ford Ranger (36071)
- 61035 RX-7 all models '79 thru '85
- 61036 RX-7 all models '86 thru '91
- 61040 626 (rear-wheel drive) all models '79 thru '82
- 61041 626/MX-6 (front-wheel drive) '83 thru '92
- 61042 626, MX-6/Ford Probe '93 thru '02
- 61043 Mazda6 '03 thru '11

MERCEDES-BENZ
- 63012 123 Series Diesel '76 thru '85
- 63015 190 Series four-cyl gas models, '84 thru '88
- 63020 230/250/280 6 cyl sohc models '68 thru '72
- 63025 280 123 Series gasoline models '77 thru '81
- 63030 350 & 450 all models '71 thru '80
- 63040 C-Class: C230/C240/C280/C320/C350 '01 thru '07

MERCURY
- 64200 Villager & Nissan Quest '93 thru '01
- All other titles, see FORD Listing.

MG
- 66010 MGB Roadster & GT Coupe '62 thru '80
- 66015 MG Midget, Austin Healey Sprite '58 thru '80

MINI
- 67020 Mini '02 thru '11

MITSUBISHI
- 68020 Cordia, Tredia, Galant, Precis & Mirage '83 thru '93
- 68030 Eclipse, Eagle Talon & Ply. Laser '90 thru '94
- 68031 Eclipse '95 thru '05, Eagle Talon '95 thru '98
- 68035 Galant '94 thru '10
- 68040 Pick-up '83 thru '96 & Montero '83 thru '93

NISSAN
- 72010 300ZX all models including Turbo '84 thru '89
- 72011 350Z & Infiniti G35 all models '03 thru '08
- 72015 Altima all models '93 thru '06
- 72016 Altima '07 thru '10
- 72020 Maxima all models '85 thru '92
- 72021 Maxima all models '93 thru '04
- 72025 Murano '03 thru '10
- 72030 Pick-ups '80 thru '97 Pathfinder '87 thru '95
- 72031 Frontier Pick-up, Xterra, Pathfinder '96 thru '04
- 72032 Frontier & Xterra '05 thru '11
- 72040 Pulsar all models '83 thru '86
- Quest - see MERCURY Villager (64200)
- 72050 Sentra all models '82 thru '94
- 72051 Sentra & 200SX all models '95 thru '06
- 72060 Stanza all models '82 thru '90
- 72070 Titan pick-ups '04 thru '10 Armada '05 thru '10

OLDSMOBILE
- 73015 Cutlass V6 & V8 gas models '74 thru '88
- For other OLDSMOBILE titles, see BUICK, CHEVROLET or GENERAL MOTORS listing.

PLYMOUTH
- For PLYMOUTH titles, see DODGE listing.

PONTIAC
- 79008 Fiero all models '84 thru '88
- 79018 Firebird V8 models except Turbo '70 thru '81
- 79019 Firebird all models '82 thru '92
- 79025 G6 all models '05 thru '09
- 79040 Mid-size Rear-wheel Drive '70 thru '87
- Vibe '03 thru '11 - see TOYOTA Matrix (92060)
- For other PONTIAC titles, see BUICK, CHEVROLET or GENERAL MOTORS listing.

PORSCHE
- 80020 911 except Turbo & Carrera 4 '65 thru '89
- 80025 914 all 4 cyl models '69 thru '76
- 80030 924 all models including Turbo '76 thru '82
- 80035 944 all models including Turbo '83 thru '89

RENAULT
- Alliance & Encore - see AMC (14020)

SAAB
- 84010 900 all models including Turbo '79 thru '88

SATURN
- 87010 Saturn all S-series models '91 thru '02
- 87011 Saturn Ion '03 thru '07
- 87020 Saturn all L-series models '00 thru '04
- 87040 Saturn VUE '02 thru '07

SUBARU
- 89002 1100, 1300, 1400 & 1600 '71 thru '79
- 89003 1600 & 1800 2WD & 4WD '80 thru '94
- 89100 Legacy all models '90 thru '99
- 89101 Legacy & Forester '00 thru '06

SUZUKI
- 90010 Samurai/Sidekick & Geo Tracker '86 thru '01

TOYOTA
- 92005 Camry all models '83 thru '91
- 92006 Camry all models '92 thru '96
- 92007 Camry, Avalon, Solara, Lexus ES 300 '97 thru '01
- 92008 Toyota Camry, Avalon and Solara and Lexus ES 300/330 all models '02 thru '06
- 92009 Camry '07 thru '11
- 92015 Celica Rear Wheel Drive '71 thru '85
- 92020 Celica Front Wheel Drive '86 thru '99
- 92025 Celica Supra all models '79 thru '92
- 92030 Corolla all models '75 thru '79
- 92032 Corolla all rear wheel drive models '80 thru '87
- 92035 Corolla all front wheel drive models '84 thru '92
- 92036 Corolla & Geo Prizm '93 thru '02
- 92037 Corolla models '03 thru '11
- 92040 Corolla Tercel all models '80 thru '82
- 92045 Corona all models '74 thru '82
- 92050 Cressida all models '78 thru '82
- 92055 Land Cruiser FJ40, 43, 45, 55 '68 thru '82
- 92056 Land Cruiser FJ60, 62, 80, FZJ80 '80 thru '96
- 92060 Matrix & Pontiac Vibe '03 thru '11
- 92065 MR2 all models '85 thru '87
- 92070 Pick-up all models '69 thru '78
- 92075 Pick-up all models '79 thru '95
- 92076 Tacoma, 4Runner, & T100 '93 thru '04
- 92077 Tacoma all models '05 thru '09
- 92078 Tundra '00 thru '06 & Sequoia '01 thru '07
- 92079 4Runner all models '03 thru '09
- 92080 Previa all models '91 thru '95
- 92081 Prius all models '01 thru '08
- 92082 RAV4 all models '96 thru '10
- 92085 Tercel all models '87 thru '94
- 92090 Sienna all models '98 thru '09
- 92095 Highlander & Lexus RX-330 '99 thru '07

TRIUMPH
- 94007 Spitfire all models '62 thru '81
- 94010 TR7 all models '75 thru '81

VW
- 96008 Beetle & Karmann Ghia '54 thru '79
- 96009 New Beetle '98 thru '10
- 96016 Rabbit, Jetta, Scirocco & Pick-up gas models '75 thru '92 & Convertible '80 thru '92
- 96017 Golf, GTI & Jetta '93 thru '98, Cabrio '95 thru '02
- 96018 Golf, GTI, Jetta '99 thru '05
- 96019 Jetta, Rabbit, GTI & Golf '05 thru '11
- 96020 Rabbit, Jetta & Pick-up diesel '77 thru '84
- 96023 Passat '98 thru '05, Audi A4 '96 thru '01
- 96030 Transporter 1600 all models '68 thru '79
- 96035 Transporter 1700, 1800 & 2000 '72 thru '79
- 96040 Type 3 1500 & 1600 all models '63 thru '73
- 96045 Vanagon all air-cooled models '80 thru '83

VOLVO
- 97010 120, 130 Series & 1800 Sports '61 thru '73
- 97015 140 Series all models '66 thru '74
- 97020 240 Series all models '76 thru '93
- 97040 740 & 760 Series all models '82 thru '88
- 97050 850 Series all models '93 thru '97

TECHBOOK MANUALS
- 10205 Automotive Computer Codes
- 10206 OBD-II & Electronic Engine Management
- 10210 Automotive Emissions Control Manual
- 10215 Fuel Injection Manual '78 thru '85
- 10220 Fuel Injection Manual '86 thru '99
- 10225 Holley Carburetor Manual
- 10230 Rochester Carburetor Manual
- 10240 Weber/Zenith/Stromberg/SU Carburetors
- 10305 Chevrolet Engine Overhaul Manual
- 10310 Chrysler Engine Overhaul Manual
- 10320 Ford Engine Overhaul Manual
- 10330 GM and Ford Diesel Engine Repair Manual
- 10333 Engine Performance Manual
- 10340 Small Engine Repair Manual, 5 HP & Less
- 10341 Small Engine Repair Manual, 5.5 - 20 HP
- 10345 Suspension, Steering & Driveline Manual
- 10355 Ford Automatic Transmission Overhaul
- 10360 GM Automatic Transmission Overhaul
- 10405 Automotive Body Repair & Painting
- 10410 Automotive Brake Manual
- 10411 Automotive Anti-lock Brake (ABS) Systems
- 10415 Automotive Detailing Manual
- 10420 Automotive Electrical Manual
- 10425 Automotive Heating & Air Conditioning
- 10430 Automotive Reference Manual & Dictionary
- 10435 Automotive Tools Manual
- 10440 Used Car Buying Guide
- 10445 Welding Manual
- 10450 ATV Basics
- 10452 Scooters 50cc to 250cc

SPANISH MANUALS
- 98903 Reparación de Carrocería & Pintura
- 98904 Manual de Carburador Modelos Holley & Rochester
- 98905 Códigos Automotrices de la Computadora
- 98906 OBD-II & Sistemas de Control Electrónico del Motor
- 98910 Frenos Automotriz
- 98913 Electricidad Automotriz
- 98915 Inyección de Combustible '86 al '99
- 99040 Chevrolet & GMC Camionetas '67 al '87
- 99041 Chevrolet & GMC Camionetas '88 al '98
- 99042 Chevrolet & GMC Camionetas Cerradas '68 al '95
- 99043 Chevrolet/GMC Camionetas '94 al '04
- 99048 Chevrolet/GMC Camionetas '99 al '06
- 99055 Dodge Caravan & Plymouth Voyager '84 al '95
- 99075 Ford Camionetas y Bronco '80 al '94
- 99076 Ford F-150 '97 al '09
- 99077 Ford Camionetas Cerradas '69 al '91
- 99088 Ford Modelos de Tamaño Mediano '75 al '86
- 99089 Ford Camionetas Ranger '93 al '10
- 99091 Ford Taurus & Mercury Sable '86 al '95
- 99095 GM Modelos de Tamaño Grande '70 al '90
- 99100 GM Modelos de Tamaño Mediano '70 al '88
- 99106 Jeep Cherokee, Wagoneer & Comanche '84 al '00
- 99110 Nissan Camioneta '80 al '96, Pathfinder '87 al '95
- 99118 Nissan Sentra '82 al '94
- 99125 Toyota Camionetas y 4Runner '79 al '95

Over 100 Haynes motorcycle manuals also available

7-12

Haynes North America, Inc., 859 Lawrence Drive, Newbury Park, CA 91320-1514 • (805) 498-6703 • http://www.haynes.com